Fundamentals of
MULTIVARIABLE
CALCULUS

Fundamentals of
MULTIVARIABLE
CALCULUS

Leonid P Lebedev
Universidad Nacional de Colombia, Colombia

Michael J Cloud
Professor Emeritus
Lawrence Technological University, USA

World Scientific

NEW JERSEY · LONDON · SINGAPORE · BEIJING · SHANGHAI · HONG KONG · TAIPEI · CHENNAI

Published by

World Scientific Publishing Co. Pte. Ltd.

5 Toh Tuck Link, Singapore 596224

USA office: 27 Warren Street, Suite 401-402, Hackensack, NJ 07601

UK office: 57 Shelton Street, Covent Garden, London WC2H 9HE

Library of Congress Control Number: 2024044556

British Library Cataloguing-in-Publication Data
A catalogue record for this book is available from the British Library.

FUNDAMENTALS OF MULTIVARIABLE CALCULUS

ISBN 978-981-98-0001-8 (hardcover)
ISBN 978-981-98-0002-5 (ebook for institutions)
ISBN 978-981-98-0003-2 (ebook for individuals)

For any available supplementary material, please visit
https://www.worldscientific.com/worldscibooks/10.1142/14020#t=suppl

Desk Editors: Soundararajan Raghuraman /Lai Fun Kwong

Typeset by Stallion Press
Email: enquiries@stallionpress.com

Preface

Our age is revolutionary. Just as the appearance of hand-held calculators buried the art of mental arithmetic and logarithmic slide rules, without which a century ago one could scarcely imagine a practicing engineer, we can expect many other arts to recede. These may include cumbersome algebraic and trigonometric transformations, evaluation of complicated integrals, or the analytic solution of differential equations. A good portion of modern mathematical textbooks is repetition from 19th century textbooks, despite the inclusion of modern pseudo-applications. But computers have already taken over the tasks of solving various equations — both algebraic and differential — and provide numerical, analytic, and graphical results. The study of tricks required to evaluate awkward integrals, and so on, will be left to professional mathematicians and theoretical physicists who need to develop new mathematical studies. One of the authors of this book asked a masters-degreed civil engineer "What kinds of mathematics do you use in your practice?" The answer was, "A calculator!" With the continued development of handheld electronics, this answer will surely become more and more typical.

What should the modern engineering or science student learn? Certainly not how to solve artificial exercises they will never apply. Rather they should come to understand what is done in various parts of mathematics, and why. Students can be given tools, but without understanding their proper use is impossible.

We cannot say how the mathematical curriculum for engineering or science students should be changed, except that (a) it should include a good portion of numerical methods, and (b) the cumbersome tricks of the 18th and 19th centuries should be eliminated. It is not clear how to achieve the main goal of education in anything, which is *understanding* of the matter, but it should be done.

This book is an introduction to multivariable calculus (or "vector calculus"). The prerequisites are limited to single-variable calculus, linear algebra, and analytic geometry.

So what should a student know about multivariable calculus, and how is this knowledge to be acquired? The second question is by far the easier one: the presentation must be clear and well motivated, and exercises must be worked. Real conceptual grasp cannot come without problem solving activity, though the exercises themselves need not be technically burdensome.

Our list of topics is also quite conventional: limits, continuity, partial derivatives, extrema, the nabla operator, multiple integrals, line integrals, and surface integrals. However, our classroom experience with engineering students has led to certain choices regarding emphasis and approach. Reasons for a few of these are noted in the following list.

(1) The first obstacle is the relative complexity of multiple limits in comparison with ordinary limits. An iterated limit does not always represent a multiple limit (which may not even exist).

(2) Attention should be drawn to the usefulness of the total differential. In particular the tangent plane helps us understand the behavior of a function at a point.

(3) To avoid difficulty with implicit differentiation, students should bear firmly in mind that the formula $y'(x) = -f_x(x, y)/f_y(x, y)$ works only on the curve $f(x, y) = c$ and only if $y = y(x)$.

(4) While we acknowledge that extrema of multivariable functions must be sought largely through numerical approaches, a solid grasp of the theory is still required for those who will go on to study extremal principles in continuum physics.

(5) The standard definition of the Riemann integral presents an unnecessary hurdle for students at the sophomore level. An expedient alternative is to define the integral as an ordinary limit

$$\lim_{n \to \infty} \sum_{k=1}^{n} f(x_k) \frac{b-a}{n}$$

for a bounded and continuous integrand function f. It can then be asserted that the same limit is obtained for various partitions and for arbitrary selection of intermediate points in the segments Δx_k, even for functions having discontinuities.

(6) The study of each type of integral — single or multiple — can commence with the problem of finding the mass of an object. This shows students that the integral is not an abstract tool but something really useful. Again, the exercises should not be very hard or artificial, as the tasks can all be handled via machine computation.

(7) It is well to stress that the main theorems of vector calculus are steps toward real applications in continuum physics, mechanics, and other field theoretical areas.

In this book we have covered only standard topics in multivariable calculus, leaving numerical calculation of multiple integrals (which is a sophisticated problem even with modern computers) for specialized books and articles.

The authors predict that applied mathematics textbooks will have to shift their emphasis from cookbook procedures toward a real understanding of mathematics. We hope the program realized in the present book is a step in this transition.

We are grateful to our wives, Natasha Lebedeva and Beth Lannon-Cloud, for their understanding and support. Special thanks are due to our colleague Niam Shandi; we are indebted for Niam's kind help in reading the manuscript and for the valuable suggestions she contributed. It remains to express our sincere appreciation to the staff of World Scientific Publishing, including engineering editor Amanda Yun and mathematics editor Lai Fun Kwong.

Leonid P. Lebedev
Universidad Nacional de Colombia, Bogotá, Colombia

Michael J. Cloud
Okemos, Michigan, USA

About the Authors

Michael J. Cloud received a Ph.D. in Electrical Engineering from Michigan State University in 1987. He is Professor Emeritus in the Department of Electrical and Computer Engineering at Lawrence Technological University. Originally from Grand Rapids, Michigan (USA), he now resides in Okemos, Michigan. Dr. Cloud has co-authored 11 books on mathematical engineering.

Leonid P. Lebedev is a Professor at the Department of Mathematics at the National University of Colombia and holds a Doctor of Science (D.Sc.) degree in Physics and Mathematics. With over 40 years of academic and research experience, his expertise lies in applied mathematics and the mechanics of solids. Lebedev earned both his Ph.D. and D.Sc. from Rostov State University, where he specialized in the mechanics of solids and served as a Professor in the Division of Elasticity until 2001.

Throughout his career, Lebedev has made significant contributions to the mathematical challenges of nonlinear mechanics. He has published more than 50 research articles and co-authored 13 books, covering topics such as functional analysis in mechanics, elasticity, and engineering applications of mechanics.

Contents

Chapter 1

Preliminaries: Lines, Trajectories, Surfaces, and 3D Figures

This chapter reviews some material needed for the theory of functions in many variables. It is concerned with the description, through relatively simple formulas, of various spatial or planar figures.

1.1 Straight Lines and Segments in Space

Much of the present book deals with assorted geometric figures and their properties in two- and three-dimensional Euclidean spaces of points.

We shall follow what has been the practice in mechanics since Newton: the use of an absolute space. Euclid's fifth axiom — the parallel postulate — is assumed.[1] Next, in the space we artificially appoint a system of reference coordinates: we fix an origin O and erect a right-handed[2] orthonormal basis $(\mathbf{i}, \mathbf{j}, \mathbf{k})$. Within this system we identify a spatial point A using coordinates (x, y, z). We also introduce the *position vector* of A, denoted $\mathbf{r}_A$, extending from O to A. By identifying the point A with the numerical triple (x, y, z) and thereby with its position vector

$$\mathbf{r} = x\,\mathbf{i} + y\,\mathbf{j} + z\,\mathbf{k}, \tag{1.1}$$

---

[1]This of course posits the existence in the plane of a unique straight line parallel to a given straight line L, and passing through a given point P not on L. We could also suppose that such a line does not exist, obtaining an axiom of Riemann's geometry, or that infinitely many "parallel" straight lines exist as assumed in Lobachevsky's geometry. We do not know which geometry precisely describes the space we live in; it is a difficult question. However, Euclidean geometry is certainly an excellent approximation in the reference frames ordinarily used in engineering and in much of continuum physics.

[2]The handedness of the basis determines the direction of the cross product of two vectors, and hence the formulations of various physical results presented by means of the cross product.

we get a space $\mathbb{R}^3$ that we identify with the initial absolute space. This permits us to draw a picture where the point, the coordinate triple, and the position vector appear together (Fig. 1.1).

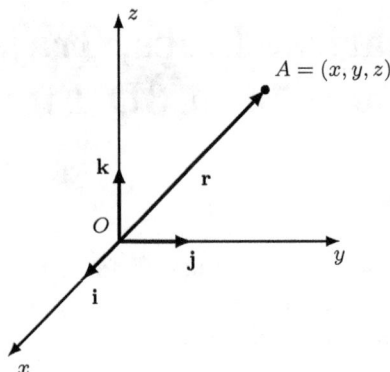

Fig. 1.1   A point A along with its Cartesian coordinates (x, y, z) and position vector $\mathbf{r}$. The basis vectors $\mathbf{i}, \mathbf{j}, \mathbf{k}$ are mutually perpendicular and have unit magnitudes.

Remark 1.1. The space $\mathbb{R}^3$ of linear algebra differs from the space $\mathbb{R}^3$ used here: our position vectors are defined with respect to the absolute space origin, whereas no spatial points exist in the space $\mathbb{R}^3$ of linear algebra. The situation with the initial two-dimensional space, the plane, and its identification with $\mathbb{R}^2$ is analogous.

It should be emphasized that in picturing points and vectors together, we combine two very different objects. One is the space with introduced coordinates; the other is the set of vectors we attach to points. It is only for convenience that we picture points and vectors simultaneously. The vectors are external, additional structures with respect to the point space.   □

From the various possibilities for writing the equations of curves and surfaces, we choose *parametric representation* because of its usefulness in calculating curvilinear and multiple integrals. In the calculation of line and surface integrals, it is the only option for their representation.

A segment in space is defined by its initial point A, with coordinates (x_0, y_0, z_0), and its final point B, with coordinates (x_1, y_1, z_1). See Fig. 1.2. The corresponding endpoint position vectors are

$$\mathbf{r}_A = x_0\,\mathbf{i} + y_0\,\mathbf{j} + z_0\,\mathbf{k} \quad \text{and} \quad \mathbf{r}_B = x_1\,\mathbf{i} + y_1\,\mathbf{j} + z_1\,\mathbf{k},$$

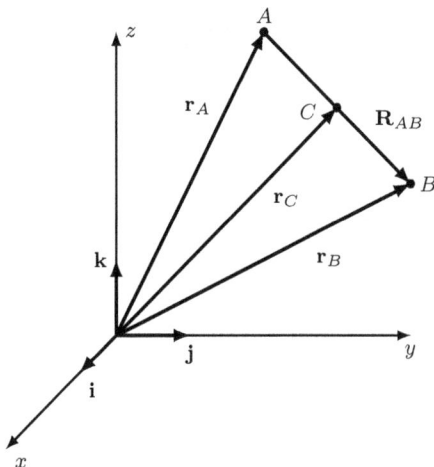

Fig. 1.2   A segment AB in three-dimensional space, along with position vectors identifying its endpoints and a general point C on the segment.

respectively; they define a displacement vector

$$\mathbf{R}_{AB} = (x_1 - x_0)\,\mathbf{i} + (y_1 - y_0)\,\mathbf{j} + (z_1 - z_0)\,\mathbf{k},$$

running parallel to the segment. For an arbitrary point C of the segment, we have position vector

$$\mathbf{r}_C = x\,\mathbf{i} + y\,\mathbf{j} + z\,\mathbf{k}$$

(and so the coordinates (x, y, z)), obtained by adding to $\mathbf{r}_A$ a vector $t\mathbf{R}_{AB}$ for some $t \in [0, 1]$. This yields

$$\begin{cases} x = x_0 + t(x_1 - x_0), \\ y = y_0 + t(y_1 - y_0), \\ z = z_0 + t(z_1 - z_0). \end{cases} \tag{1.2}$$

In vector form these are

$$\mathbf{r}(t) = \mathbf{r}_0 + t\mathbf{R}_{AB}. \tag{1.3}$$

Note that $\mathbf{r}(0) = \mathbf{r}_A$ and $\mathbf{r}(1) = \mathbf{r}_B$. Taking $t \notin [0, 1]$, we access other points on the straight line of which the given segment is a part.

Example 1.1.   Consider the line segment $P_1 P_2$ of Fig. 1.3. Its total length is $d = d_1 + d_2$, and we say that point P divides the segment in the ratio η if $d_1/d_2 = \eta$. An old problem in geometry is how to find P for given endpoints and a desired value of η. We can solve this problem easily using

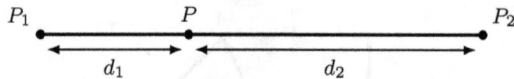

Fig. 1.3   Dividing a line segment.

the vector equation of the segment

$$\mathbf{r}(t) = (1 - t)\,\mathbf{r}_1 + t\,\mathbf{r}_2 \qquad (0 \le t \le 1)$$

where $\mathbf{r}_1$ and $\mathbf{r}_2$ are the positions of P_1 and P_2, respectively. Let us start with the proportion

$$\frac{d_1}{d_1 + d_2} = \frac{t}{1},$$

take reciprocals to get

$$\frac{1}{t} = \frac{d_1 + d_2}{d_1} = 1 + \frac{d_2}{d_1},$$

and set d_1/d_2 equal to the desired ratio η:

$$\frac{1}{t} = 1 + \frac{1}{\eta}.$$

Solving for t, we get the position of P:

$$\mathbf{r}_P = \frac{1}{\eta + 1}\,\mathbf{r}_1 + \frac{\eta}{\eta + 1}\,\mathbf{r}_2.$$

For instance the point P that lies $3/4$ of the way from $\mathbf{r}_1$ to $\mathbf{r}_2$ is found by putting $\eta = 3$. But the most important case is $\eta = 1$, which gives the classical midpoint formula. $\square$

Figure 1.2 does not show the parameter t of (1.2). However, the vector $\mathbf{r} = \mathbf{r}(t)$ represents the position of a point moving with constant velocity in space, and we can regard t as the time variable. Then the segment from A to B becomes the *trajectory* of the moving point.

In the general case, at instant t the coordinates of a point moving through the space $\mathbb{R}^3$ will depend on t. That is, we shall have $x = f(t)$, $y = g(t)$, $z = k(t)$, with some functions f, g, k also denoted by

$$x = x(t), \qquad y = y(t), \qquad z = z(t). \tag{1.4}$$

By marking in $\mathbb{R}^3$ the points given by (1.4) over an interval $[t_0, t_1]$, we get the trajectory of the moving point. This trajectory is a space curve

(Fig. 1.4). For an arbitrary space curve, the variable t should not have temporal meaning; rather, t is called a *parameter* and the representation (1.4) is the *parametric representation* of the curve.

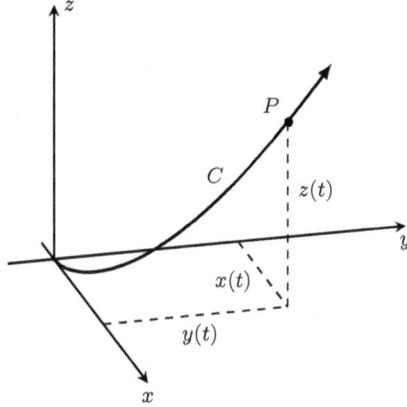

Fig. 1.4   Space curve C as the trajectory of a moving particle P.

Equations (1.4) also define the vectorial representation of the curve:

$$\mathbf{r} = \mathbf{r}(t) \equiv x(t)\,\mathbf{i} + y(t)\,\mathbf{j} + z(t)\,\mathbf{k} \quad \text{for } t \in [t_0, t_1]. \tag{1.5}$$

To each value $t \in [t_0, t_1]$ there corresponds a vector $\mathbf{r}(t) \in \mathbb{R}^3$. By analogy with the case of ordinary functions, we call $\mathbf{r}$ a *vector-valued function* or simply a *vector function*. The domain of definition for a vector function is a segment or other subset of the real axis, and its range is a curve in $\mathbb{R}^3$. (For an abstract "curve," an image of segment $[t_0, t_1]$, the range can lie in $\mathbb{R}^k$ with any positive integer k.) As an image of a one-dimensional set (again, some subset of the real axis), the curve is said to be a *one-dimensional object*. The parametric equations for a curve depend on just one parameter. Later we discuss differentiation and integration of such functions, but let us continue with a few functions of practical importance.

To get the equations for a straight line in the xy-plane, we simply omit the equation for z in (1.2):

$$x = x_0 + t(x_1 - x_0), \qquad y = y_0 + t(y_1 - y_0).$$

This trick is often used to reduce formulas obtained in $\mathbb{R}^3$ to corresponding formulas in $\mathbb{R}^2$.

When a plane curve is given in the form $y = f(x)$, it can also be represented in parametric form as

$$x = t, \qquad y = f(t).$$

Example 1.2. The parabola $y = ax^2$ has parametric and position vector representations

$$\begin{cases} x = t \\ y = at^2 \end{cases} \qquad \mathbf{r}(t) = t\,\mathbf{i} + at^2\,\mathbf{j} \qquad (-\infty < t < \infty).$$

See Fig. 1.5. □

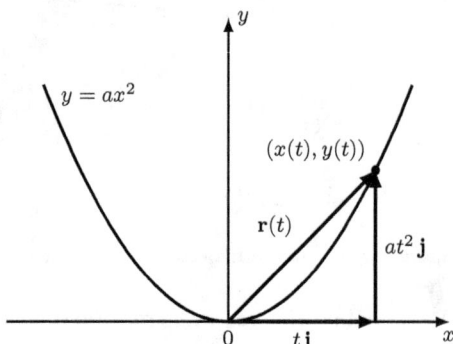

Fig. 1.5   Example 1.2, parameterizing the parabola $y = ax^2$, $a > 0$. Note the "head-to-tail" vector addition that produces the vector $\mathbf{r}(t)$ from its components $t\,\mathbf{i}$ and $at^2\,\mathbf{j}$.

However this practice is unnecessary, as x is a natural parameter here. Usually we write only $y = f(x)$, bearing in mind the first equation $x = x$.

1.2   Circle, Ellipse, Helix

The Circle

Consider the equations for a circle of radius R with center at the origin. In terms of the polar coordinates R and θ they are

$$x = R\cos\theta, \qquad y = R\sin\theta,$$

and give rise to the vector representation

$$\mathbf{r} = R\cos\theta\,\mathbf{i} + R\sin\theta\,\mathbf{j} \tag{1.6}$$

with parameter θ. For an entire circle the angle θ usually runs from 0 to 2π in the counterclockwise sense. But at times we may prefer to keep θ in an interval $[\theta_0, \theta_0 + 2\pi]$ for some real θ_0, preserving one-to-one correspondence between the points of the circle and those of the interval.

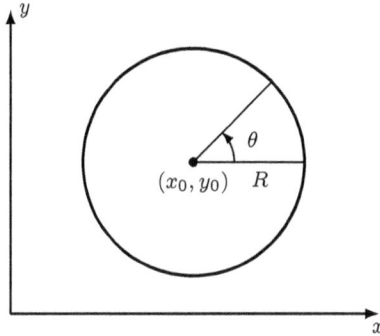

Fig. 1.6   Circle with center (x_0, y_0) and radius R in the xy-plane.

Note that in calculus θ is always expressed in radians, as for other measures the derivative of $\sin\theta$ is not $\cos\theta$. If the center of the circle is shifted to (x_0, y_0) as in Fig. 1.6, the formulas become

$$\left. \begin{array}{l} x = x_0 + R\cos\theta \\ y = y_0 + R\sin\theta \end{array} \right\} \quad \text{or} \quad \mathbf{r} = \mathbf{r}_0 + R\cos\theta\,\mathbf{i} + R\sin\theta\,\mathbf{j}.$$

In certain cases it is possible to deduce simple equations for the shifted circle in terms of radius R and angle θ if we work in polar coordinates.

Example 1.3. For a circle with center $(0, R)$ and radius R, the equation in Cartesian coordinates is $x^2 + (y - R)^2 = R^2$ or $x^2 + y^2 - 2yR = 0$. Setting $x = r\cos\theta$ and $y = r\sin\theta$, we reduce the equation of the circle to $r^2 - 2rR\sin\theta = 0$ or $r = 2R\sin\theta$. □

The Ellipse

From the parametric representation

$$x = a\cos\theta, \qquad y = b\sin\theta \qquad (0 \le \theta < 2\pi) \tag{1.7}$$

with positive constants a, b, we easily establish that

$$\frac{x^2}{a^2} + \frac{y^2}{b^2} = 1. \tag{1.8}$$

Thus equations (1.7) with fixed positive values a, b represent an ellipse centered at the origin and having semiaxes a and b. The ellipse centered at (x_0, y_0) has the parametric equations

$$x = x_0 + a \cos \theta, \qquad y = y_0 + b \sin \theta \qquad (0 \le \theta < 2\pi).$$

See Fig. 1.7. But, as for the circle, θ can occupy any interval of length 2π. Note that when $b = a$, the ellipse is a circle.

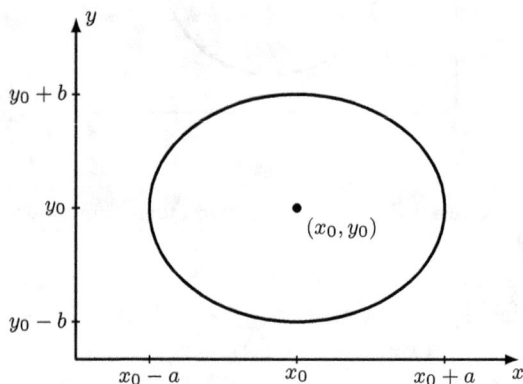

Fig. 1.7   Ellipse with center (x_0, y_0) and semiaxes a and b in the xy-plane.

Example 1.4. Consider the equation

$$9x^2 + 4y^2 - 18x - 16y = 11. \tag{1.9}$$

Factoring and completing the square, we can rewrite (1.9) as

$$9[(x-1)^2 - 1] + 4[(y-2)^2 - 4] = 11$$

or

$$\frac{(x-1)^2}{4} + \frac{(y-2)^2}{9} = 1.$$

Hence a parametric form of (1.9) is

$$x = 1 + 2 \cos \theta, \qquad y = 2 + 3 \cos \theta. \qquad \square$$

The Helix

We obtain the representation for this curve by supplementing (1.6) with a uniform motion along the z-direction:

$$\mathbf{r} = R\cos\theta\,\mathbf{i} + R\sin\theta\,\mathbf{j} + c\theta\,\mathbf{k}. \tag{1.10}$$

So the Cartesian coordinates of a point on the curve are $(R\cos\theta, R\sin\theta, c\theta)$, where c characterizes the *pitch* of the helix. See Fig. 1.8.

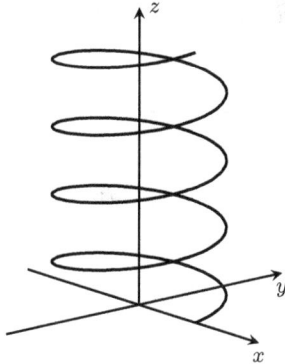

Fig. 1.8   Circular helix coaxial with the z-axis.

Quadratic Curves

A general class of plane curves is represented by the equation

$$a_1 x^2 + a_2 xy + a_3 y^2 + a_4 x + a_5 y = a_6 \tag{1.11}$$

provided at least one of the coefficients a_1, a_2, a_3 is nonzero. In certain cases (e.g., $x^2 + y^2 = -1$) the equation fails to describe a curve; otherwise the result is an ellipse, hyperbola, or parabola. For simplicity we shall set $a_2 = 0$ and consider curves defined by quadratic coefficients a_1 and a_3. If $a_1 a_3 > 0$, the curve is an ellipse; if $a_1 a_3 < 0$, it is a hyperbola; if $a_1 = 0$ or $a_3 = 0$, it is a parabola. The remaining coefficients a_4 and a_5 determine the position of the curve relative to the coordinate origin. Our interest will lie in the parametric representation of selected quadratic curves.

Example 1.5. Let us parameterize the curves (a) $x = 2y^2 - y + 3$, (b) $4x^2 + y^2 - 8x + 2y = 11$, and (c) $4x^2 - y^2 - 8x - 2y = 13$. From the coefficients of x^2 and y^2 we conclude at once that curve (a) is a parabola, curve (b) is an ellipse, and curve (c) is a hyperbola. Parameterization of the parabola is trivial; we can simply set $y = t$ and write $x = 2t^2 - t + 3$. For the ellipse, we first derive a canonical equation, commencing with a few algebraic transformations. The first step is to rewrite the given equation as

$$(4x^2 - 8x) + (y^2 + 2y) = 11.$$

Completing the squares in the parenthetical expressions, we obtain

$$4(x - 1)^2 + (y + 1)^2 = 16.$$

Dividing through by 16 we have

$$\frac{(x - 1)^2}{2^2} + \frac{(y + 1)^2}{4^2} = 1.$$

Comparison with the trigonometric identity $\cos^2\theta + \sin^2\theta = 1$ prompts us to set

$$\frac{x - 1}{2} = \cos\theta, \qquad \frac{y + 1}{4} = \sin\theta.$$

So the parametric equations are

$$x = 1 + 2\cos\theta, \qquad y = -1 + 4\sin\theta.$$

The center of the ellipse is located at $(1, -1)$, and its semiaxes have lengths 2 and 4. Similar techniques can be applied to the hyperbola. The given equation is rewritten as

$$4(x - 1)^2 - (y + 1)^2 = 16$$

which yields

$$\frac{(x - 1)^2}{2^2} - \frac{(y + 1)^2}{4^2} = 1.$$

Comparison with the identity $\cosh^2 t - \sinh^2 t = 1$ prompts us to set

$$\frac{x - 1}{2} = \cosh t, \qquad \frac{y + 1}{4} = \sinh t,$$

and obtain finally

$$x = 1 + 2\cosh t, \qquad y = -1 + 4\sinh t. \qquad \square$$

Note that when considered in $\mathbb{R}^3$, the quadratic equations (1.11) define cylindrical surfaces with axes parallel to the z-axis.

1.3   Derivative of $\mathbf{r} = \mathbf{r}(t)$

Now we wish to discuss the differentiation of vector functions in one variable, and some characteristics of a curve in space.

If a point undergoes rectilinear motion and we know its distance $s(t)$ along the line from a fixed point O on the line as a function of time t, we can obtain its speed as $v(t) = ds(t)/dt$. More generally we can define the *velocity*

$$\mathbf{v}(t) = \frac{d\mathbf{r}(t)}{dt}$$

of a particle moving along trajectory $\mathbf{r} = \mathbf{r}(t)$. It is given by

$$\mathbf{v}(t) = \lim_{\Delta t \to 0} \frac{\mathbf{r}(t + \Delta t) - \mathbf{r}(t)}{\Delta t}$$

$$= \lim_{\Delta t \to 0} \left[\frac{x(t + \Delta t) - x(t)}{\Delta t} \mathbf{i} + \frac{y(t + \Delta t) - y(t)}{\Delta t} \mathbf{j} + \frac{z(t + \Delta t) - z(t)}{\Delta t} \mathbf{k} \right]$$

$$= x'(t)\,\mathbf{i} + y'(t)\,\mathbf{j} + z'(t)\,\mathbf{k}.$$

So now we consider the curve as a trajectory of a moving point; in its parametric representation, time t is a parameter. By applying a limiting idea to Fig. 1.9 (similar to finding the tangent to the graph of an ordinary function $y = f(x)$ at point x), we conclude that the vector $\mathbf{v}(t)$ is tangent to the trajectory $\mathbf{r} = \mathbf{r}(t)$ at time instant t.

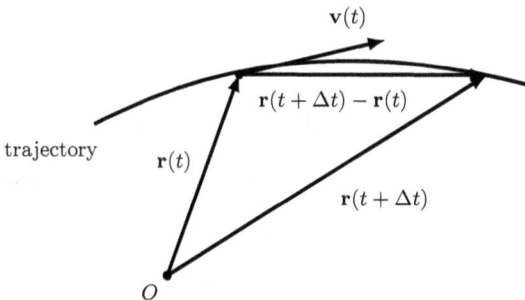

Fig. 1.9   Velocity of a particle at time t.

Clearly $d\mathbf{r}/dt$ is still tangent to curve $\mathbf{r} = \mathbf{r}(t)$ if t is an arbitrary parameter. This fact will be of fundamental importance to us: *the derivative*

$\mathbf{v}$ of $\mathbf{r} = \mathbf{r}(t)$ *is tangent to the curve* $\mathbf{r} = \mathbf{r}(t)$, *independent of the physical meanings of* t *and* $\mathbf{r}(t)$. Recall that if $\mathbf{a}$ is a nonzero vector, then $\mathbf{a}/|\mathbf{a}|$ is a unit vector in the direction of $\mathbf{a}$. Hence the unit tangent $\mathbf{t}(t)$ to the curve at t is

$$\mathbf{t}(t) = \mathbf{r}'(t)/|\mathbf{r}'(t)|. \tag{1.12}$$

Example 1.6. We find the angle between the parabolas $y = x^2$ and $x = y^2$ at point $(1,1)$. See Fig. 1.10. Let us label and parameterize the two curves

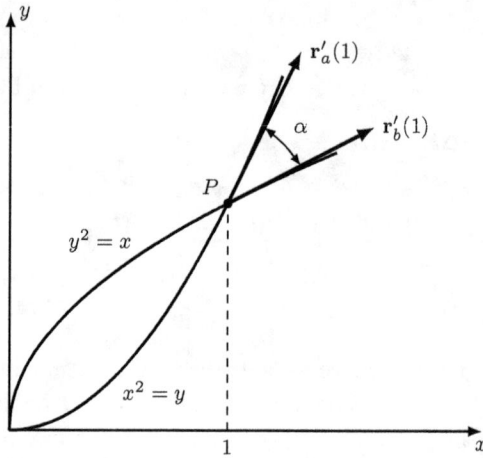

Fig. 1.10   Example 1.6.

as follows:

(a) $y = x^2$ is represented by $\mathbf{r}_a(t) = t\,\mathbf{i} + t^2\,\mathbf{j}$ for $0 \le t \le 1$,

(b) $x = y^2$ is represented by $\mathbf{r}_b(t) = t^2\,\mathbf{i} + t\,\mathbf{j}$ for $0 \le t \le 1$.

The angle α between the curves at P will be obtained from the dot product formula

$$\cos\alpha = \frac{\mathbf{r}_a'(1) \cdot \mathbf{r}_b'(1)}{|\mathbf{r}_a'(1)||\mathbf{r}_b'(1)|}.$$

For both curves the intersection point corresponds to $t = 1$. We have $\mathbf{r}_a'(t) = \mathbf{i} + 2t\,\mathbf{j}$ and $\mathbf{r}_b'(t) = 2t\,\mathbf{i} + \mathbf{j}$ so that $\mathbf{r}_a'(1) = \mathbf{i} + 2\,\mathbf{j}$ and $\mathbf{r}_b'(1) = 2\,\mathbf{i} + \mathbf{j}$. Then

$$\cos\alpha = \frac{(\mathbf{i} + 2\,\mathbf{j}) \cdot (2\,\mathbf{i} + \mathbf{j})}{|\mathbf{i} + 2\,\mathbf{j}|\,|2\,\mathbf{i} + \mathbf{j}|} = \frac{4}{5}.$$

Therefore $\alpha = \cos^{-1}(4/5) = \tan^{-1}(3/4)$. □

We shall follow an older tradition and not show explicitly the domains of functions or the smoothness classes to which the functions belong, unless required for a particular development. Rather, we proceed under the assumption that the functions possess all the properties needed to justify the transformations we perform.

Because differentiating $\mathbf{r}$ with respect to t in Cartesian coordinates reduces to simply differentiating the components of $\mathbf{r}$, we can use the properties of the derivatives for an ordinary function in one variable. In the same way we can introduce the second derivative of $\mathbf{r}$ with respect to t, which is the *acceleration* $\mathbf{a}$ of the point along the trajectory if t is the time variable:

$$\mathbf{a}(t) = \frac{d^2\mathbf{r}(t)}{dt^2} = x''(t)\,\mathbf{i} + y''(t)\,\mathbf{j} + z''(t)\,\mathbf{k}.$$

Example 1.7. Suppose the position of a moving point is given by

$$\mathbf{r} = \mathbf{r}(t) = \mathbf{i} + 4t\,\mathbf{j} - 6t^2\,\mathbf{k},$$

where t is the time variable. Then the velocity and speed of the point are, respectively,

$$\mathbf{v} = d\mathbf{r}/dt = 4\,\mathbf{j} - 12t\,\mathbf{k}, \qquad |\mathbf{v}| = 4(1 + 9t^2)^{1/2},$$

and the acceleration is the constant value $\mathbf{a} = d\mathbf{v}/dt = -12\,\mathbf{k}$. □

It is important to note that, with respect to acceleration, the case of a time-dependent basis set is not so simple. Suppose, for instance, that a point moves on the Earth's surface. The unit basis vectors, being attached to the Earth, depend on the Earth's motion (its rotation about its axis as well as its revolution about the Sun). So the formulas for $\mathbf{v}$ and $\mathbf{a}$ must include the derivatives of basis vectors $\mathbf{i}(t)$, $\mathbf{j}(t)$, and $\mathbf{k}(t)$. This complicates the solution of practical problems in mechanics.

We can successively find higher-order derivatives of $\mathbf{r}$:

$$\frac{d^3\mathbf{r}(t)}{dt^3} = \frac{d}{dt}\frac{d^2\mathbf{r}(t)}{dt^2} = x'''(t)\,\mathbf{i} + y'''(t)\,\mathbf{j} + z'''(t)\,\mathbf{k},$$

and so on.

Derivatives of the Dot and Cross Products

Let $\mathbf{r}_i = x_i\,\mathbf{i} + y_i\,\mathbf{j} + z_i\,\mathbf{k}$ for $i = 1, 2$. Then

$$\mathbf{r}_1 \cdot \mathbf{r}_2 = x_1 x_2 + y_1 y_2 + z_1 z_2 \tag{1.13}$$

and we obtain

$$\frac{d}{dt}[\mathbf{r}_1(t) \cdot \mathbf{r}_2(t)] = x_1'x_2 + y_1'y_2 + z_1'z_2 + x_1x_2' + y_1y_2' + z_1z_2'$$

$$= \mathbf{r}_1'(t) \cdot \mathbf{r}_2(t) + \mathbf{r}_1(t) \cdot \mathbf{r}_2'(t). \tag{1.14}$$

Similarly

$$\frac{d}{dt}[\mathbf{r}_1(t) \times \mathbf{r}_2(t)] = \mathbf{r}_1'(t) \times \mathbf{r}_2(t) + \mathbf{r}_1(t) \times \mathbf{r}_2'(t). \tag{1.15}$$

Let $\mathbf{d}(t)$ be a unit vector for all t so that $\mathbf{d}(t) \cdot \mathbf{d}(t) \equiv 1$. We can of course differentiate (or integrate) an identity. So

$$\frac{d}{dt}[\mathbf{d}(t) \cdot \mathbf{d}(t)] = 0$$

and (1.14) implies

$$\mathbf{d}(t) \cdot \mathbf{d}'(t) = 0.$$

This means $\mathbf{d}'(t)$ and $\mathbf{d}(t)$ are perpendicular.

In particular, if we differentiate the unit tangent $\mathbf{t}(t) = \mathbf{r}'(t)/|\mathbf{r}'(t)|$ to the curve $\mathbf{r} = \mathbf{r}(t)$, we get a vector $\mathbf{N}(t)$ normal to $\mathbf{t}(t)$. The vector

$$\mathbf{n}(t) = \mathbf{N}(t)/|\mathbf{N}(t)| \tag{1.16}$$

is called the *principal unit normal* to the curve at t. There are infinitely many vectors normal to $\mathbf{t}$. From them we select

$$\mathbf{b}(t) = \mathbf{t}(t) \times \mathbf{n}(t), \tag{1.17}$$

called the *binormal* of the curve. The unit vectors $\mathbf{t}(t), \mathbf{n}(t), \mathbf{b}(t)$ constitute a local orthonormal basis at point $\mathbf{r}(t)$, which plays an important role in the theory of curves.

When the parameter t is the length s of the curve measured from a fixed point, the value

$$\kappa(s) = \left| \frac{d\mathbf{t}(s)}{ds} \right| \tag{1.18}$$

is the *curvature* of the curve at a point. A detailed treatment of length calculations will come later. Here we simply mention that for a curve $\mathbf{r} = \mathbf{r}(t)$, the differential of the length function $s = s(t)$ is $ds = |\mathbf{r}'(t)|\, dt$. In particular, for the circle of radius R centered at the origin the length parameter is $s = R\theta$.

For an arbitrary parameter t, the chain rule gives

$$\frac{dt}{dt} = \frac{dt}{ds}\frac{ds}{dt} = \frac{dt}{ds}\,|\mathbf{r}'(t)|$$

so that

$$\kappa(t) = \frac{|\mathbf{t}'(t)|}{|\mathbf{r}'(t)|}. \tag{1.19}$$

In terms of $\mathbf{r} = \mathbf{r}(t)$ it can be shown also that

$$\kappa(t) = \frac{|\mathbf{r}'(t) \times \mathbf{r}''(t)|}{|\mathbf{r}'(t)|^3}. \tag{1.20}$$

The reader is encouraged to supply the proof.

For a plane curve, $\mathbf{n}$ lies in the plane. For a space curve, the vectors $\mathbf{t}$ and $\mathbf{n}$ lie in the *osculating plane*. Intuitively, this is the plane which most neatly "hugs" the curve in a small neighborhood of a given point on the curve.

Example 1.8. For a circle of radius R (equation (1.4)), we find that $R = 1/\kappa$. In fact, a circle of radius $R = 1/\kappa$ provides the best (quadratic) fit to any plane or space curve in a small neighborhood of a point. □

Example 1.9. Let us find the tangent vector, normal, and curvature at a point on a circular helix with parameters R and c. The helix has the equation $\mathbf{r} = R\cos\theta\,\mathbf{i} + R\sin\theta\,\mathbf{j} + c\theta\,\mathbf{k}$. The tangent vector is

$$\mathbf{r}'(\theta) = -R\sin\theta\,\mathbf{i} + R\cos\theta\,\mathbf{j} + c\,\mathbf{k}.$$

The unit tangent is $\mathbf{t}(\theta) = \mathbf{r}'(\theta)/(R^2 + c^2)^{1/2}$. The normal $\mathbf{N}$ is given by

$$\mathbf{N}(\theta) = \mathbf{t}'(\theta) = -R(\cos\theta\,\mathbf{i} + \sin\theta\,\mathbf{j})/(R^2 + c^2)^{1/2}.$$

Finally, the curvature is $\kappa = (R^2 + c^2)^{-1/2}$. If $c = 0$ the helix reduces to a circle of radius R and its curvature to $1/R$. □

1.4 Polar Coordinate System and Coordinate Lines

Polar coordinates should be familiar to the reader but we offer a brief review. For any point $P = (x, y)$ of the plane we introduce the distance from the origin O to P, which is

$$r = (x^2 + y^2)^{1/2},$$

and the angle θ between the x-axis and OP in the counterclockwise sense. The pair (r, θ) gives us the polar coordinates of P. Here $x = r\cos\theta$ and $y = r\sin\theta$. When $r \geq 0$ and $0 \leq \theta < 2\pi$, we observe a one-to-one correspondence between the pairs (x, y) and (r, θ) (except at the origin, but this presents no difficulties in calculating double integrals in polar coordinates).

To get the one-to-one correspondence, the requirement $r \geq 0$ is obligatory. So in drawing the graph of $r = \cos 3\theta$ in Cartesian coordinates, we should be careful when $\cos 3\theta$ is negative: we should not draw the corresponding part of the graph, as r cannot be negative. See Fig. 1.11.

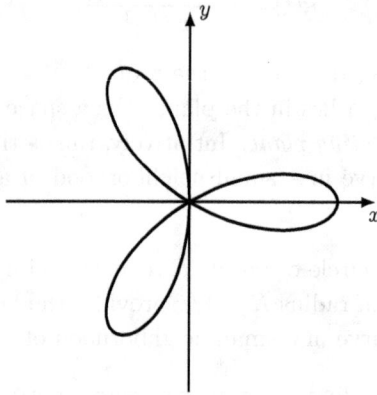

Fig. 1.11 A three-leaved rose.

It is useful to remember that for a one-to-one correspondence between Cartesian and polar coordinates, it suffices to confine θ to any interval of length 2π. So when working in the right half-plane $x \geq 0$, it may be convenient to have $\theta \in [-\pi/2, \pi/2]$ instead of $\theta \in [0, \pi/2] \cup [3\pi/2, 2\pi)$ as would be the case for $0 \leq \theta < 2\pi$.

By fixing one of the polar coordinates, we produce the *coordinate lines* in the polar system: an *r-line* when θ is constant, which is a radial line from the origin, and a *θ-line* when r is constant, which is a circle. See Fig. 1.12. We have said that a plane curve is given by a pair of equations $x = x(t)$ and $y = y(t)$, or in vector form by $\mathbf{r} = \mathbf{r}(t)$. So when we fix $\theta = \theta_0$ or $r = r_0$, we have the description of a corresponding coordinate line $\mathbf{r} = \mathbf{r}(r, \theta_0)$ or $\mathbf{r} = \mathbf{r}(r_0, \theta)$. The derivatives of these vector functions with respect to the parameters $t = r$ or $t = \theta$, with the other coordinate being held fixed, are tangent to the corresponding coordinate lines. We denote them by $\mathbf{r}_r$ and

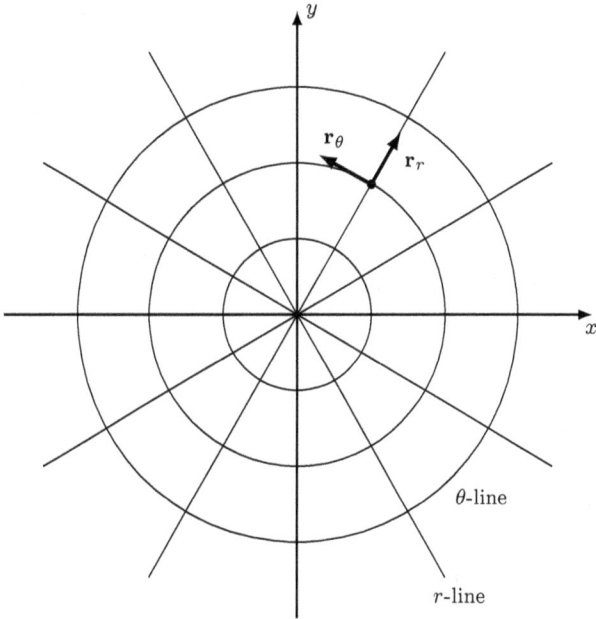

Fig. 1.12 Coordinate lines in polar system.

$\mathbf{r}_\theta$. Later we shall also write them as

$$\mathbf{r}_r = \frac{\partial \mathbf{r}(r,\theta)}{\partial r}, \qquad \mathbf{r}_\theta = \frac{\partial \mathbf{r}(r,\theta)}{\partial \theta},$$

and call them the partial derivatives of $\mathbf{r}$ with respect to r or θ, as they are each taken when the other argument is fixed. The notations $\mathbf{r}_r$ and $\mathbf{r}_\theta$ are also commonly used for the corresponding partial derivatives.

It is easy to calculate the tangent vectors to the coordinate lines:

$$\mathbf{r}_r = \cos\theta\,\mathbf{i} + \sin\theta\,\mathbf{j}, \quad r = \text{constant}, \quad 0 \le \theta < 2\pi,$$

$$\mathbf{r}_\theta = -r\sin\theta\,\mathbf{i} + r\cos\theta\,\mathbf{j}, \quad r \ge 0, \quad \theta = \text{constant}.$$

We see that $\mathbf{r}_r$ is a unit vector codirected with $\mathbf{r} = r(\cos\theta\,\mathbf{i} + \sin\theta\,\mathbf{j})$. Moreover, $\mathbf{r}_r \cdot \mathbf{r}_\theta = 0$ so $\mathbf{r}_r$ and $\mathbf{r}_\theta$ are orthogonal. This is well known from elementary geometry: a tangent at a point of a circle is orthogonal to the radius from the center to the point. So at any point of $\mathbb{R}^2$ the vectors $\mathbf{r}_r$ and $\mathbf{r}_\theta$ constitute a local coordinate basis in which any vector can be spanned. Such a representation of a vector field, using the local basis at each point, is widely used in areas such as hydrodynamics, the theory of elasticity, and electrodynamics.

It is also useful to consider small cells delineated by the r- and $(r+dr)$-lines and the θ- and $(\theta+d\theta)$-lines. As the area of a circular sector of radius R and angle ϕ is $\frac{1}{2}R^2\phi$, the area of the cell is

$$\Delta A = \tfrac{1}{2}(r+dr)^2\, d\theta - \tfrac{1}{2}r^2\, d\theta = r\, dr\, d\theta + \tfrac{1}{2}\, dr^2 d\theta.$$

If dr and $d\theta$ are very small, then the last term on the right is of higher smallness than the first term, and we have approximately

$$\Delta A \approx r\, dr\, d\theta.$$

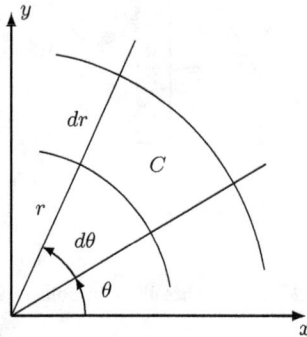

Fig. 1.13   Cell C bounded by four coordinate lines.

The same area value ΔA for small dr and $d\theta$ can be obtained in another way (Fig. 1.13). The cell C for $r \neq 0$ is nearly a rectangle; here we replace the part of the circumference $r\, d\theta$ by the perpendicular to the radius r of length $r\, d\theta$ and get a rectangular area $r\, dr\, d\theta$. This quantity we shall denote by

$$dA = r\, dr\, d\theta. \tag{1.21}$$

It will be used to calculate double integrals in polar coordinates. Note that, unlike Cartesian coordinates in which dA, the area of an elementary rectangular cell with sides dx and dy, is $dx\, dy$, in polar coordinates we have an additional factor r depending on the position of the point. Such a factor appears for dA, the area of an elementary cell constituted by coordinate lines (and for elementary cell volumes constituted by coordinate surfaces in $\mathbb{R}^3$), in any non-Cartesian coordinate system.

1.5 Cylindrical and Spherical Coordinate Systems and Their Coordinate Lines

Cylindrical Coordinates

It is easy to extend polar coordinates to cylindrical coordinates in $\mathbb{R}^3$. In the xy-plane we introduce polar coordinates, then bring in the third coordinate by means of the z-axis with its unit directional vector $\mathbf{k}$. So between Cartesian and cylindrical coordinates we have the relations

$$x = r \cos \theta, \qquad y = r \sin \theta, \qquad z = z, \qquad (1.22)$$

which determine the cylindrical coordinates (r, θ, z) of the point with Cartesian coordinates (x, y, z). The restrictions on r and θ in polar coordinates hold in cylindrical coordinates: $r \geq 0$, and θ is confined to a range of length 2π, usually to $[0, 2\pi)$. The third coordinate $z \in (-\infty, \infty)$.

In cylindrical coordinates the position vector is given by

$$\mathbf{r} = r \cos \theta \, \mathbf{i} + r \sin \theta \, \mathbf{j} + z \, \mathbf{k}. \qquad (1.23)$$

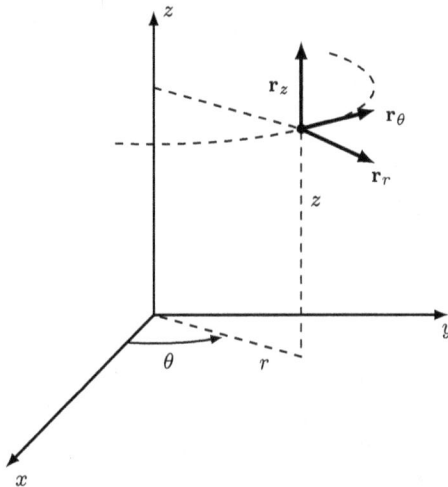

Fig. 1.14 Cylindrical coordinate system.

We get the coordinate lines of the cylindrical coordinate system by fixing two of three coordinates. The tangent vectors to the coordinate lines at each space point are

$$\mathbf{r}_r = \cos\theta\,\mathbf{i} + \sin\theta\,\mathbf{j}, \qquad \mathbf{r}_\theta = -r\sin\theta\,\mathbf{i} + r\cos\theta\,\mathbf{j}, \qquad \mathbf{r}_z = \mathbf{k}.$$

The local coordinate vectors $(\mathbf{r}_r, \mathbf{r}_\theta, \mathbf{r}_z)$ constitute a right-handed orthogonal basis trihedron (Fig. 1.14).

By fixing just one cylindrical coordinate we produce a *coordinate surface*. For a fixed r it is a circular cylinder $x^2 + y^2 = r^2$. The surface $\theta = $ constant is a half-plane with the z-axis as its edge. For $z = $ constant we get a plane parallel to the xy-plane. These surfaces bound elementary cells, and by taking such a cell at a point with coordinates (r, θ, z) and having infinitesimal sides we can evaluate triple integrals in the cylindrical system. It is clear that the cell volume is

$$dV = r\,dr\,d\theta\,dz. \qquad (1.24)$$

Typically we use the cylindrical coordinate system to calculate various characteristics of a body if its shape contains cylindrical elements like a portion of a circular cylinder between two planes parallel to the xy-plane.

Sometimes a cylindrical body will have its main axis along the x-axis or the y-axis. In this case we should transform the z-axis of the cylindrical system to one of these two axes. A cyclic change of coordinate axes

$$(x, y, z) \to (z, x, y) \to (y, z, x)$$

will preserve right-hand orientation in the new cylindrical system. For example, if we take y as the cylinder axis the formulas relating Cartesian and the new cylindrical coordinates (r, θ, y) are

$$z = r\cos\theta, \qquad x = r\sin\theta, \qquad y = y.$$

Spherical Coordinates

The familiar spherical coordinate system is shown in Fig. 1.15. A point is determined by the distance ρ from the origin and two angles (ϕ, θ). By fixing ρ we obtain a sphere $\rho = $ constant. If furthermore one of the two angles is fixed, we generate lines analogous to the geographic parallels and meridians. However, in spherical coordinates the angles are measured in radians and the "zero point" for the parallels is the "North pole."

Let us consider this more precisely. First we show that the relation between (x, y, z) and (ρ, ϕ, θ) is given by the equations

$$x = \rho\sin\phi\cos\theta, \qquad y = \rho\sin\phi\sin\theta, \qquad z = \rho\cos\phi. \qquad (1.25)$$

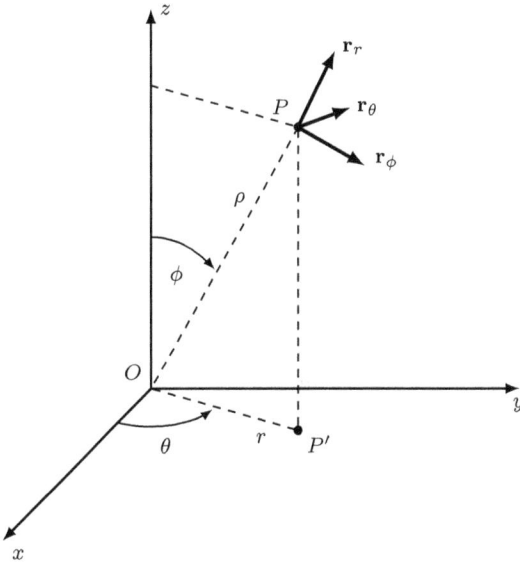

Fig. 1.15 Spherical coordinate system.

To get these, we draw the sphere through the point $P = (x, y, z)$, centered at the origin and having radius

$$\rho = (x^2 + y^2 + z^2)^{1/2}. \tag{1.26}$$

This provides the first spherical coordinate ρ of P. Fixing ρ, we get a spherical coordinate surface. Next, the angle between the z-axis and OP we denote by ϕ. Note that $z = \rho \cos \phi$. The values of ϕ lie in the interval $[0, \pi]$. The distance from P to Oz is $r = (x^2 + y^2)^{1/2} = \rho \sin \phi$. If we draw the projection P' of P onto the xy-plane, then using the xy-plane with the polar coordinate system we get the third spherical coordinate of P, the angle θ from Ox to OP', and

$$x = r \cos \theta = \rho \sin \phi \cos \theta, \qquad y = r \sin \theta = \rho \sin \phi \sin \theta.$$

We have noted the restrictions $\rho \geq 0$ and $0 \leq \phi \leq \pi$. For the values of θ we can use any interval of length 2π; it is usually taken as $[0, 2\pi)$, but this is not obligatory.

By fixing two of three coordinates (ρ, ϕ, θ), we produce a coordinate line. Fixing $\rho = \rho_0$ and $\phi = \phi_0$ on the sphere $\rho = \rho_0$, we get a latitude line. Fixing $\rho = \rho_0$ and $\theta = \theta_0$, we get a longitude line. Fixing two angles, we initiate a ray from the origin. At each point, the tangent vectors to the coordinate lines constitute an orthogonal trihedron $\mathbf{r}_\rho, \mathbf{r}_\phi, \mathbf{r}_\theta$. Its orthogonality is clear from elementary geometry but can of course be checked analytically. For $\rho \neq 1$ the trihedron is not orthonormal but it constitutes a local right-handed basis which changes from point to point. In many application problems the order of $(\mathbf{r}_r, \mathbf{r}_\phi, \mathbf{r}_\theta)$ is important.

We can also produce the coordinate surfaces by fixing one of the three coordinates at a time. For $\rho =$ constant it is a sphere, for $\phi =$ constant it is a circular cone with vertex at O and aligned on the z-axis, and for $\theta =$ constant it is a semiplane with edge on the z-axis.

The volume of an "infinitesimal" cell bounded by the spheres $\rho = \rho_0$ and $\rho = \rho_0 + d\rho$, cones $\phi = \phi_0$ and $\phi = \phi_0 + d\phi$, and half-planes $\theta = \theta_0$ and $\theta = \theta_0 + d\theta$ is given by

$$dV = \rho_0^2 \sin \phi_0 \, d\rho \, d\phi \, d\theta.$$

We leave the deduction of this for Section 4.5.

1.6   The Plane

In mathematics, certain familiar objects are formally left undefined. These include points, lines, and planes. Mathematicians did state axioms pertaining to these objects: two straight lines on a plane can intersect at a point or coincide; through two distinct points there can pass only one straight line, etc. Upon combining this situation with the hurdles involved with a space that is itself logically undefinable, we arrive at the scenario in which the ancient Greeks found themselves. They responded by declaring that certain things are obvious. But modern mathematicians have sought complete abstractness, insisting that geometry should be presented in such a way that intuitions regarding points, straight lines, planes — and space itself — are unnecessary. The famous David Hilbert even wrote a textbook based on this approach, though it may not be suitable for beginners. Isaac Newton addressed the problem of "defining" space and time by terming them *absolute*. This of course did not resolve the issue of definition either. But, like the Greeks, modern engineers can still address many real-life problems without access to rigorous mathematical definitions.

In $\mathbb{R}^3$ we can write the equation for a plane if we know the coordinates (x_0, y_0, z_0) of a point on the plane and the components (a, b, c) of a vector normal to the plane. Indeed, let (x, y, z) be the coordinates of an arbitrary point on the plane. Then the vector (a, b, c) and the displacement vector

$(x - x_0, y - y_0, z - z_0)$ are orthogonal:

$$a(x - x_0) + b(y - y_0) + c(z - z_0) = 0. \tag{1.27}$$

This is the equation of the plane through (x_0, y_0, z_0) with normal (a, b, c); see Fig. 1.16. There are various equations for the same plane, but the coefficients (a, b, c), the components of a normal, are defined to within a constant factor.

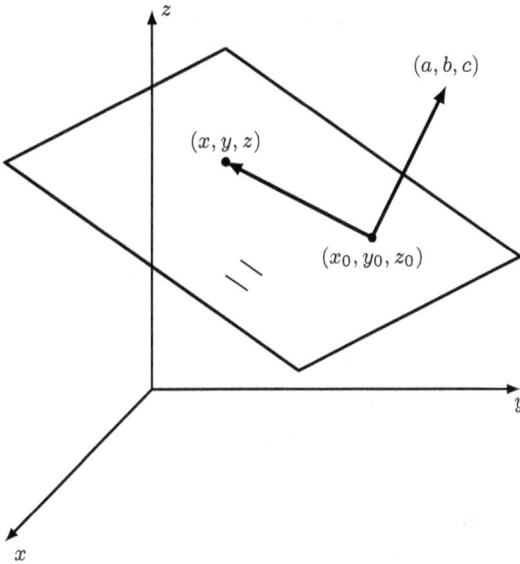

Fig. 1.16   A plane in 3D-space, determined by a single point (x_0, y_0, z_0) and a normal vector (a, b, c).

We have noted that, in vector calculus, parametric forms of representation are preferred. For a surface this requires the introduction of two parameters. For the plane these can be two of the three coordinates (x, y, z), such that for the third coordinate, the corresponding normal component has nonzero value. If $c \neq 0$, say, then the parametric equations with two parameters u, v are

$$x = u, \quad y = v, \quad z = z_0 - c^{-1}[a(u - x_0) + b(v - y_0)].$$

Typically the same letters x and y are used instead:

$$x = x, \quad y = y, \quad z = z_0 - c^{-1}[a(x - x_0) + b(y - y_0)].$$

More generally, for a surface given by the equation $z = f(x, y)$ one of its parametric representations is

$$x = x, \qquad y = y, \qquad z = f(x, y). \qquad (1.28)$$

Here we should not place trivial equations $x = x$, $y = y$ but leave them in mind as they are necessary for the parametric representation.

In some problems we must write the equation of a plane knowing three of its points (x_0, y_0, z_0), (x_1, y_1, z_1), and (x_2, y_2, z_2), not all on the same straight line. Three such points constitute two non-collinear vectors

$$(x_1 - x_0, y_1 - y_0, z_1 - z_0) \quad \text{and} \quad (x_2 - x_0, y_2 - y_0, z_2 - z_0).$$

Let (x, y, z) be an arbitrary point on the plane. Then the vectors

$$(x - x_0, y - y_0, z - z_0),$$
$$(x_1 - x_0, y_1 - y_0, z_1 - z_0),$$
$$(x_2 - x_0, y_2 - y_0, z_2 - z_0),$$

of Fig. 1.17 lie in the plane. But the volume of the parallelepiped spanned by three coplanar vectors must vanish:

$$\begin{vmatrix} x - x_0 & y - y_0 & z - z_0 \\ x_1 - x_0 & y_1 - y_0 & z_1 - z_0 \\ x_2 - x_0 & y_2 - y_0 & z_2 - z_0 \end{vmatrix} = 0.$$

This is the equation of the plane containing points (x_0, y_0, z_0), (x_1, y_1, z_1), and (x_2, y_2, z_2). We note again that the coefficients of x, y, z after expansion of the determinant are the components of the normal vector. They are

$$a = \begin{vmatrix} y_1 - y_0 & z_1 - z_0 \\ y_2 - y_0 & z_2 - z_0 \end{vmatrix}, \qquad b = - \begin{vmatrix} x_1 - x_0 & z_1 - z_0 \\ x_2 - x_0 & z_2 - z_0 \end{vmatrix},$$

and

$$c = \begin{vmatrix} x_1 - x_0 & y_1 - y_0 \\ x_2 - x_0 & y_2 - y_0 \end{vmatrix}.$$

Sometimes we can determine the equation of a plane from the geometrical sense of its coefficients.

Example 1.10. Let us find the equation of the plane containing points $(2, 0, 0)$, $(0, 2, 0)$, and $(0, 0, 2)$. All planes passing through $(2, 0, 0)$ have equations of the form

$$a(x - 2) + b(y - 0) + c(z - 0) = 0,$$

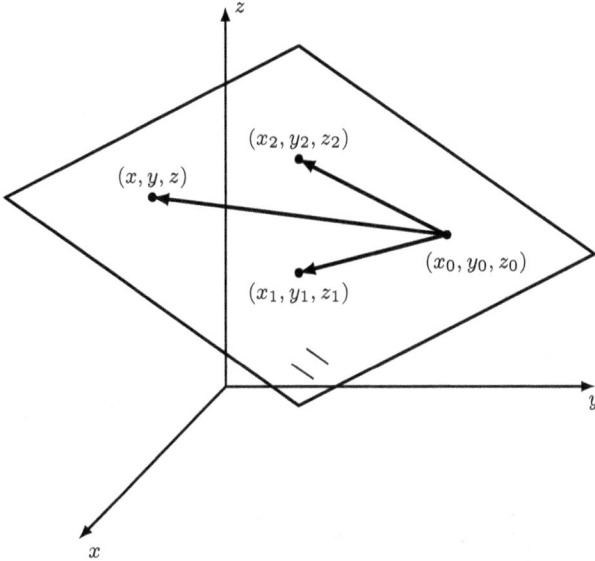

Fig. 1.17   A plane determined by three non-collinear points in 3-space.

where (a, b, c) are components of a normal vector. By the symmetrical way in which the given plane crosses the three coordinate axes, the components of the normal must be equal and we can take $a = b = c = 1$. Hence the equation is $(x - 2) + y + z = 0$. □

1.7   Surfaces and the Functions in Two Variables

If a plane is not parallel to the z-axis, its equation can be written in the form

$$z = a_1 x + a_2 y + a_3$$

with constants a_k; this fits the more general form

$$z = f(x, y).$$

Hence we introduce a function f dependent on two variables x and y. This means, of course, that to each point (x, y) of the domain D of f there corresponds a unique numerical value $f(x,y)$. The set R of all values $f(x,y)$ corresponding to $(x,y) \in D$ is the range of f.[3]

The set of points $(x,y,f(x,y))$ for $(x,y) \in D$ constitutes the graph of f in space and is of course a *surface*. Its equation in vector form,

$$\mathbf{r}(x,y) = x\,\mathbf{i} + y\,\mathbf{j} + f(x,y)\,\mathbf{k}, \tag{1.29}$$

yields a vector-valued function $\mathbf{r}(x,y)$ assigning to each $(x,y) \in D$ a unique vector value given by the right-hand side. More generally surfaces are described by relations

$$x = x(u,v), \qquad y = y(u,v), \qquad z = z(u,v), \tag{1.30}$$

assigning to each point (u,v) in some planar domain of the parameters u, v a unique point $(x(u,v), y(u,v), z(u,v))$ of the surface S in space (Fig. 1.18). In vectorial form,

$$\mathbf{r} = \mathbf{r}(u,v) = x(u,v)\,\mathbf{i} + y(u,v)\,\mathbf{j} + z(u,v)\,\mathbf{k}. \tag{1.31}$$

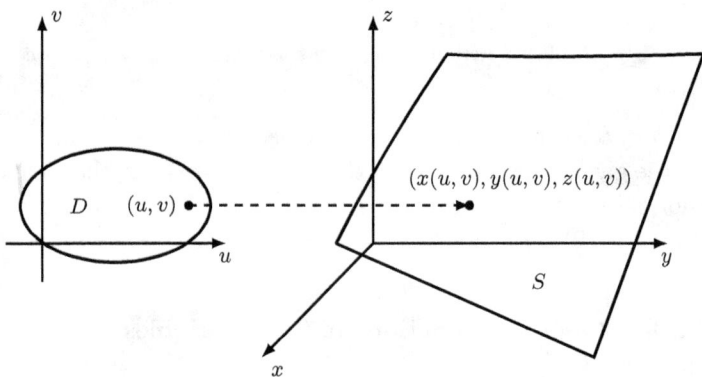

Fig. 1.18   Surface S as the image of a domain D in the uv-plane.

As S is defined by two parameters u and v, a surface is said to be a *two-dimensional (2D) object* in space. The parameters (u,v) are also called the *inner coordinates* in S. When one parameter is fixed, say $v = v_0$, we have the vector equation of the curve

$$\mathbf{r}(u,v_0) = x(u,v_0)\,\mathbf{i} + y(u,v_0)\,\mathbf{j} + z(u,v_0)\,\mathbf{k}$$

---

[3] A real-valued function in n variables $x_1, \ldots, x_n$ is often denoted $f \colon D \subset \mathbb{R}^n \to \mathbb{R}$ but we shall avoid this notation to the extent possible.

with variable u, called a *coordinate curve* on S. Familiar examples are the parallels and meridians on the globe, a nearly spherical surface representing the Earth.

Example 1.11. The elliptic cylinder

$$\frac{x^2}{a^2} + \frac{y^2}{b^2} = 1$$

can be parameterized as $x = a\cos u$, $y = b\sin u$, $z = v$, hence has position vector representation

$$\mathbf{r}(u,v) = a\cos u\,\mathbf{i} + b\sin u\,\mathbf{j} + v\,\mathbf{k}.$$

The two types of coordinate curves on the surface are

$$\mathbf{r}(u,v_0) = a\cos u\,\mathbf{i} + b\sin u\,\mathbf{j} + v_0\,\mathbf{k},$$
$$\mathbf{r}(u_0,v) = a\cos u_0\,\mathbf{i} + b\sin u_0\,\mathbf{j} + v\,\mathbf{k},$$

which are ellipses and vertical lines, respectively. □

We have seen that many definitions and ideas from single-variable calculus can be transferred to multivariable calculus with evident changes. This will occur in the following section as well.

1.8 Quadric Surfaces

Among the surfaces used in engineering practice are the quadric surfaces. The conic sections — circles, ellipses, hyperbolas, and parabolas — were known to the ancient Greeks. The "Mathematica" of Pythagoras consisted of the theory of numbers, musical harmony, and astronomy. Afterwards mathematics as a science lost music and astronomy but gained algebra. A grand step in connecting algebra with geometry was produced by René Descartes who showed that the conic sections all have equations expressed as quadratics: we have

$$x^2 + y^2 = R^2, \qquad \frac{x^2}{a^2} + \frac{y^2}{b^2} = 1, \qquad \frac{x^2}{a^2} - \frac{y^2}{b^2} = 1, \qquad y = ax^2$$

for a circle, ellipse, hyperbola, and parabola respectively. These equations, extended to three dimensions, produce quadric surfaces such as spheres

$$x^2 + y^2 + z^2 = R^2, \tag{1.32}$$

ellipsoids

$$\frac{x^2}{a^2} + \frac{y^2}{b^2} + \frac{z^2}{c^2} = 1, \tag{1.33}$$

hyperboloids

$$\frac{x^2}{a^2} - \frac{y^2}{b^2} \pm \frac{z^2}{c^2} = 1, \tag{1.34}$$

and paraboloids

$$z = ax^2 + by^2. \tag{1.35}$$

Another example of a quadric surface is the elliptic cone

$$z^2 = \frac{x^2}{a^2} + \frac{y^2}{b^2}. \tag{1.36}$$

With $a^2 = b^2 = c$ it becomes a circular cone $z^2 = c(x^2 + y^2)$. Clearly these are all special cases of a general quadric equation

$$\sum_{i,j=1}^{3} a_{ij}x_i x_j + \sum_{i=1}^{3} b_i x_i = c$$

whose quadratic part, depending on the coefficients, can be reduced to one of the main types of quadric surfaces (1.32)–(1.36) by a linear transformation and whose linear terms determine the position of the figure with respect to the origin.

The equations of many surfaces used in this book have a simplified form

$$ax^2 + by^2 + cz^2 + a_1 x + b_1 y + c_1 z = d \tag{1.37}$$

with coefficients $a, b, c, a_1, b_1, c_1, d$. The values a, b, c define the type of surface, and all the coefficients determine the surface shape and position in space. Note that some equations (e.g., $x^2 + y^2 + z^2 = -1$) do not define a surface.

Here is a useful classification of some quadric surfaces (we do not mention the cases when the surface does not exist).

(a) $a = b = c$. Equation (1.37) describes a sphere.

(b) a, b, c are nonzero of the same sign. Equation (1.37) describes an ellipsoid.

(c) a, b, c are nonzero and have different signs. Equation (1.37) describes a hyperboloid (or a cone for appropriate coefficients).

(d) One of a, b, c is zero. Equation (1.37) describes a paraboloid.

(e) Equation (1.37) does not contain a variable x, y or z. It then describes a cylinder with axis parallel to the missing variable.

Example 1.12. Parametric equations for the ellipsoid (1.33) are $z = c \cos \phi$, $x = a \sin \phi \cos \theta$, and $y = b \sin \phi \sin \theta$. □

Example 1.13. A hyperboloid of one sheet can be represented parametrically as

$$\mathbf{r} = a \cosh u \cos v \, \mathbf{i} + b \cosh u \sin v \, \mathbf{j} + c \sinh u \, \mathbf{k}$$

where a, b, c are constants and u, v are the parameters. □

1.9 Sketching Surfaces

Simple drawings by hand can be of great assistance in thinking and problem solving. While photographic realism is certainly not required, a sketch must capture the essence of the situation under consideration. The following examples illustrate a few of the many techniques available.

Example 1.14. It is often necessary to envision how two given surfaces intersect. Figure 1.19 indicates how the first-octant portion of an ellipsoid is cut by two planes. Some intuition can be developed by envisioning the

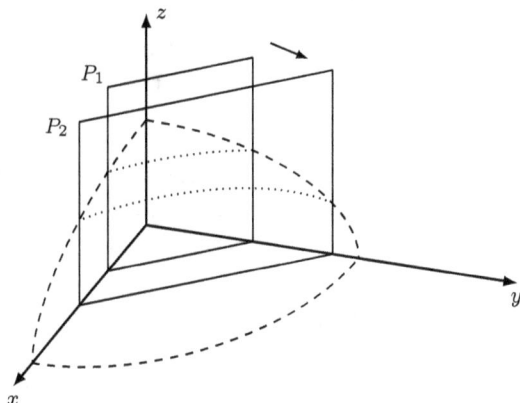

Fig. 1.19 Portion of an ellipsoid (dashed lines), cut by two vertical planes P_1 and P_2 (solid lines) in two curves (dotted lines). The ellipsoid $x^2/9 + y^2/4 + z^2 = 1$ was chosen for illustration, along with the planes $x + y = 1$ and $x + y = 7/4$, respectively.

intersection curve as the first plane slides toward the second. This idea is
encouraged by the arrow at upper right.   □

Example 1.15. The intersections of a surface with a sequence of regularly-
spaced parallel planes can help us envision the surface. Figure 1.20 shows a
stack of ellipses produced by taking cross sections of the elliptic paraboloid

$$2cz = \frac{x^2}{a^2} + \frac{y^2}{b^2} \tag{1.38}$$

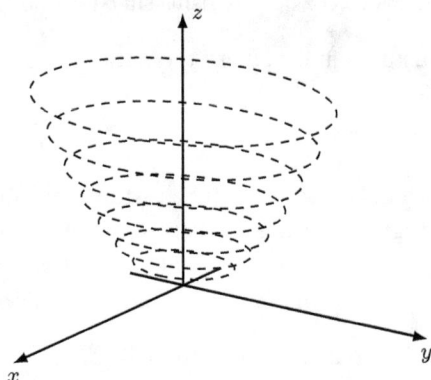

Fig. 1.20   Planar cuts through the surface (1.38).

with planes of the form $z = $ constant. From this quickly sketched "framework," we can get a reasonable idea of what the elliptic paraboloid looks
like. Some practice with this technique is provided in Problem 1.47.   □

Example 1.16. It may be enough to indicate the edges of some chosen
portion of the surface. In Fig. 1.21 this is done for the portion of the
surface $z = (x+y)^2$ over the ranges $0 \le x \le 1$ and $0 \le y \le 1$.   □

1.10   Surfaces of Revolution

Consider the surface generated by revolving the graph of a nonnegative
function $z = h(y)$ about the y-axis (Fig. 1.22). Because a section through
y parallel to the zx-plane yields a circle of radius $h(y)$, the surface has
equation

$$(x^2 + z^2)^{1/2} = h(y). \tag{1.39}$$

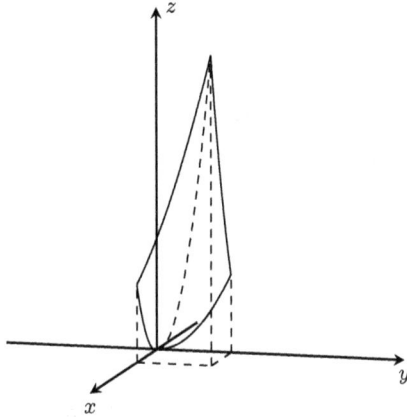

Fig. 1.21 Example 1.16: the surface $z = (x + y)^2$.

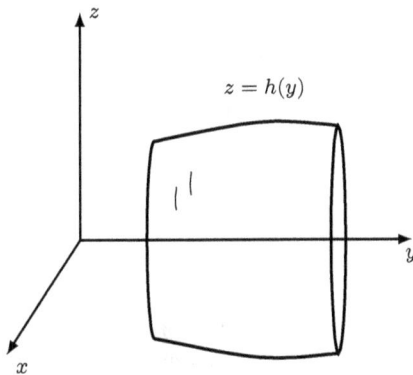

Fig. 1.22 Surface of revolution with respect to y-axis generated by the curve $z = h(y)$ in yz-plane.

In parametric form this is

$$z = h(y)\cos\theta, \qquad x = h(y)\sin\theta, \qquad y = y \qquad (1.40)$$

with two parameters y and θ.

Similarly, for a surface of revolution with respect to the x-axis we start with a curve $z = h(x)$. Rotating it about x-axis we get the equation

$$(y^2 + z^2)^{1/2} = h(x) \qquad (1.41)$$

or in parametric form

$$y = h(x)\cos\theta, \qquad z = h(x)\sin\theta, \qquad x = x. \qquad (1.42)$$

Finally, a surface of revolution with respect to the z-axis (the hyperboloid of Example 1.13, for instance) starts with a curve $y = h(z)$ in the yz-plane. This leads to the equation

$$(x^2 + y^2)^{1/2} = h(z) \qquad (1.43)$$

and the parametric forms

$$x = h(z)\cos\theta, \qquad y = h(z)\sin\theta, \qquad z = z \qquad (1.44)$$

with parameters z and θ. A general equation of a surface of revolution about the z-axis takes the form

$$F((x^2 + y^2)^{1/2}, z) = 0. \qquad (1.45)$$

Similarly, the general forms for the other types are

$$F((x^2 + z^2)^{1/2}, y) = 0, \qquad F((y^2 + z^2)^{1/2}, x) = 0. \qquad (1.46)$$

Having finished this introductory chapter, we can proceed to vector calculus.

1.11   Problems

3-Space and Vectors

1.1 In this book we often use *Cartesian products* to denote regions of space. Recall that the Cartesian product of two nonempty sets A and B is defined as

$$A \times B = \{(a, b) : a \in A, \ b \in B\}.$$

The Cartesian product $A \times B \times C$ of three sets is defined analogously. Sketch the following regions:

(a) the square $[0, 1]^2 = [0, 1] \times [0, 1]$ in the xy-plane

(b) the rectangle $[-1, 0] \times [1, 3]$ in the xy-plane

(c) the cube $[0, 1]^3 = [0, 1] \times [0, 1] \times [0, 1]$ in 3-space

1.2 A point $P = (x, y)$ is given in Cartesian coordinates in the xy-plane. Find its reflection through (a) the x-axis, (b) the y-axis, (c) the origin, and (d) the line $y = x$.

1.3 Sketch in three space dimensions the graph of each equation:

(a) $z = 0$

(b) $y = 1$

(c) $x = \pi$

(d) $y = x$

(e) $y = x^2$

(f) $z = -x$

(g) $y = z - 1$

(h) $x + 2y - z = 1$

(i) $z = x^2 + y^2$

(j) $z = x^2 - y^2$

(k) $x^2 + y^2/4 + z^2/9 = 1$

(l) $(x - 1)^2 + (y + 2)^2 + (z - 2)^2 = 4$

1.4 Given the Cartesian coordinates of each point, convert to both cylindrical coordinates and spherical coordinates:

(a) $(1, 0, 0)$

(b) $(0, 1, 0)$

(c) $(0, 1, 1)$

(d) $(1, 0, 1)$

(e) $(-1, 0, 1)$

(f) $(1, 1, 1)$

1.5 Given the cylindrical coordinates of each point, convert to spherical coordinates:

(a) $P(r, \theta, z) = (2, \pi/2, 1)$

(b) $P(r, \theta, z) = (1, \pi/4, -1)$

1.6 Given the spherical coordinates of each point, convert to cylindrical coordinates:

(a) $P(\rho, \phi, \theta) = (3, \pi/2, \pi)$

(b) $P(\rho, \phi, \theta) = (1, \pi/4, \pi/2)$

1.7 Derive an expression for the distance between two points in terms of their cylindrical coordinates.

1.8 Write down the position vector of each point:

(a) $P(x, y, z) = (1, -1, 2)$

(b) $P(r, \theta, z) = (1, \pi/4, -1)$

(c) $P(r, \theta, z) = (2, 3\pi/2, 2)$

(d) $P(\rho, \phi, \theta) = (1, \pi/2, \pi/2)$

1.9 For each given vector $\mathbf{A}$, find $|\mathbf{A}|$ and a unit vector in the direction of $\mathbf{A}$:

(a) $\mathbf{A} = \mathbf{i} + 2\mathbf{j}$

(b) $\mathbf{A} = 2\mathbf{i} - 3\mathbf{k}$

(c) $\mathbf{A} = 4\mathbf{i} + 2\mathbf{j} + \mathbf{k}$

(d) $\mathbf{A} = -\mathbf{i} + 3\mathbf{j} - \mathbf{k}$

1.10 For each given pair of vectors $\mathbf{A}$ and $\mathbf{B}$, calculate $-\mathbf{A}$, $2\mathbf{B}$, $\mathbf{A} + \mathbf{B}$, $\mathbf{A} - \mathbf{B}$, $\mathbf{A} \cdot \mathbf{B}$, and $\mathbf{A} \times \mathbf{B}$:

(a) $\mathbf{A} = \mathbf{i} + 2\mathbf{j}, \mathbf{B} = \mathbf{i} - 2\mathbf{j}$ (c) $\mathbf{A} = \mathbf{i} + \mathbf{j} - 2\mathbf{k}, \mathbf{B} = -\mathbf{i} + \mathbf{j} + \mathbf{k}$

(b) $\mathbf{A} = 2\mathbf{i} - 3\mathbf{k}, \mathbf{B} = \mathbf{j} - 2\mathbf{k}$ (d) $\mathbf{A} = 2\mathbf{i} + \mathbf{j} + \mathbf{k}, \mathbf{B} = 2\mathbf{i} - \mathbf{j} - \mathbf{k}$

1.11 Give a geometric meaning of the cross product of two vectors, each of which carries units of length.

1.12 For the given vectors $\mathbf{A}, \mathbf{B}, \mathbf{C}$, calculate $\mathbf{A} \cdot (\mathbf{B} \times \mathbf{C})$ and $\mathbf{A} \times (\mathbf{B} \times \mathbf{C})$:

(a) $\mathbf{A} = \mathbf{i} + 2\mathbf{j}, \mathbf{B} = \mathbf{i} - 2\mathbf{j}, \mathbf{C} = 2\mathbf{i} - 3\mathbf{j} + \mathbf{k}$

(b) $\mathbf{A} = \mathbf{i} + 2\mathbf{j} + \mathbf{k}, \mathbf{B} = \mathbf{i} + 2\mathbf{j}, \mathbf{C} = 2\mathbf{i} + \mathbf{j} + \mathbf{k}$

(c) $\mathbf{A} = -\mathbf{i} + \mathbf{j} - \mathbf{k}, \mathbf{B} = \mathbf{i} + 2\mathbf{j} + \mathbf{k}, \mathbf{C} = -2\mathbf{i} + \mathbf{k}$

1.13 Is it true in general that $(\mathbf{A} \times \mathbf{B}) \times \mathbf{C} = \mathbf{A} \times (\mathbf{B} \times \mathbf{C})$? If not, state a counterexample.

1.14 In each case decide whether $\mathbf{A}$ and $\mathbf{B}$ are parallel, perpendicular, or neither:

(a) $\mathbf{A} = 2\mathbf{i} - 2\mathbf{j} + 5\mathbf{j}$ and $\mathbf{B} = 3\mathbf{i} + 8\mathbf{j} + 2\mathbf{k}$

(b) $\mathbf{A} = 2\mathbf{i} + \mathbf{j} + \mathbf{k}$ and $\mathbf{B} = 2\mathbf{i} - \mathbf{j} - \mathbf{k}$

(c) $\mathbf{A} = \mathbf{i} - \mathbf{j} + \mathbf{k}$ and $\mathbf{B} = -2\mathbf{i} + 2\mathbf{j} - 2\mathbf{k}$

1.15 The *direction cosines* of a vector $\mathbf{A}$ are the cosines of the angles between $\mathbf{A}$ and the Cartesian coordinate axes. Find these for each of the following vectors:

(a) $\mathbf{A} = \mathbf{j}$ (c) $\mathbf{A} = \mathbf{i} + \mathbf{j} + \mathbf{k}$

(b) $\mathbf{A} = \mathbf{i} + \mathbf{j}$ (d) $\mathbf{A} = -\mathbf{i} + \mathbf{j} - \mathbf{k}$

1.16 What are the geometric meanings of the following inequalities? Under what conditions does equality hold in each?

(a) $|\mathbf{A} + \mathbf{B}| \le |\mathbf{A}| + |\mathbf{B}|$ (*triangle inequality*)

(b) $|\mathbf{A} \cdot \mathbf{B}| \le |\mathbf{A}||\mathbf{B}|$ (*Cauchy–Schwarz inequality*)

1.17 Prove each statement for arbitrary vectors $\mathbf{A}$ and $\mathbf{B}$:

(a) $\mathbf{A} \cdot \mathbf{A} = |\mathbf{A}|^2$

(b) $(\mathbf{A} + \mathbf{B}) \cdot (\mathbf{A} \cdot \mathbf{B}) = \mathbf{A} \cdot \mathbf{A} + \mathbf{B} \cdot \mathbf{B} + 2\mathbf{A} \cdot \mathbf{B}$

(c) if $\mathbf{A}$ and $\mathbf{B}$ are perpendicular then $|\mathbf{A} + \mathbf{B}|^2 = |\mathbf{A}|^2 + |\mathbf{B}|^2$

Plane Curves

1.18 Write a set of parametric equations for the line $y = mx + b$.

1.19 Sketch each given curve, which is either an ellipse or at least one branch of a hyperbola:

(a) $5x^2 + 2y^2 = 10$

(b) $2x^2 - 5y^2 = 10$

(c) $x^2 - 2x + 4y^2 - 16y = -13$

(d) $x^2 - 2x - 4y^2 + 16y = 19$

(e) $x = 3\cos t,\ y = 2\cos 4\ (0 \le t < 2\pi)$

(f) $xy = 8$

(g) $x = \cosh t,\ y = \sinh t\ (-\infty < t < \infty)$

1.20 The equation

$$\mathbf{r} = (a\cos t + at\sin t)\,\mathbf{i} + (a\sin t - at\cos t)\,\mathbf{j} \qquad (t \ge 0)$$

is a parametric representation for an *involute of a circle*. Plot this curve for $a = 2$ and $0 \le t \le 2$.

1.21 Identify the conic sections represented parametrically by (a) $x = k_1 t^2$, $y = k_2 t$ where k_1, k_2 are constants, and (b) $x = a\sec t,\ y = b\tan t$ where a, b are constants.

1.22 Plot the following plane curves:

(a) the cycloid $(r > 0)$

$$x(t) = r(t - \sin t)$$
$$y(t) = r(1 - \cos t)$$

(b) the epicycloid $(R > r > 0$; experiment with the ratio $R/r)$

$$x(t) = (R + r)\cos t - r\cos\left(\frac{R + r}{r}t\right)$$

$$y(t) = (R + r)\sin t - r\sin\left(\frac{R + r}{r}t\right)$$

(c) the hypocycloid $(R > r > 0)$

$$x(t) = (R - r)\cos t + r\cos\left(\frac{R - r}{r}t\right)$$

$$y(t) = (R - r)\sin t - r\sin\left(\frac{R - r}{r}t\right)$$

(d) the semicubic parabola $(r > 0)$

$$x(t) = t^2$$
$$y(t) = rt^3$$

1.23 Plot each of the following parametric plane curves:

(a) $x = -t$, $y = t$ $(0 \leq t \leq 1)$

(b) $x = 1/(1+t)$, $y = 1-t$ $(0 \leq t \leq 1)$

(c) $x = t$, $y = 1/t$ $(1 \leq t \leq 2)$

(d) $x = \ln t$, $y = e^{-t}$ $(1 \leq t \leq 2)$

1.24 Find the distance from the point (x_1, y_1) to the line $y = mx + b$.

1.25 Assuming a is a positive constant, graph each of the following polar equations:

(a) $r = a(1 + \cos \theta)$ (e) $r = a \sin 2\theta$

(b) $r = a(1 + \sin \theta)$ (f) $r = a\theta$

(c) $r = a(1 - \sin \theta)$ (g) $r = a/\theta$

(d) $r = a \cos \theta$ (h) $r = a^\theta$

1.26 Convert each given equation to polar coordinates and graph the resulting polar equation:

(a) $(x^2 + y^2)^2 = x^2 y$ (b) $(x^2 + y^2)^3 = x^2$

1.27 Find the points of intersection between the curve $r = 2a \cos \theta$ and (a) the circle $r = a$, (b) the line $x = b$ where $0 < b < 2a$, (c) the ray $\theta = \pi/4$, and (d) the line $y = b$ where $-a < b < a$.

1.28 A circle has radius a and center located by the polar coordinates (r_0, θ_0). Find its polar equation.

1.29 A curve is given in polar coordinates by

$$r = \frac{\sin \theta + (\sin^2 \theta + 4ab \cos^2 \theta)^{1/2}}{2a \cos^2 \theta} \qquad (a, b > 0).$$

Find its Cartesian coordinate description.

1.30 A curve is given parametrically by the equations $x = a \cosh t$, $y = b \sinh t$ $(-\infty < t < \infty)$. Eliminate the parameter and identify the type of curve.

1.31 Identify the curve given by $3x^2 + 6x + 2y^2 - 8y + 5 = 0$.

1.32 Eliminate the parameter t from each of the following pairs of equations:

(a) $x = at^2 + b$ and $y = ct^2 + d$ (b) $x = at^2$ and $y = bt$

1.33 Find a parametric representation of

(a) the curve $(x/a)^{1/2} + (y/b)^{1/2} = 1$ from $(a, 0)$ to $(0, b)$,

(b) the whole curve $(y - x)^2 = a^2 - x^2$, and

(c) the whole curve $x^{2/3} + y^{2/3} = 1$.

1.34 Find the angle between each given curve and the line $y = x$ at the point $(1, 1)$:

(a) $y = 2 - x$ (c) $x = y^2$

(b) $y = x^2$ (d) $y = x^3$

1.35 Find the angle between the graphs of $y = \sin x$ and $y = \cos x$ at $x = \pi/4$.

Space Curves

1.36 Write down a set of three parametric equations representing the line segment running from

(a) the coordinate origin to the point $(1, 3, 2)$,

(b) the point $(1, -1, 1)$ to the coordinate origin, and

(c) the point $(1, 1, 1)$ to the point (a, b, c).

1.37 Sketch each curve:

(a) $\mathbf{r} = t \cos t\, \mathbf{i} + t \sin t\, \mathbf{j} + t\, \mathbf{k}$ for $t \geq 0$

(b) $\mathbf{r} = t\, \mathbf{i} + t^2\, \mathbf{j} + t\, \mathbf{k}$ for $0 \leq t \leq 1$

(c) $\mathbf{r} = \sin t\, \mathbf{i} + \sin^2 t\, \mathbf{j} + \sin^3 t\, \mathbf{k}$ for $0 \leq t \leq \pi/2$

(d) $\mathbf{r} = \cosh t\, \mathbf{i} + \sinh t\, \mathbf{j} + t\, \mathbf{k}$ for $0 \leq t \leq 1$

(e) $\mathbf{r} = t\, \mathbf{i} + \ln t\, \mathbf{j} + (1/t)\, \mathbf{k}$ for $1 \leq t \leq 2$

1.38 Find an expression for the unit tangent to each curve, and evaluate it at $t = 0$ and $t = 1$:

(a) $\mathbf{r}(t) = 6t\, \mathbf{i} + 3t^2\, \mathbf{j} + t^3\, \mathbf{k}$ $(0 \leq t \leq 2)$

(b) $\mathbf{r}(t) = t^2\, \mathbf{i} + t^3\, \mathbf{j} + t^6\, \mathbf{k}$ $(0 \leq t \leq 2)$

Area Regions and Surfaces

1.39 Sketch each of the following regions in the xy-plane:

(a) $x^2/4 + y^2/2 \leq 1$ for $x \geq 0$ (c) $x \geq 0$, $y \geq 0$ for $x + y \leq 1$

(b) $y \geq x/2$, $y \leq x^{1/2}$ for $x \geq 1$ (d) $x^2/9 + y^2/4 > 1$

1.40 Sketch the regions in the xy-plane bounded by the following sets of curves:

(a) $xy = 1$, $x = y$, $x = 2$

(b) $y = x$, $y = 2x$, $y = 2(1 - x)$

(c) $y = x$, $y = 2x$, $xy = 1$, $xy = 2$

(d) $x = 1$, $x = 2$, $xy = 1$, $xy = 2$

(e) $x = 1$, $x = 2$, $y = x^2$, $y = 2x^2$

(f) $y = x^2$, $y = 2x^2$, $xy = 1$, $xy = 2$

1.41 Find the equation of the plane

(a) passing through the point $(2, -4, -2)$ and normal to the two vectors $-\mathbf{i} - 3\mathbf{j} - 7\mathbf{k}$ and $-4\mathbf{i} - \mathbf{j} - 5\mathbf{k}$;

(b) passing through the points $(a, 0, 0)$, $(0, b, 0)$, and $(0, 0, c)$.

1.42 Show that $(\mathbf{r} - \mathbf{A}) \cdot \mathbf{r} = 0$ is the vector equation of a sphere.

1.43 Write the equation of the sphere

(a) centered at (a, b, c) and passing through (x_1, y_1, z_1),

(b) having (a, b, c) and (d, e, f) as endpoints of a diametral segment.

1.44 Identify the type of quadric surface represented by each of the following equations:

(a) $2x^2 + 3y^2 + z^2 = 6$

(b) $x^2 + 2y^2 - z^2 = 0$

(c) $x^2 + 2y^2 = 2$

(d) $x^2 - y = 0$

1.45 Describe the surface given by the equation $(x^2 + y^2)^2 = 2a^2(x^2 - y^2)$.

1.46 Verify that each given surface can be parameterized as shown:

(a) elliptic cone

$$\frac{z^2}{c^2} = \frac{x^2}{a^2} + \frac{y^2}{b^2}, \qquad \mathbf{r}(u, v) = au \cos v\, \mathbf{i} + bu \sin v\, \mathbf{j} + cu\, \mathbf{k}$$

(b) elliptic paraboloid

$$2cz = \frac{x^2}{a^2} + \frac{y^2}{b^2}, \qquad \mathbf{r}(u, v) = au \cos v\, \mathbf{i} + bu \sin v\, \mathbf{j} + \frac{u^2}{2c}\, \mathbf{k}$$

(c) elliptic hyperboloid

$$2cz = \frac{x^2}{a^2} - \frac{y^2}{b^2}, \qquad \mathbf{r}(u, v) = au \cosh v\, \mathbf{i} + bu \sinh v\, \mathbf{j} + \frac{u^2}{2c}\, \mathbf{k}$$

1.47 Use the technique of Example 1.15 to sketch the elliptic cone and the elliptic hyperboloid of Problem 1.46.

1.48 What is a *degenerate conic*? Give some examples.

1.49 A surface in space is given by $x = u \cos v \sin \alpha$, $y = u \sin v \sin \alpha$, $z = u \cos \alpha$ where u, v are parameters with $-\infty < u < \infty$ and $0 \leq v < 2\pi$, and $0 < \alpha < \pi/2$ is a constant. By eliminating the parameters u, v, identify the type of surface.

1.50 Two surfaces are given by the equations

$$x^2 + y^2 + z^2 = 1, \qquad x^2 + (y-1)^2 + (z-1)^2 = 1.$$

Find the projection of their intersection onto the xy-plane.

1.51 Sketch the curves of intersection, in the first octant, between the cylinder $x^2 + z^2 = 1$ and the planes $y = ax$ and $y = bx$ for $b > a > 0$.

1.52 Parameterize the curve of intersection of

(a) the plane $ax + by + cz = d$ and the elliptic cylinder $x^2 + e^2 y^2 = e^2$;

(b) the plane $x + y = 1$ and the paraboloid $z = x^2 + y^2$, in the first octant;

(c) the cylinder $x^2 + y^2 = 1$ and the cylinder $x^2 + z^2 = 1$, in the first octant;

(d) the cylinder $x^2 + y^2 = 1$ and the surface $z = xy$, in the first octant.

1.53 Sketch the torus given by

$$\mathbf{r} = (a + b \cos u) \cos v \, \mathbf{i} + (a + b \cos u) \sin v \, \mathbf{j} + b \sin u \, \mathbf{k}$$

for $0 \leq u < 2\pi$ and $0 \leq v < 2\pi$.

1.54 A *Möbius strip* has parametric representation

$$\mathbf{r} = \left(2 \cos u + v \sin \frac{u}{2} \cos u\right) \mathbf{i} + \left(2 \sin u + v \sin \frac{u}{2} \sin u\right) \mathbf{j} + \left(v \cos \frac{u}{2}\right) \mathbf{k}$$

for $0 \leq u \leq 2\pi$ and $|v| \leq 1$. Plot it. Do some background research to learn why this surface is famous in mathematics.

1.55 Find the equation of the surface obtained by revolving

(a) the ellipse $x^2/a^2 + z^2/c^2 = 1$ about the x-axis,

(b) the ellipse of part (a) about the z-axis,

(c) the hyperbola $x^2/a^2 - z^2/c^2 = 1$ about the x-axis,

(d) the hyperbola of part (c) about the z-axis,

(e) the parabola $y^2 = 2px$ about the z-axis, and

(f) the line $y = x$ about the x-axis.

1.56 For each type of surface of revolution discussed in Section 1.10, find the inner coordinate lines, the basis vectors at a point on the surface, the angle between the coordinate lines, and the curvatures of the coordinate lines.

Some Applications

1.57 The motion of a projectile in two dimensions is described by the equations

$$x = x_0 + (v_0 \cos \theta_0)t,$$
$$y = y_0 + (v_0 \sin \theta_0)t - \tfrac{1}{2}gt^2,$$

where t is the time variable, (x_0, y_0) is the initial position, v_0 is the initial speed, θ_0 is the launch angle from the horizontal (i.e., x) direction, and g is the free-fall acceleration constant.

(a) Write down the corresponding position vector $\mathbf{r}(t)$.

(b) Find the instantaneous velocity $\mathbf{v}(t)$ and the instantaneous speed $|\mathbf{v}(t)|$.

(c) Find the instantaneous acceleration $\mathbf{a}(t)$.

(d) Taking $(x_0, y_0) = (0, 0)$, eliminate t from the given equations to find the equation of the trajectory.

1.58 (a) An electrically charged particle travels through an electrostatic field region such that its position coordinates are described by

$$x(t) = u_0 t, \qquad y(t) = \tfrac{1}{2}\eta E_0 t^2, \qquad z(t) = 0,$$

where u_0, η, E_0 are constants and t is time. Find the equation of the particle's trajectory. (b) Repeat for a particle whose position coordinates are

$$x(t) = \tfrac{1}{2}\eta E_0 t^2, \qquad y(t) = 0, \qquad z(t) = \tfrac{1}{2}gt^2,$$

where E_0, η, and g are constants.

1.59 An electrically charged particle travels through a magnetostatic field region such that its position coordinates are described by

$$x(t) = \frac{u_{x0}}{\omega_c} \sin \omega_c t, \qquad y(t) = \frac{u_{x0}}{\omega_c} \cos \omega_c t - \frac{u_{x0}}{\omega_c}, \qquad z(t) = u_{z0} t,$$

where u_{x0}, u_{z0}, ω_c are constants. What kind of motion does this represent?

Chapter 2

Differential Calculus

2.1 A Couple of Remarks on Convergence

An old, somewhat naive approach to the limit of a function in one variable x was to say that

$$\lim_{x \to a} f(x) = A \text{ if } f(x) \text{ approaches } A \text{ when } x \text{ approaches } a.$$

Many calculus textbooks still use this as a first introduction to the concept of the limit of a function at a point. To explain the phrase "x approaches a," older books turned to the notion of sequence limit, reminding the reader that $x_k \to a$ if for any $\varepsilon > 0$ there exists N (depending on ε) such that $k > N$ implies $|x_k - a| < \varepsilon$. The next step is Cauchy's ε-δ definition: we say that A is the limit of $f(x)$ as $x \to a$ if for any $\varepsilon > 0$ there exists $\delta > 0$ such that $|f(x) - A| < \varepsilon$ whenever $0 < |x - a| < \delta$. Here we do not need to know *how* x tends to a. So newer calculus books can avoid reliance on the notion of sequence limit, although it is still needed in order to present the definite integral.

Similarly, modern textbooks can introduce limits of multivariable functions $f = f(x_1, \ldots, x_n) = f(\mathbf{x})$ without having to talk about limits of vector sequences $\{\mathbf{x}_k\}$. But when we must prove that a certain limit does *not* exist, we usually select a particular sequence or a path along which $\mathbf{x} \to \mathbf{a}$. So it makes sense to present

Definition 2.1. The vector sequence $\{\mathbf{x}_k\}$ tends to $\mathbf{a}$ if and only if each component of $\mathbf{x}_k$ tends to the corresponding component of $\mathbf{a}$.

The approach taken in single-variable calculus to introduce limits, derivatives, differentials, and applications to extremum problems is here broadly repeated for functions of several variables. But the theory is not a simple repetition or extension, as functions in many variables present certain nontrivial difficulties that are absent from ordinary calculus. Let us start with some essential notions regarding point sets.

2.2   Open and Closed Sets in $\mathbb{R}^n$

By analogy with balls in $\mathbb{R}^3$, we speak of *open* and *closed balls* in $\mathbb{R}^n$, centered at $(x_1^*, \ldots, x_n^*)$ and having radii R. These are the sets of points $(x_1, \ldots, x_n) \in \mathbb{R}^n$ satisfying the inequalities

$$\sum_{i=1}^n (x_i - x_i^*)^2 < R^2, \qquad \sum_{i=1}^n (x_i - x_i^*)^2 \le R^2,$$

respectively. Note that the closed ball contains both the corresponding open ball and its boundary, which of course is the sphere described by the equation

$$\sum_{i=1}^n (x_i - x_i^*)^2 = R^2.$$

A *neighborhood* of a point is a set that contains an open ball centered at the point and having nonzero radius.

Example 2.1. In $\mathbb{R}^2$, the ball consisting of the points (x, y) satisfying

$$[(x - a)^2 + (y - b)^2]^{1/2} < \varepsilon$$

is a *circular ε-neighborhood* of the point (a, b). A *square ε-neighborhood* of (a, b) is the set of points satisfying $|x - a| < \varepsilon$ and $|y - b| < \varepsilon$ (the reader should sketch this). Note that any square ε-neighborhood of (a, b) contains a circular ε-neighborhood of (a, b), and a circular ε-neighborhood contains a square ε_1-neighborhood with $\varepsilon_1 = \varepsilon/\sqrt{2}$; see Problem 2.11. □

A general set in $\mathbb{R}^n$ is *open* if for each point there is a ball of nonzero radius centered at the point and contained in the set. Any union of open sets in $\mathbb{R}^n$ is open.

Example 2.2. In $\mathbb{R}^2$, the set of points (x, y) satisfying $x > 0$ and $y > 0$ is open. Indeed, the point (x, y) is at the center of a ball of radius $\frac{1}{2}\min(x, y)$ that is contained in the set. □

A point of a set in $\mathbb{R}^n$ is an *accumulation* (or *limit*) *point* if each neighborhood of the point contains a point of the set different from the point in question. A set that contains all its accumulation points is said to be *closed*. It can be shown that a set is closed if and only if it contains the limits of all its convergent sequences: S is closed if and only if $\{x_n\} \subset S$ and $x_n \to x_0$ together imply that $x_0 \in S$. A finite union of closed sets is closed, as is the intersection of any collection of closed sets. The proofs are elementary and are left to the reader.

For the relatively simple shapes encountered in engineering practice (and in textbooks), open sets do not contain their boundaries, and supplementing an open set with its boundary results in a closed set. Sometimes it is important to know whether the domain of a function is open or closed.

We consider functions f in n variables:

$$f = f(x_1, \ldots, x_n),$$

where n is usually 2 or 3, namely $z = f(x, y)$ or $f = f(x, y, z)$. Because the position vector $\mathbf{r}$ is given by $\mathbf{r} = x\,\mathbf{i} + y\,\mathbf{j} + z\,\mathbf{k}$, we shall employ the notation $f(x, y, z) \equiv f(\mathbf{r})$ although formally it is less than correct. The definition of a function in n variables, along with its domain and range, involves mere reformulation of the definition for a function in one variable.

2.3 Limits

We first consider the notion of limit in the context of a function $z = f(x, y)$ in two variables. Intuitively, A is the limit of f at (a, b) if $f(x, y)$ approaches A as (x, y) approaches (a, b). The ancient Greeks used this concept to calculate various limits, like the circumference of the circle. In fact it was used naively for many centuries. The idea is that we construct various (and all possible!) paths in the xy-plane through, but not including, the point (a, b), obtaining along *any* such curve the same value A as the limit of $f(x, y)$. See Fig. 2.1. The paths can be continuous or discrete, so they include sequences tending to (a, b).

But checking this condition directly for each of infinitely many paths is of course impossible. So when Augustin-Louis Cauchy stated his ε-δ definition of limit (which, by Heine's theorem, turned out to be equivalent to the definition considering all the paths through (a, b)), it began to appear in textbooks. This definition mimics the limit definition for a function of one variable, with evident changes.

Definition 2.2. The value A is the *limit* of f at (a, b) if to each preassigned $\varepsilon > 0$ there corresponds a $\delta > 0$ such that $|f(x, y) - A| < \varepsilon$ whenever

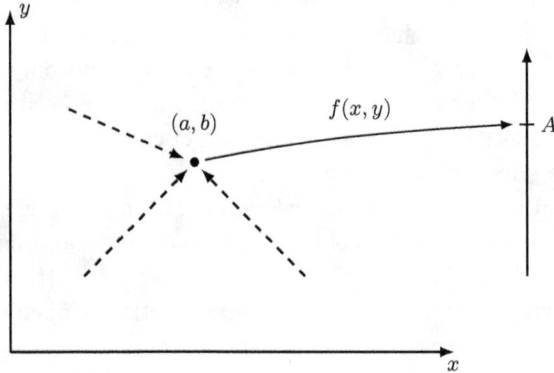

Fig. 2.1 Various paths (dashed) in xy-plane along which, as the point (x, y) tends to the point (a, b), the value $f(x, y)$ tends to the value A.

$0 < [(x-a)^2 + (y-b)^2]^{1/2} < \delta.$ We write

$$A = \lim_{(x,y) \to (a,b)} f(x, y)$$

and sometimes refer to A as a *double limit*.

It is important to stress that we must try all positive values for ε, exclude the point (a, b) from consideration, and select the value A — the value suspected as the limit — in advance. Clearly we should consider relatively small values for ε; if for some $\varepsilon_0 > 0$ the needed $\delta > 0$ exists, then this δ serves for any $\varepsilon > \varepsilon_0$.

We should mention that a typical beginner's mistake is to say "the limit tends to A." But a limit is a fixed number; it cannot "tend" to anything.

Why is this seemingly cumbersome definition used, while the previous and much clearer path definition is relegated to showing only that certain limits do *not* exist? The answer is that by Cauchy's definition, for many functions we can establish whether A is the limit by using well developed techniques for the solution of inequalities. To prove that A is a limit of f at (a, b), we start with an arbitrary $\varepsilon > 0$ and the inequality

$$|f(x, y) - A| < \varepsilon.$$

By the ε-δ definition, we should prove that its solution contains a "punctured disk" of some radius $\delta > 0$:

$$0 < [(x-a)^2 + (y-b)^2]^{1/2} < \delta.$$

This is easier than trying to find the limit over *every* path through (a, b).

Example 2.3. We use the ε-δ definition to show that $x^2 + y^2$ tends to 5 as $(x, y) \to (1, 2)$. (We can take the value of the given function at point $(2, 1)$ as a candidate for the limit value.) So we take $\varepsilon > 0$ and consider the inequality

$$|x^2 + y^2 - 5| < \varepsilon. \tag{2.1}$$

We should show that its solution set contains a punctured disk

$$0 < [(x - 1)^2 + (y - 2)^2]^{1/2} < \delta \tag{2.2}$$

for some $\delta > 0$. Letting $x = 1 + s$ and $y = 2 + t$, we transform (2.1) into the inequality

$$|s(s + 2) + t(t + 4)| < \varepsilon \tag{2.3}$$

and (2.2) into the inequality

$$(s^2 + t^2)^{1/2} < \delta. \tag{2.4}$$

We can restrict consideration to $|s| < 1$ and $|t| < 1$. Using the elementary inequality

$$|a| + |b| \leq [2(a^2 + b^2)]^{1/2} \tag{2.5}$$

let us estimate

$$\begin{aligned}
|s(s + 2) + t(t + 4)| &\leq |s(s + 2)| + |t(t + 4)| \\
&= |s||s + 2| + |t||t + 4| \\
&\leq 3|s| + 5|t| \\
&\leq 5(|s| + |t|) \\
&\leq 5\sqrt{2}\,(s^2 + t^2)^{1/2}.
\end{aligned}$$

So we can take $\delta = \varepsilon/(5\sqrt{2})$, and all the pairs (s, t) satisfying (2.4) also satisfy (2.3). □

Remark 2.1. If the limit A of f at (a, b) exists and f is continuous at (a, b) (see Section 2.3) then obviously there exist two ordinary limits

$$A = \lim_{x \to a} f(x, b) = \lim_{y \to b} f(a, y),$$

one of which is often used as a starting point to find a possible value A for the double limit. As this A is a limit of a function in one variable x or y, we may apply l'Hôpital's rule in order to find it. Let $\lim_{x \to a} f(x, y)$, depending on y, exist for all y in some vicinity of $y = b$, and let $\lim_{y \to b} f(x, y)$ exist as a function of x in a vicinity of $x = a$. The *iterated limits*

$$\lim_{y \to b} \left[\lim_{x \to a} f(x, y) \right], \qquad \lim_{x \to a} \left[\lim_{y \to b} f(x, y) \right],$$

both exist and equal A provided that A exists. However, existence of the four limits mentioned above does not guarantee that the double limit of f exists at (a, b). $\qquad \square$

More generally we can bring in the notion of limit for a function of n variables using ball-type neighborhoods. We leave the formalities to the reader. The reader should already be aware of the properties of the limit for a function in one variable, and how to prove these. The proofs are easily extended to multivariable functions with evident changes.

As is the case in single-variable calculus, we often seek the limit of a two-variable function at some point (a, b) on the boundary of a closed set S. Rather than using a δ-ball $B_\delta(a, b)$ for the neighborhood of (a, b), we should use the intersection of $B_\delta(a, b)$ with S as reflected in the next definition.

Definition 2.3. A is the limit of f at a point (a, b) lying on the boundary of S if for any $\varepsilon > 0$ there exists $\delta > 0$ such that $|f(x, y) - A| < \varepsilon$ whenever $(x, y) \in \overline{S}$ and $0 < [(x - a)^2 + (y - b)^2]^{1/2} < \delta$. Here $\overline{S}$ is the closure of S, obtained by adding to S all of its accumulation points.

The extension of this idea to n variables is straightforward. An important property of the limit is stated in

Theorem 2.1. *The limit value* $\lim_{(x,y) \to (a,b)} f(x, y)$ *is unique.*

Proof. Suppose there are two limits

$$\lim_{(x,y) \to (a,b)} f(x, y) = A_1 \quad \text{and} \quad \lim_{(x,y) \to (a,b)} f(x, y) = A_2$$

where $A_2 - A_1 = 3c > 0$. Taking $\varepsilon = c$, for the limit A_1 we find $\delta_1 > 0$ such that for any (x, y) satisfying $0 < [(x - a)^2 + (y - b)^2]^{1/2} < \delta_1$ we have

$$|f(x, y) - A_1| < c. \tag{2.6}$$

Similarly, taking $\varepsilon = c$ again, for the limit A_2 we find $\delta_2 > 0$ such that for any (x, y) satisfying $0 < [(x - a)^2 + (y - b)^2]^{1/2} < \delta_2$ we have

$$|f(x, y) - A_2| < c. \tag{2.7}$$

Take $\delta = \min(\delta_1, \delta_2)$. Then (x, y) satisfying $0 < [(x - a)^2 + (y - b)^2]^{1/2} < \delta$ is inside both circles of radii δ_1 and δ_2 respectively, hence for any (x, y) in the ball B_δ both (2.6) and (2.7) hold simultaneously. This is impossible because the distance between A_1 and A_2 is $3c$, and $f(x, y)$ cannot be in both the c-neighborhood of A_1 and the c-neighborhood of A_2. □

As we see, this uniqueness proof for the limit mimics the corresponding proof for a function in one variable. In a similar way we can prove other elementary properties of the limit for a multivariable function.

Theorem 2.2. *If f, g, h have limits at point (a, b) and*

$$\lim_{(x,y)\to(a,b)} h(x, y) \neq 0,$$

then for any constants α, β we have

$$\lim_{(x,y)\to(a,b)} \alpha f(x, y) \pm \lim_{(x,y)\to(a,b)} \beta g(x, y)$$

$$= \alpha \lim_{(x,y)\to(a,b)} f(x, y) \pm \beta \lim_{(x,y)\to(a,b)} g(x, y),$$

and

$$\lim_{(x,y)\to(a,b)} f(x, y)g(x, y) = \lim_{(x,y)\to(a,b)} f(x, y) \lim_{(x,y)\to(a,b)} g(x, y),$$

and

$$\lim_{(x,y)\to(a,b)} \frac{f(x, y)}{h(x, y)} = \frac{\lim_{(x,y)\to(a,b)} f(x, y)}{\lim_{(x,y)\to(a,b)} h(x, y)}.$$

It is easy to understand that if the double limit

$$\lim_{(x,y)\to(a,b)} f(x, y) = A$$

exists, then for any path *through* (a, b), for example $y = \phi(x)$ where $b = \phi(a)$, there exists

$$\lim_{x\to a} f(x, \phi(x)) = A. \tag{2.8}$$

In particular, there exist iterated limits:

$$\lim_{x\to a}\lim_{y\to b} f(x,y) = \lim_{y\to b}\lim_{x\to a} f(x,y) = A.$$

Remark 2.2. It is crucial to remember that even if we take 10^{100} different paths and all of them yield the same limit in (2.8), this does not prove that the double limit exists. Too often students find that $\lim_{x\to a} f(x,b) = \lim_{y\to b} f(a,y) = A$ and decide that the double limit $\lim_{(x,y)\to(a,b)} f(x,y)$ exists and equals A. $\qquad\square$

Note that *if for two different paths* $y = \phi_k(x)$ *the limits* (2.8) *differ, we conclude at once that* $\lim_{(x,y)\to(a,b)} f(x,y)$ *does not exist.* This is used to prove non-existence of a double limit.

Example 2.4. The limit

$$\lim_{(x,y)\to(0,0)} \frac{xy}{x^2+y^2}$$

does not exist. Indeed, for two paths $x = 0$ and $y = x$ we get respectively

$$\lim_{y\to 0} \frac{0y}{0^2+y^2} = 0 \quad\text{and}\quad \lim_{x\to 0} \frac{xx}{x^2+x^2} = \frac{1}{2}.$$

As these limits differ we have proved non-existence of the double limit. Note that the iterated limits

$$\lim_{x\to 0}\lim_{y\to 0} \frac{xy}{x^2+y^2} = 0, \qquad \lim_{y\to 0}\lim_{x\to 0} \frac{xy}{x^2+y^2} = 0,$$

both exist and are equal. $\qquad\square$

The first step in finding a double limit is to establish a candidate value for the limit using a particular path in conjunction with (2.8). This is usually done by finding the limit along a path $y = h(x)$ such that $b = h(a)$, that is $\lim_{x\to a} f(x, h(x)) = A$. In particular we can use the limit $\lim_{x\to a} f(x,b) = A$. However, to prove that A is really the value of the double limit, we must follow through with the ε-δ procedure.

In some cases it makes sense to make the change of variables $x = a+t$, $y = b+s$ and to treat

$$\lim_{(t,s)\to(0,0)} f(a+t,b+s) \quad\text{which is equal to}\quad \lim_{(x,y)\to(a,b)} f(x,y).$$

Some rearrangement may aid in the evaluation of a limit.

Example 2.5.

$$\lim_{(x,y)\to(0,0)} \frac{\sqrt{xy+1}-1}{xy} = \lim_{(x,y)\to(0,0)} \frac{\sqrt{xy+1}-1}{xy} \cdot \frac{\sqrt{xy+1}+1}{\sqrt{xy+1}+1}$$

$$= \lim_{(x,y)\to(0,0)} \frac{1}{\sqrt{xy+1}+1}$$

$$= \frac{1}{2}. \qquad \square$$

One tool for establishing the existence of a limit is the following simple theorem. Its analogue for functions in one variable is known as the *sandwich* or *squeeze theorem*.

Theorem 2.3. *Let* $|f(x,y) - A| \le h(x,y)$ *in some circle with center* (a,b) *and* $h(x,y) \to 0$ *as* $(x,y) \to (0,0)$. *Then* $f(x,y) \to A$ *as* $(x,y) \to (0,0)$.

Proof. The condition

$$\lim_{(x,y)\to(a,b)} h(x,y) = 0$$

means that corresponding to any $\varepsilon > 0$ there is a $\delta > 0$ such that for any (x,y) satisfying $0 < [(x-a)^2 + (y-b)^2]^{1/2} < \delta$ we have $|h(x,y)| < \varepsilon$. The other condition implies that $|f(x,y) - A| < \varepsilon$ holds for any such pair (x,y). $\qquad \square$

Corollary 2.1. *Assume that* $f(x,y) - A = g(x,y)h(x,y)$, *where* $g(x,y)$ *is bounded in a neighborhood of* (a,b) *excluding point* (a,b), *and that* $h(x,y) \to 0$ *as* $(x,y) \to (a,b)$. *Then* $f(x,y) \to A$ *as* $(x,y) \to (a,b)$.

Proof. Let $|g(x,y)| \le c = $ constant. Then $|g(x,y)h(x,y)| \le c|h(x,y)| \to 0$ as $(x,y) \to (a,b)$, and Theorem 2.3 applies. $\qquad \square$

The reader can extend Theorem 2.3 to functions in three variables.

Example 2.6. Consider

$$\lim_{(x,y)\to(0,0)} \frac{x^2 y}{x^2 + y^2}.$$

Taking a path limit for $y = 0$, we identify a candidate for A as

$$\lim_{x\to 0} \frac{x^2 \cdot 0}{x^2 + 0^2} = 0.$$

Here $f(x,y) = x^2 y/(x^2 + y^2) = g(x,y)h(x,y)$, where $g(x,y) = x^2/(x^2 + y^2)$ and $h(x,y) = y$. But $|x^2/(x^2 + y^2)| \leq 1$ for $(x,y) \neq (0,0)$ and $h(x,y) = y$ has the limit 0 as $(x,y) \to (0,0)$. Thus the limit exists and is 0. □

For multivariable functions we also can use various classical limits like

$$\lim_{x \to 0} \frac{\sin x}{x} = 1, \quad \lim_{x \to 0} \frac{1 - \cos x}{x^2} = \frac{1}{2}, \quad \text{or} \quad \lim_{x \to 0} (1 + x)^{1/x} = e.$$

In such cases we can use equivalences between various limit expressions such as that between $\sin x$ and x as $x \to 0$. In the multidimensional version it could be the equivalence of $\sin(xy)$ with xy as $(x,y) \to 0$ and so

$$\lim_{(x,y) \to (0,0)} \frac{x \sin(xy)}{x^2 + y^2} = \lim_{(x,y) \to (0,0)} \frac{x(xy)}{x^2 + y^2} = 0$$

by the previous example.

Next we turn to non-existence of the limit at a point. Understanding that the agreement of even a hundred limits of a function at a point (a, b) along particular paths does not imply existence of the double limit, we realize we must use the ε-δ definition to prove existence. But to prove that a *limit does not exist, it is enough to find two paths along which the limiting values disagree.* Example 2.4 illustrated this point. Let us consider some additional examples.

Example 2.7. (1) Consider

$$\lim_{(x,y) \to (1,1)} \frac{x}{y}.$$

As the denominator is not zero, we obtain the limit by substituting the values for x and y. The result is 1. (2) Nonexistence of the limit

$$\lim_{(x,y) \to (0,0)} \frac{x}{y}$$

can be established by considering the limit along the single path $y = 0$. Because the limit $\lim_{x \to 0} \frac{x}{0}$ does not exist, neither does the double limit. (3) The limit

$$\lim_{(x,y) \to (0,0)} \frac{x}{x - y}$$

does not exist, as we can see by taking the path $y = x$. □

Example 2.8. The nonexistence of

$$\lim_{(x,y)\to(0,0)} \frac{2xy^2}{4x^2 + y^4}$$

can be shown in various ways. One is to bring in new variables $t = 2x$ and $s = y^2$, getting

$$\lim_{(x,y)\to(0,0)} \frac{2xy^2}{4x^2 + y^4} = \lim_{(t,s)\to(0,0)} \frac{ts}{t^2 + s^2}$$

and so reducing the problem to Example 2.4. Another is to identify two paths along which the limits differ. The limit taken along the path $x = 0$ is

$$\lim_{y\to0} \frac{2(0)y^2}{4x^2 + y^4} = 0.$$

On the other hand, the limit taken along the path described by $2x = y^2$ is

$$\lim_{y\to0} \frac{(y^2)(y^2)}{y^4 + y^4} = \frac{1}{2}. \qquad \square$$

Example 2.9. Consider the limits

$$(1) \quad \lim_{(x,y)\to(0,0)} \frac{\sin(x^2 + y^2)}{x^2 + y^2} \quad \text{and} \quad (2) \quad \lim_{(x,y)\to(0,0)} \frac{\sin(x + y)}{x + y}.$$

Using the equivalence $\sin x \sim x$ as $x \to 0$, we obtain for (1)

$$\lim_{(x,y)\to(0,0)} \frac{\sin(x^2 + y^2)}{x^2 + y^2} = \lim_{(x,y)\to(0,0)} \frac{x^2 + y^2}{x^2 + y^2} = 1.$$

We could reduce the fraction because $x^2 + y^2 \neq 0$ for $(x,y) \neq (0,0)$. We cannot repeat this procedure for (2), as along the path $y = -x$ we get

$$\lim_{x\to0} \frac{\sin(x - x)}{x - x} = \lim_{x\to0} \frac{0}{0}$$

which does not exist. Hence the double limit (2) does not exist.

For the case of (2), it can be noted that were we to define

$$f(x) = \begin{cases} \dfrac{\sin x}{x}, & x \neq 0, \\ 1, & x = 0, \end{cases}$$

which is continuous at any point, we should have existence of the double limit $\lim_{(x,y)\to(0,0)} f(x - y) = 1.$ $\qquad \square$

Example 2.10. We are required to show nonexistence of the double limit

$$\lim_{(x,y)\to(1,2)} \frac{(x-1)(y-2)}{(x-1)^2 + (y-2)^2}.$$

A common error involves taking two paths of the form $y = kx$, which fail to pass through the point $(a,b) = (1,2)$. Utilizing the paths $x = 1$ and $x - 1 = y - 2$, one finds that the limits differ so the double limit does not exist. Alternatively, the change of variables $s = x - 1$ and $t = y - 2$ will reduce the problem to Example 2.4.   □

Let us summarize the things to be learned from the above theorems and examples. *To prove that the limit of a function at a point **exists** we can*

(a) *use the ε-δ limit definition directly, or*

(b) *use an analogue of Theorem 2.3, or*

(c) *exploit classical limits for functions in one variable to which the multivariable function can be reduced.*

*To prove that the limit of a function $f(x,y)$ as $(x,y) \to (a,b)$ **does not exist** we can*

(a) *find two paths through (a,b) along which the limits are unequal, or*

(b) *find a path through (a,b) along which the limit does not exist, or*

(c) *prove that there exists $\varepsilon > 0$ such that no $\delta > 0$ satisfies the definition of the limit (often a difficult approach).*

Touching on the question whether we can perform a change of variables under the limit sign, we should mention the possibility of transformation to polar coordinates. Consider a limit of the form

$$\lim_{(x,y)\to(0,0)} f(x,y) \quad \text{with } x = r\cos\theta \text{ and } y = r\sin\theta.$$

Clearly $r \to 0$ here, but the angle θ is not fixed and we face computation of the limit $\lim_{r\to 0} f(r\cos\theta, r\sin\theta)$ knowing only that $0 \le \theta < 2\pi$. This strange situation arises because $(0,0)$ is a singular point of the transformation between the Cartesian and polar systems. But sometimes the approach is fruitful, as is illustrated next.

Example 2.11. Let us compute

$$A = \lim_{(x,y)\to(0,0)} \frac{x^3 + y^3}{x^2 + y^2}.$$

A change of variables yields

$$A = \lim_{r \to 0} \frac{r^3 \cos^3 \theta + r^3 \sin^3 \theta}{r^2} = \lim_{r \to 0} r(\cos^3 \theta + \sin^3 \theta) = 0$$

since $\cos^3 \theta + \sin^3 \theta$ is bounded and $r \to 0$. ☐

We close this section by noting that the principles above can be applied to functions of more than two variables.

Example 2.12. We can show that

$$\lim_{(x,y,z) \to (0,0,0)} \frac{xyz}{x^2 + y^2 + z^2} = 0$$

by applying a straightforward extension of Theorem 2.3 to three dimensions. If $f(x, y, z) = g(x, y, z)h(x, y, z)$ where g is bounded and $h(x, y, z) \to 0$ as $(x, y, z) \to (0, 0, 0)$, then

$$\lim_{(x,y,z) \to (0,0,0)} f(x, y, z) = 0.$$

We may select $g(x, y, z) = xy/(x^2 + y^2 + z^2)$ and $h(x, y, z) = z$. Indeed g is bounded because

$$\left| \frac{xy}{x^2 + y^2 + z^2} \right| = \frac{|x||y|}{x^2 + y^2 + z^2} \leq \frac{x^2 + y^2}{2(x^2 + y^2 + z^2)} \leq \frac{1}{2}. \qquad \square$$

The problem of multiple limits is a tricky one in applications. Too often a practitioner will evaluate iterated limits, discover their equality, and forget that this is not sufficient for existence of a multiple limit.

Remark 2.3. In single-variable calculus they traditionally introduce limits starting with the limit of a numerical sequence $\{a_n\}$. In multivariable calculus they start with the limit of a function at a point, rather than with sequence limits of the form

$$\lim_{\substack{i \to \infty \\ j \to \infty}} a_{ij}.$$

But this type of double limit will be encountered in the way we use Riemann sums to treat the double integral, so a few words about it are warranted here.

First of all, the double limit, if it exists, should not depend on the way in which the two indices i and j tend to infinity. This is achieved through the following, clearly a two-dimensional analogue to Cauchy's definition of limit.

Definition 2.4. We call A the limit of the sequence $\{a_{ij}\}$ if for any $\varepsilon > 0$ there is a number N (dependent on ε) such that $|a_{ij} - A| < \varepsilon$ for any pair (i, j) such that $i > N$ and $j > N$.

If the double limit of a sequence exists, it is equal to the iterated limits:

$$A = \lim_{\substack{i \to \infty \\ j \to \infty}} a_{ij} = \lim_{i \to \infty} \lim_{j \to \infty} a_{ij} = \lim_{j \to \infty} \lim_{i \to \infty} a_{ij}.$$

As a candidate for A we usually take one of the two iterated limits. But we still must prove that the double limit is equal to A (if it exists).

The double sequence limit possesses some properties similar to those of the single sequence limit.

(1) If the double limit of a sequence exists, it is unique. (This property is used to prove non-existence of the double limit: in this case for some sequences $\{(i_k, j_k)\}$ tending to infinity, there are particular unequal ordinary limits $\lim\limits_{k \to \infty} a_{i_k j_k}$, or one of the particular limits does not exist.)

(2) If the limits

$$\lim_{\substack{i \to \infty \\ j \to \infty}} a_{ij}, \quad \lim_{\substack{i \to \infty \\ j \to \infty}} b_{ij}, \quad \text{and} \quad \lim_{\substack{i \to \infty \\ j \to \infty}} c_{ij} \neq 0$$

exist, then

(a)

$$\lim_{\substack{i \to \infty \\ j \to \infty}} (\alpha\, a_{ij} + b_{ij}) = \alpha \lim_{\substack{i \to \infty \\ j \to \infty}} a_{ij} + \lim_{\substack{i \to \infty \\ j \to \infty}} b_{ij}$$

for any constant α;

(b)

$$\lim_{\substack{i \to \infty \\ j \to \infty}} a_{ij} b_{ij} = \lim_{\substack{i \to \infty \\ j \to \infty}} a_{ij} \cdot \lim_{\substack{i \to \infty \\ j \to \infty}} b_{ij};$$

(c)

$$\lim_{\substack{i \to \infty \\ j \to \infty}} \frac{a_{ij}}{c_{ij}} = \frac{\lim_{\substack{i \to \infty \\ j \to \infty}} a_{ij}}{\lim_{\substack{i \to \infty \\ j \to \infty}} c_{ij}}.$$

The proofs are practically repetitions of those for ordinary one-index sequences.

As in the theory of limits for ordinary sequences, the most interesting cases concern indeterminate forms such as $0 \cdot \infty$, $0/0$, and ∞/∞.

We do not present examples of finding double limits for sequences, as normally we can reduce the double limit for $\{a_{ij}\}$ to a double limit for a function $f(x, y)$ as $(x, y) \to (0, 0)$ in two variables by setting $x = 1/i$ and $y = 1/j$ in the expressions for a_{ij}. □

2.4 Continuity

Definition 2.5. A function f is *continuous* at point (a, b) if

$$\lim_{(x,y)\to(a,b)} f(x, y) = f(a, b).$$

We say that f is continuous on an open set if it is continuous at each point of the set.

We can rephrase the definition in ε-δ terms. A function f is continuous at point (a, b) if for any $\varepsilon > 0$ there exists $\delta > 0$ such that $|f(x, y) - f(a, b)| < \varepsilon$ for any (x, y) such that $0 < [(x - a)^2 + (y - b)^2]^{1/2} < \delta$. This definition is used when (a, b) is a point of an open set but it also holds at a boundary point (a, b) of the closure $\overline{S}$ when we invoke Definition 2.3. If the notion of limit for f is expressed in terms of paths through (a, b) on the boundary of the domain S, then all the paths should lie inside $\overline{S}$.

A continuous function defined on an open domain need not be bounded. This is illustrated by the function $f(x, y) = 1/x$ defined on the open square $(0, 1) \times (0, 1)$. But for a function continuous on a compact set, we have the following crucial result.

Theorem 2.4 (Weierstrass). *A function continuous on a bounded and closed domain $D \subset \mathbb{R}^2$ is bounded and takes its infimum and supremum values on D.*

To prove this theorem we need a simple

Lemma 2.1. *A bounded sequence in a closed subset D of $\mathbb{R}^2$ contains a Cauchy subsequence convergent to a point of D.*

Proof. Let D be inside a square S with side length a and sides parallel to the coordinate axes (Fig. 2.2).

Divide the square in half by a cut parallel to the x-axis. One of the rectangular portions contains infinitely many of the sequence terms. Choose from it a term to be labeled (x_1, y_1). Divide the rectangular portion in half through the longer side and select a half that contains infinitely many terms

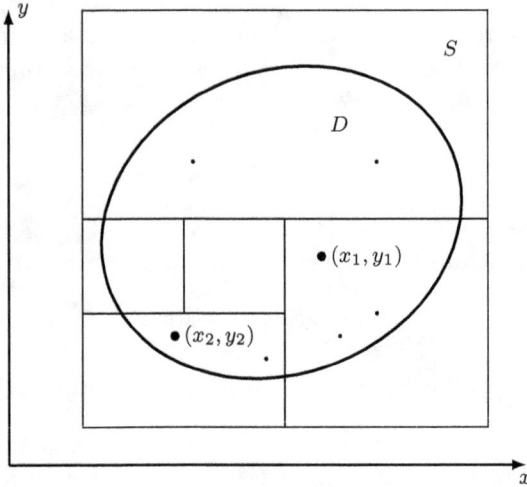

Fig. 2.2   Halving procedure for square S in the proof of Weierstrass' theorem. The larger points are the selected points of a sequence.

of the previous subsequence. From these terms select a point, to be labeled (x_2, y_2), in such a way that its index in the initial sequence exceeds that of (x_1, y_1). Now repeat the halving procedure with the latter portion: take a rectangle containing infinitely many terms of the last subsequence, and select (x_3, y_3) having a further position in the initial sequence. Repeat this procedure ad infinitum to get a subsequence $\{(x_k, y_k)\}$ with the needed properties. As all its terms starting with index k are inside the rectangle obtained in the kth step, and all rectangles of the subsequent steps are inside that rectangle with their sides tending to zero, the subsequence is a Cauchy sequence. Its limit lies in D since D is closed. $\square$

Now we can establish Theorem 2.4.

Proof (of Weierstrass' theorem). First we show that f is bounded on D. Suppose to the contrary that f is unbounded from above on D. Then there is a sequence $\{(x_m, y_m)\}$ such that $f(x_m, y_m) \to \infty$ as $m \to \infty$. By Lemma 2.1 we select a subsequence from the sequence, which we denote again by $\{(x_m, y_m)\}$, convergent to a point $(a, b) \in D$. Continuity of f in D implies

$$\lim_{m \to \infty} f(x_m, y_m) = f(a, b)$$

which is impossible since $f(x_m, y_m) \to \infty$ as $m \to \infty$. So there is a finite supremum of f on D equal to M. Clearly in D there is a sequence of points $\{(x_m, y_m)\}$ such that $f(x_m, y_m) \to M$. By Lemma 2.1 this sequence contains a Cauchy subsequence $\{(x_{m_k}, y_{m_k})\}$ having a limit point belonging to D on which the limit of f exists; by continuity of f, this limit is equal to M. Similarly proved is the assertion regarding the infimum of f.

Weierstrass' theorem can be extended to closed and bounded (i.e., compact) domains D in $\mathbb{R}^n$ for any finite n.

2.5 Partial Derivatives

In discussing the coordinate lines in $\mathbb{R}^2$, we encountered partial derivatives along the trajectories. Now we generalize this. Let $f \colon D \subset \mathbb{R}^n \to \mathbb{R}$ with D an open set. Fixing the $n-1$ variables $x_1 = x_1{}^*, \ldots, x_n = x_n{}^*$ and letting x_k vary, we get a function

$$y = g(x_k) = f(x_1{}^*, \ldots, x_k, \ldots x_n{}^*)$$

in one variable. We can introduce the derivative

$$\lim_{h \to 0} \frac{f(x_1{}^* \ldots, x_k{}^* + h, \ldots, x_n{}^*)}{h}$$

if it exists. It is called the partial derivative of f with respect to x_k. The various ways in which it can be denoted in terms of f include

$$\frac{\partial f(x_1{}^*, \ldots, x_n{}^*)}{\partial x_k}, \qquad f_{x_k}(x_1{}^*, \ldots, x_n{}^*), \qquad f_k(x_1{}^*, \ldots, x_n{}^*),$$

and also

$$D_k f(x_1{}^*, \ldots, x_n{}^*), \qquad D_{x_k} f, \qquad D^{(0, \ldots, 1, \ldots, 0)} f,$$

where in the last form the subscript 1 stands in the kth position. Rather than showing the arguments in parentheses, we can show them as

$$f_{x_k}\Big|_{(x_1, \ldots, x_n) = (x_1{}^*, \ldots, x_n{}^*)}.$$

Sometimes the arguments are not indicated explicitly if they are clear from the context.

For partial derivatives we use the symbol ∂, reserving d for a "complete derivative" when the other variables can depend on the differentiation variable. For example, many authors would write

$$\frac{df(x, y(x))}{dx} = \frac{\partial f}{\partial x} + \frac{\partial f}{\partial y} \frac{dy}{dx},$$

where y is a function of x.

So the procedure for finding the partial derivative $\partial/\partial x$ of a function $f = f(x, y)$ is

(1) fix y as a constant, and then

(2) take the ordinary derivative of f with respect to x.

The partial derivative with respect to y is taken analogously.

Example 2.13. For $f(x, y) = x^2 + xy + y^2 + \sin xy$, we have

$$f_x(x, y) = \frac{\partial f(x, y)}{\partial x} = (x^2 + xy + y^2 + \sin xy)_x = 2x + y + 0 + y \cos xy,$$

$$f_y(x, y) = \frac{\partial f(x, y)}{\partial y} = (x^2 + xy + y^2 + \sin xy)_y = 0 + x + 2y + x \cos xy.$$

$\square$

Example 2.14. Letting $f(x, y) = x^5 - x^4 y - y^3$, we find $f_x(x, y)$ at the point $(2, 1)$:

$$f_x(2, 1) = (5x^4 - 4x^3 y)\Big|_{(x,y)=(2,1)} = (5x^4 - 4x^3 y)\Big|_{\substack{x=2 \\ y=1}} = 48. \qquad \square$$

In a similar way we can introduce partial derivatives of higher order. We get second-order derivatives by differentiating first-order partial derivatives. There are small peculiarities in notation. The derivatives

$$\frac{\partial^2 f}{\partial y\, \partial x} = \frac{\partial}{\partial y}\left(\frac{\partial f}{\partial x}\right) = \frac{\partial(f_x)}{\partial y} = f_{xy},$$

$$\frac{\partial^2 f}{\partial x\, \partial y} = \frac{\partial}{\partial x}\left(\frac{\partial f}{\partial y}\right) = \frac{\partial(f_y)}{\partial x} = f_{yx},$$

are said to be *mixed* partial derivatives. The non-mixed derivatives are

$$\frac{\partial^2 f}{\partial x^2} = \frac{\partial}{\partial x}\left(\frac{\partial f}{\partial x}\right) = f_{xx}, \qquad \frac{\partial^2 f}{\partial y^2} = \frac{\partial}{\partial y}\left(\frac{\partial f}{\partial y}\right) = f_{yy}.$$

For higher-order partial derivatives one may also use the following notation. Let the "multi-index" $\alpha = (\alpha_1, \ldots, \alpha_n)$ denote the order of the differentiation and write $|\alpha| = \alpha_1 + \cdots + \alpha_n$. Then

$$\frac{\partial^{|\alpha|} f}{\partial x_1^{\alpha_1} \ldots \partial x_n^{\alpha_n}} = D^\alpha f = \partial_\alpha f.$$

Such a notation is permissible only when the order in which one takes mixed partial derivatives is immaterial. For mixed derivatives, we should calculate repeated limits. From the theory of limits we know that repeated limits taken in different orders can be unequal. We are interested to learn when a mixed derivative does not depend on the order of differentiation. An answer is given by

Theorem 2.5 (Clairaut). *Let f be defined on open set $D \subset \mathbb{R}^2$ with the functions f_{xy} and f_{yx} continuous on D. Then at each $(a,b) \in D$ we have $f_{xy}(a,b) = f_{yx}(a,b)$; that is, the mixed derivative does not depend on the order in which the derivatives are taken.*

The proof is a bit lengthy and we omit it.

Note that in order to have f_{xy} and f_{yx} continuous we should have continuous first partial derivatives f_x and f_y.

Example 2.15. For $f(x,y) = e^{xy}$ we obtain

$$f_{xy} = f_{yx} = (1 + xy)e^{xy}.$$ □

Clairaut's theorem extends to mixed derivatives of higher order. Say, if f_{xxy}, f_{xyx}, and f_{yxx} are continuous in D, they are equal at each point of D. Here f and all its first and second derivatives must also be continuous in D; this is necessary for the existence of continuous third-order mixed derivatives of f.

To grasp the significance of Clairaut's theorem, it suffices to note that this result directly supports a large portion of elementary thermodynamics.

Some Important Partial Differential Equations

The entire theory of partial differential equations is based on the notion of partial derivative. This theory is the most important aspect of those mathematical techniques used in continuum physics and, in particular, in continuum mechanics. Of course it lies outside the official goals of the present book. However, it is worth pausing to examine a few of the central partial differential equations. One that is encountered in practically all phases of continuum physics is *Laplace's equation*

$$u_{xx} + u_{yy} = 0$$

for an unknown function u in two variables x and y. This is frequently written as

$$\nabla^2 u = 0 \quad \text{or} \quad \Delta u = 0, \tag{2.9}$$

where ∇ is the nabla operator to be treated in Section 2.6, and Δ is called the Laplacian operator. Laplace's equation belongs to the class of linear elliptic equations. Taken together with boundary conditions, it constitutes a boundary value problem. The most common conditions on the boundary ∂D are

$$\left.(u - \phi)\right|_{\partial D} = 0 \qquad \textit{Dirichlet's condition}$$

and

$$\left.\left(\frac{\partial u}{\partial n} - \psi\right)\right|_{\partial D} = 0 \qquad \textit{Neumann's condition}$$

with functions ϕ and ψ given on ∂D. The Dirichlet boundary value problem has a unique solution for a wide class of functions ϕ, while the Neumann problem determines a solution only up to an indefinite constant. Unfortunately further details regarding existence and uniqueness must be left to specialized texts on boundary value problems. Laplace's equation is also formulated in three space dimensions, where it becomes $u_{xx} + u_{yy} + u_{zz} = 0$ but is still written as (2.9).

The two-dimensional Laplace equation is a special case of another elliptic equation, *Poisson's equation*

$$u_{xx} + u_{yy} = -f, \tag{2.10}$$

which describes the equilibrium of a flat membrane under given load $f = f(x, y)$; here the unknown function $u = u(x, y)$ is the transverse membrane deflection. This equation is encountered in other areas of continuum physics, where u and f have other meanings. To describe the equilibrium problem for a flat membrane, Poisson's equation should be accompanied by Dirichlet, Neumann, or other types of boundary conditions.

Also important in classical mathematical physics are the wave equation

$$u_{tt} - c^2 u_{xx} = 0$$

and the diffusion or heat equation

$$u_t = a^2 u_{xx}$$

where t is time and x is a spatial coordinate. In two or three space dimensions the term u_{xx} is replaced by $\nabla^2 u$, yielding

$$u_{tt} - c^2 \nabla^2 u = 0, \qquad u_t - a^2 \nabla^2 u = 0,$$

respectively. The wave equation is classed as a linear hyperbolic equation, while the diffusion equation is classed as a linear parabolic equation. For uniqueness of solution they should be equipped with Dirichlet or Neumann type boundary conditions and, additionally, with initial conditions at some time instant $t = t_0$.

These partial differential equations are important enough to occupy entire books. The reader of a typical calculus text will merely be asked to verify that given functions are solutions or that they satisfy certain special properties.

Example 2.16. To prove that $u(x, y) = e^x \sin y$ is a solution of Laplace's equation $u_{xx} + u_{yy} = 0$, we can use direct substitution. Indeed, calculation shows that

$$u_{xx} + u_{yy} = e^x \sin y - e^x \sin y$$

as required. □

2.6 Total Differential

The First Differential: Brief Review

Before proceeding to the topic of this section, let us review the notion of first differential for a function of just one variable. Consider a function $y = f(x)$ at point $x \in (a, b)$. If Δx denotes a change (assumed positive) in x, and f is continuous and differentiable on $[x, x + \Delta x]$, then by Lagrange's mean value theorem there is a point $\xi \in [x, x + \Delta x]$ such that

$$\Delta y = \Delta f(x, \Delta x) \equiv f(x + \Delta x) - f(x) = f'(\xi) \Delta x.$$

If, moreover, f' is continuous at x, we have

$$f'(\xi) = f'(x) + \varepsilon \quad \text{with } \varepsilon \to 0 \text{ as } \Delta x \to 0$$

so that

$$\Delta f(x, \Delta x) = f'(x) \Delta x + \varepsilon \Delta x.$$

Since $\varepsilon \to 0$ as $\Delta x \to 0$, the term $f'(x) \Delta x$ approximates $\Delta f(x, \Delta x)$ near the point x. As the best linear approximation of $\Delta f(x, \Delta x)$ with respect to Δx at point x, it is called the *first differential* of f at x and is denoted dy. For Δx we also use the two-letter notation dx, and hence obtain by definition

$$dy \equiv df = f'(x) \, dx. \tag{2.11}$$

Note that we have introduced two variables, dx and dy, each denoted by two letters. Despite the impression of some students that dx and dy are something "infinitesimal," which comes from the symbol dy/dx for the derivative, dx is an ordinary variable that can take any value. The symbol dy denotes a new dependent variable; the value of the function dy depends on two variables, x through $y'(x)$, and dx. Note that dy depends on dx linearly. Again, neither of these quantities should be regarded as "infinitesimal." We know that $f'(x) = \tan\alpha$, where α is the angle between the x-axis and the tangent to the curve $y = f(x)$ at x. Since

$$f(x + \Delta x) \approx f(x) + f'(x)\,\Delta x,$$

the right-hand side gives the tangent line at point a with $\Delta x = x - a$:

$$y = f(a) + f'(a)(x - a).$$

Total Differential for a Function in Two Variables

Let us take the idea reviewed above and extend it to a function $z = f(x, y)$ having partial derivatives at each point of some open domain D. The desired analogue, also known as the *total differential*, is given by

$$dz = f_x(x, y)\,dx + f_y(x, y)\,dy. \tag{2.12}$$

Note that it does reduce to (2.11) if f does not depend on y. An alternate notation is

$$dz = z_x(x, y)\,dx + z_y(x, y)\,dy.$$

As before, dx and dy represent changes in x and y, respectively, which can be denoted also by Δx and Δy.

Now we deduce (2.12). The change in z corresponding to the changes in x and y by Δx and Δy is

$$\Delta z = f(x + \Delta x, y + \Delta y) - f(x, y).$$

Let us transform this expression by adding and subtracting $f(x, y + \Delta y)$:

$$\Delta z = [f(x + \Delta x, y + \Delta y) - f(x, y + \Delta y)] + [f(x, y + \Delta y) - f(x, y)].$$

Each bracketed term on the right can be considered as Δf obtained via a change in just one variable. Two applications of Lagrange's mean value theorem yield

$$f(x + \Delta x, y + \Delta y) - f(x, y + \Delta y) = f_x(\xi, y + \Delta y)\,\Delta x$$

and

$$f(x, y + \Delta y) - f(x, y) = f_y(x, \eta)\, \Delta y$$

for some $\xi \in [x, x + \Delta x]$ and $\eta \in [y, y + \Delta y]$. Now suppose the partial derivatives f_x, f_y are both continuous at the point (x, y). This implies that

$$f_x(\xi, y + \Delta y) = f_x(x, y) + \varepsilon_1 \quad \text{and} \quad f_y(x, \eta) = f_y(x, y) + \varepsilon_2$$

with $\varepsilon_1, \varepsilon_2$ tending to zero when $(\Delta x, \Delta y) \to (0,0)$. Substituting into Δz, we obtain

$$\Delta z = [f_x(x, y)\, \Delta x + f_y(x, y)\, \Delta y] + (\varepsilon_1\, \Delta x + \varepsilon_2\, \Delta y). \qquad (2.13)$$

It follows that

$$\frac{|\Delta z - [f_x(x, y)\, \Delta x + f_y(x, y)\, \Delta y]|}{(\Delta x^2 + \Delta y^2)^{1/2}} = \frac{|\varepsilon_1\, \Delta x + \varepsilon_2\, \Delta y|}{(\Delta x^2 + \Delta y^2)^{1/2}} \to 0$$

as $(\Delta x, \Delta y) \to (0,0)$. So $f_x(x, y)\, \Delta x + f_y(x, y)\, \Delta y$ is the best linear approximation to Δz with respect to $(\Delta x, \Delta y)$ in a small neighborhood of (x, y). We denote it by dz. Writing $\Delta x = dx$ and $\Delta y = dy$, we get (2.12).

We have deduced the expression for dz under some conditions placed on the first partial derivatives. But the same dz holds for a function that is differentiable at a point; this notion is defined as follows.

Definition 2.6. A function f is *differentiable* at point (x, y) if its increment takes the form (2.13). We call f *continuously differentiable* at (x, y) if f_x and f_y are continuous at (x, y). The portion of the increment (2.13) that is linear in Δx and Δy, namely the quantity $f_x(x, y)\, \Delta x + f_y(x, y)\, \Delta y$, is the *differential* of f and is denoted df.

Example 2.17. Consider the function $f(x, y) = xy^2$. We have

$$\Delta z = f(x + \Delta x, y + \Delta y) - f(x, y)$$
$$= (x + \Delta x)(y + \Delta y)^2 - xy^2$$
$$= y^2\, \Delta x + 2xy\, \Delta y + x\, \Delta y^2 + 2y\, \Delta x\, \Delta y + \Delta x\, \Delta y^2.$$

But

$$\lim_{(\Delta x, \Delta y) \to (0,0)} \frac{x\, \Delta y^2 + 2y\, \Delta x\, \Delta y + \Delta x\, \Delta y^2}{(\Delta x^2 + \Delta y^2)^{1/2}} = 0$$

since no term in the numerator is of degree less than two in $\Delta x, \Delta y$. Hence f is differentiable at any point (x, y) and $df = y^2\, \Delta x + 2xy\, \Delta y$. □

It is worth emphasizing that the right member of (2.12) contains four independent variables x, y, dx, dy, while the left member is the value of a function dz that depends on these. There is nothing "infinitesimal" about dx, dy, or dz; they are ordinary variables.

Example 2.18. Let us find df for $f(x,y) = x + 2y + \sin(xy)$ at the point $(1,2)$. Since $f_x = 1 + y\cos(xy)$ and $f_y = 2 + x\cos(xy)$, we have

$$df(x,y,dx,dy) = [1 + y\cos(xy)]\,dx + [2 + x\cos(xy)]\,dy,$$

$$df(1,2,dx,dy) = (1 + 2\cos 2)\,dx + (2 + \cos 2)\,dy. \qquad \square$$

Let us rewrite (2.13) with $x = a$, $y = b$, $\Delta x = x - a$, and $\Delta y = y - b$:

$$\Delta z = f(x,y) - f(a,b)$$
$$= f_x(a,b)(x-a) + f_y(a,b)(y-b) + [\varepsilon_1(x-a) + \varepsilon_2(y-b)].$$

Denoting by z the linear approximation to $f(x,y)$ in both $x - a$ and $y - b$, i.e., $\Delta z = z - f(a,b)$, we get the equation of a plane:

$$z = f(a,b) + f_x(a,b)(x-a) + f_y(a,b)(y-b). \qquad (2.14)$$

This plane is tangent to the graph of $z = f(x,y)$ at the point (a,b). Indeed, at (a,b) we have $z(a,b) = f(a,b)$ while $z_x(a,b) = f_x(a,b)$ and $z_y(a,b) = f_y(a,b)$. So the tangents to the intersection of the planes parallel to the xz- and yz-planes with graph $z = f(x,y)$ which pass through $(a,b,f(a,b))$ lie in the plane described by (2.14).

Similarly defined is the total differential for a function in any number of variables. For a function $f = f(x,y,z)$, the total differential

$$df(x,y,z,dx,dy,dz) = f_x(x,y,z)\,dx + f_y(x,y,z)\,dy + f_z(x,y,z)\,dz \quad (2.15)$$

gives the best linear approximation to Δf in terms of dx, dy, dz in a small vicinity of the point $(x,y,z) \in \mathbb{R}^3$.

It is worth noting that in the theory of asymptotic expansions, (2.14) is called the first-order asymptotic expansion of f at point (a,b). This is because in the vicinity of (a,b) the quantity $z - f(x,y)$, with z given by formula (2.14), is of higher order of smallness than $[(x-a)^2 + (y-b)^2]^{1/2}$.

The fact that approximation by the differential is linear with respect to the differentials of the independent variables is used in the approximate theory of errors in engineering calculations. The linearity of the total differential allows us to quickly estimate small errors; moreover, it allows us to estimate the maximum error in view of the linearity of Δf with respect to the errors Δx, Δy, Δz, etc., as the maximum value occurs on the boundary of the domain of possible errors of the parameters.

Example 2.19. The area of an ellipse is given by $S(a,b) = \pi a b$ where a, b are the lengths of the semi-axes. We can approximate the error in the area due to small errors in a and b. Since the derivatives of S with respect to the two length parameters are $S_a(a,b) = \pi b$ and $S_b(a,b) = \pi a$, the absolute error of S is

$$|\Delta S| \approx |\pi b \, \Delta a + \pi a \, \Delta b|$$

and the relative error is

$$\left| \frac{S(a+\Delta a, b+\Delta b) - S(a,b)}{S(a,b)} \right| \approx \left| \frac{\pi(b \, \Delta a + a \, \Delta b)}{\pi a b} \right| = \left| \frac{\Delta a}{a} + \frac{\Delta b}{b} \right|.$$

The maximum errors are estimated by substituting the maximum errors of Δa and Δb into the formula. $\square$

Example 2.20. A circular right cylinder has radius R, altitude h, and volume V. We seek the relative error in V, produced by small errors in R and h. From $V = \pi R^2 h$ we obtain $V_R = 2\pi R h$ and $V_h = \pi R^2$, hence

$$\left| \frac{\Delta V}{V} \right| = \left| \frac{V_R \Delta R}{V} + \frac{V_h \Delta h}{V} \right| = \left| \frac{2}{R} \Delta R + \frac{1}{h} \Delta h \right|.$$

If the absolute values of maximum errors in both R and h are the same value $\varepsilon > 0$, then the maximum error in V is

$$|\Delta V_{\max}| = \pi(2Rh + R^2)\varepsilon. \qquad \square$$

We end this section with a

Theorem 2.6. *If f is differentiable at (x,y), it is continuous at (x,y).*

Proof. By definition, if f is differentiable at (x,y) then

$$f(x+\Delta x, y+\Delta y) - f(x,y) = [f_x(x,y)\,\Delta x + f_y(x,y)\,\Delta y] + (\varepsilon_1 \, \Delta x + \varepsilon_2 \, \Delta y)$$

where $\varepsilon_1 \to 0$ and $\varepsilon_2 \to 0$ as $(\Delta x, \Delta y) \to (0,0)$. This means that

$$\lim_{(\Delta x, \Delta y) \to (0,0)} f(x+\Delta x, y+\Delta y) = f(x,y)$$

which in turn means that f is continuous at (x,y). $\square$

It follows that if f is not continuous at (x, y), then it is not differentiable there.

Example 2.21. The function

$$f(x, y) = \begin{cases} 1, & xy \neq 0, \\ 0, & xy = 0, \end{cases}$$

is zero on the coordinate axes and unity elsewhere. It is discontinuous at $(0, 0)$, although its partial derivatives f_x and f_y both exist there. Hence it is not differentiable at $(0, 0)$. □

Of course, mere continuity does not imply differentiability. Even a function of a single variable, such as

$$f(x) = \begin{cases} x \sin(1/x), & x \neq 0, \\ 0, & x = 0, \end{cases}$$

is enough to demonstrate this.

2.7   Vector Form of Total Differential; Nabla Operator

We can cast the total differential into vector form, starting with the position vector $\mathbf{r} = x\,\mathbf{i} + y\,\mathbf{j} + z\,\mathbf{k}$. The differential of $\mathbf{r}$ is

$$d\mathbf{r} = dx\,\mathbf{i} + dy\,\mathbf{j} + dz\,\mathbf{k}. \tag{2.16}$$

Let us introduce a vector quantity

$$\nabla f = f_x\,\mathbf{i} + f_y\,\mathbf{j} + f_z\,\mathbf{k} \quad \text{or} \quad \nabla f = (f_x, f_y, f_z). \tag{2.17}$$

Then

$$df = f_x\,dx + f_y\,dy + f_z\,dz = \nabla f(x, y, z) \cdot d\mathbf{r}. \tag{2.18}$$

The quantity ∇f is called the *gradient* of f. Later we shall find that our ability to represent df in component-free vector form, using objective vectors and the dot product only, plays an important role in deducing the total differential for functions defined in non-Cartesian coordinates.

Example 2.22. Let us calculate the gradient of $f(x, y, z) = \ln(x + y + z)$ at $(1, 2, -1)$:

$$\nabla f(1, 2, -1) = \left. \frac{1}{x + y + z} \right|_{(1,2,-1)} (\mathbf{i} + \mathbf{j} + \mathbf{k}) = \frac{\mathbf{i} + \mathbf{j} + \mathbf{k}}{2}. \qquad □$$

Now we introduce an operator via the expression (2.18):

$$\nabla = \mathbf{i}\,\frac{\partial}{\partial x} + \mathbf{j}\,\frac{\partial}{\partial y} + \mathbf{k}\,\frac{\partial}{\partial z}. \tag{2.19}$$

It is known as the *nabla* or *gradient* operator. Formally, as it contains the frame vectors, it may appear at first to be a vector. In linear algebra something is a vector if it belongs to a vector space, which is a set of these entities satisfying the vector space axioms. According to physics, vectors (also known as axial vectors) have directions and magnitudes; they can be added by the parallelogram rule and can be multiplied by scalars. In this sense ∇ is not a physical vector, but we call it a vector-operator. It is an important operator in many theories dealing with objects in space (e.g., elasticity, hydrodynamics, electrodynamics).

The nabla operator can be extended to functions $f = f(x_1, \dots, x_n)$ in any finite number of variables:

$$\nabla = \mathbf{i}_1\,\frac{\partial}{\partial x_1} + \cdots + \mathbf{i}_n\,\frac{\partial}{\partial x_n}, \tag{2.20}$$

where $\mathbf{i}_1, \dots, \mathbf{i}_n$ is an orthonormal basis in $\mathbb{R}^n$. In physics it is also denoted by grad. So for $n = 3$,

$$\nabla = \mathrm{grad} = \mathbf{i}\,\frac{\partial}{\partial x} + \mathbf{j}\,\frac{\partial}{\partial y} + \mathbf{k}\,\frac{\partial}{\partial z}.$$

Although nabla is a vector only formally, we can involve it in product-type operations. The most important of these are the following.

1. Divergence. Take a vector function $\mathbf{F} = \mathbf{i}\,P + \mathbf{j}\,Q + \mathbf{k}\,R$. Dot "multiplying" ∇ by $\mathbf{F}$, we get the divergence of $\mathbf{F}$:

$$\nabla \cdot \mathbf{F} = \left(\mathbf{i}\,\frac{\partial}{\partial x} + \mathbf{j}\,\frac{\partial}{\partial y} + \mathbf{k}\,\frac{\partial}{\partial z} \right) \cdot (\mathbf{i}\,P + \mathbf{j}\,Q + \mathbf{k}\,R)$$

$$= \frac{\partial P}{\partial x} + \frac{\partial Q}{\partial y} + \frac{\partial R}{\partial z}. \tag{2.21}$$

We encounter $\nabla \cdot \mathbf{F}$ in Gauss' divergence theorem and in the law of mass conservation for an incompressible liquid. Note that $\nabla \cdot \mathbf{F}$ is a scalar quantity.

2. Rotor. Using the cross product for ∇ and $\mathbf{F}$, we define the rotor (or curl) of $\mathbf{F}$:

$$\nabla \times \mathbf{F} = \begin{vmatrix} \mathbf{i} & \mathbf{j} & \mathbf{k} \\ \dfrac{\partial}{\partial x} & \dfrac{\partial}{\partial y} & \dfrac{\partial}{\partial z} \\ P & Q & R \end{vmatrix} = \mathbf{i}\,(R_y - Q_z) + \mathbf{j}\,(P_z - R_x) + \mathbf{k}\,(Q_x - P_y). \quad (2.22)$$

At each point the result "looks like" a vector. We use this phrasing because the sign of the cross product between two vectors depends on the coordinate system orientation. Such quantities are called *pseudovectors*. The present result is for right-handed frames. We shall encounter the rot operator in Stokes' theorem. It is an important tool in electrodynamics and hydromechanics.

3. Laplacian. Produced from the dot product of two gradients, this operator is denoted

$$\nabla^2 = \Delta = \nabla \cdot \nabla = \left(\mathbf{i}\,\frac{\partial}{\partial x} + \mathbf{j}\,\frac{\partial}{\partial y} + \mathbf{k}\,\frac{\partial}{\partial z} \right) \cdot \left(\mathbf{i}\,\frac{\partial}{\partial x} + \mathbf{j}\,\frac{\partial}{\partial y} + \mathbf{k}\,\frac{\partial}{\partial z} \right)$$

$$= \frac{\partial^2}{\partial x^2} + \frac{\partial^2}{\partial y^2} + \frac{\partial^2}{\partial z^2}. \quad (2.23)$$

When applied to a function it yields

$$\nabla^2 f = f_{xx} + f_{yy} + f_{zz}. \quad (2.24)$$

We saw it appear in the partial differential equations (2.9) and (2.10).

The Laplacian can act on a vector function $\mathbf{F} = \mathbf{i}\,P + \mathbf{j}\,Q + \mathbf{k}\,R$:

$$\nabla^2 \mathbf{F} = \nabla^2 P\,\mathbf{i} + \nabla^2 Q\,\mathbf{j} + \nabla^2 R\,\mathbf{k}. \quad (2.25)$$

Again, it is encountered in nearly all the spatial continuum theories of physics.

All the above operators can be used in two-dimensional versions:

$$\text{grad} = \nabla = \mathbf{i}\,\frac{\partial}{\partial x} + \mathbf{j}\,\frac{\partial}{\partial y},$$

$$\nabla f = \mathbf{i}\,f_x + \mathbf{j}\,f_y,$$

$$\nabla \cdot \mathbf{G} = p_x + q_y \quad \text{for} \quad \mathbf{G} = p\,\mathbf{i} + q\,\mathbf{j}, \quad \mathbf{G} = \mathbf{G}(x, y),$$

$$\nabla \times \mathbf{G} = \begin{vmatrix} \mathbf{i} & \mathbf{j} & \mathbf{k} \\ \dfrac{\partial}{\partial x} & \dfrac{\partial}{\partial y} & 0 \\ p & q & 0 \end{vmatrix} = \mathbf{k}\,(q_x - p_y),$$

$$\nabla^2 f = f_{xx} + f_{yy}.$$

These operators satisfy certain identities that are used in applications. A few appear in the chapter problems.

2.8 Directional Derivative

The value of the derivative of an ordinary function tells us how the function increases at a point. Using partial derivatives for $z = f(x,y)$ (or of any number of variables), we can learn how the function changes at a point as we move parallel to the Cartesian axes. But in many circumstances we must know the change of $f(x,y)$ along an arbitrary direction as well, and find the direction along which it changes most rapidly. Hence we introduce directional derivatives. The direction at point (a,b) will be defined by a unit vector $\mathbf{v} = v_1\,\mathbf{i} + v_2\,\mathbf{j}$. To find the rate of change of f at (a,b) along the direction $\mathbf{v}$, we can use the expression (2.13) from which we deduced the total differential with $\Delta x = v_1 \Delta h$ and $\Delta y = v_2 \Delta h$, where Δh characterizes the change of the absolute value of the argument $x\,\mathbf{i} + y\,\mathbf{j}$ along $\mathbf{v}$. Now

$$\frac{\Delta f}{\Delta h} = \frac{f_x(a,b)v_1\,\Delta h + f_y(a,b)v_2\,\Delta h}{\Delta h} + \frac{\varepsilon_1 v_1\,\Delta h + \varepsilon_2 v_2\,\Delta h}{\Delta h}.$$

Taking the limit as $\Delta h \to 0$ we get

$$\lim_{\Delta h \to 0} \frac{\Delta f}{\Delta h} = f_x(a,b)v_1 + f_y(a,b)v_2.$$

We can recast the right side as $\nabla f(a,b) \cdot \mathbf{v}$. This is called the *directional derivative* of f along $\mathbf{v}$ at point (a,b). It is usually denoted

$$D_{\mathbf{v}} f(a,b) = \nabla f(a,b) \cdot \mathbf{v}. \tag{2.26}$$

For $\mathbf{v} = \mathbf{i}$ and $\mathbf{v} = \mathbf{j}$ the results are

$$D_{\mathbf{i}} f(a,b) = \nabla f(a,b) \cdot \mathbf{i} = \frac{\partial f}{\partial x},$$

$$D_{\mathbf{j}} f(a,b) = \nabla f(a,b) \cdot \mathbf{j} = \frac{\partial f}{\partial y}.$$

The vectorial form of the directional derivative is unaltered for a function $z = f(x_1, \ldots, x_n)$ in any finite number of variables, with evident changes in the gradient, the vector $\mathbf{v}$, and the point:

$$D_\mathbf{v} f(a_1, \ldots, a_n) = \nabla f(a_1, \ldots, a_n) \cdot \mathbf{v}. \tag{2.27}$$

In some textbooks the authors do not suppose that $|\mathbf{v}| = 1$, but if we wish to compare the rates of change of f in various directions at a point, we should impose this condition.

In two and three dimensions we know that the value on the right can be written as

$$|\nabla f(a_1, \ldots, a_n)| \, |\mathbf{v}| \cos \alpha,$$

where α is the angle between the vectors $\nabla f(a_1, \ldots, a_n)$ and $\mathbf{v}$. Because $|\mathbf{v}| = 1$, the maximum of $D_\mathbf{v} f$ occurs when $\cos \alpha = 1$; this is when $\mathbf{v}$ is directed along the gradient of f at the point. The minimum occurs when $\cos \alpha = -1$, i.e., when $\mathbf{v}$ is directed opposite the gradient of f at the same point. If $\mathbf{v}$ is orthogonal to $\nabla f(a_1, \ldots, a_n)$, then $D_\mathbf{v} f(a_1, \ldots, a_n) = 0$. So in the two dimensional case there are two directional vectors $\mathbf{v}$ for which the directional derivative vanishes. And in the three dimensional case, for any directional vector parallel to the plane orthogonal to $\nabla f(a_1, \ldots, a_n)$ the directional derivative is zero.

Thus at point $\mathbf{x}$, *the gradient* $\nabla f(\mathbf{x})$ *yields the direction in which the directional derivative of* f *takes its maximum value* $|\nabla f(\mathbf{x})|$. *The function* f *decreases most rapidly in the direction of* $-\nabla f(\mathbf{x})$.

Example 2.23. Let $f(x, y, z) = xyz$. We have

$$\nabla f = yz\,\mathbf{i} + xz\,\mathbf{j} + xy\,\mathbf{k}$$

and, say, $\nabla f(1, 4, 3) = 12\,\mathbf{i} + 3\,\mathbf{j} + 4\,\mathbf{k}$. At that same point, the directional derivative of f in the direction of the vector $-3\,\mathbf{j} - 6\,\mathbf{k}$ is

$$(12\,\mathbf{i} + 3\,\mathbf{j} + 4\,\mathbf{k}) \cdot \frac{-3\,\mathbf{j} - 6\,\mathbf{k}}{(9 + 36)^{1/2}} = -\frac{11}{5^{1/2}}. \qquad \square$$

Writing $\mathbf{x} = (x_1, \ldots, x_n)$, we can also write $f(x_1, \ldots, x_n) = f(\mathbf{x})$ and, for the directional derivative,

$$D_\mathbf{v} f(\mathbf{x}) = \nabla f(\mathbf{x}) \cdot \mathbf{v}. \tag{2.28}$$

Note that the directional derivative depends on the two independent vector variables $\mathbf{x}$ and $\mathbf{v}$ or, in components, on the $2n$ scalar variables $x_1, \ldots, x_n$ and $v_1, \ldots, v_n$.

We can successively produce directional derivatives of higher order. For example, the directional derivative of second order is

$$D_\mathbf{v}^2 f(\mathbf{x}) = D_\mathbf{v}[\nabla f(\mathbf{x}) \cdot \mathbf{v}]$$
$$= D_\mathbf{v}[f_{x_1}(\mathbf{x})v_1 + \cdots + f_{x_n}(\mathbf{x})v_n]$$
$$= [f_{x_1 x_1}v_1 + \cdots + f_{x_1 x_n}v_n]v_1 + \cdots + [f_{x_n x_1}v_1 + \cdots + f_{x_n x_n}v_n]v_n.$$

In terms of the components of $\mathbf{v}$, the second directional derivative of f is a quadratic form. A matrix expression for $D_\mathbf{v}^2 f(\mathbf{x})$ is

$$D_\mathbf{v}^2 f(\mathbf{x}) = (\, v_1 \; \cdots \; v_n \,) \begin{pmatrix} f_{x_1 x_1} & \cdots & f_{x_1 x_n} \\ \vdots & \ddots & \vdots \\ f_{x_n x_1} & \cdots & f_{x_n x_n} \end{pmatrix} \begin{pmatrix} v_1 \\ \vdots \\ v_n \end{pmatrix} \qquad (2.29)$$

The square matrix $(f_{x_i x_j})$, known as the *Hessian*, appears in various theoretical constructions of differential multivariable calculus.

Introduced analogously are

$$D_\mathbf{v}^3 f(\mathbf{x}) = D_\mathbf{v}[D_\mathbf{v}^2 f(\mathbf{x})], \qquad D_\mathbf{v}^4 f(\mathbf{x}) = D_\mathbf{v}[D_\mathbf{v}^3 f(\mathbf{x})],$$

and so on. The kth-order directional derivative is a kth order homogeneous form in the components of $\mathbf{v}$.

2.9 The Chain Rule and Changes of Variable

We know that for a function in one variable $y = f(x)$, the change of variable $x = g(t)$ yields for the derivative df/dt the expression

$$\frac{df}{dt} = f_x'(g(t))\, g_t'(t). \qquad (2.30)$$

This is called the *chain rule* for the substitution $x = g(t)$.

Consider the derivative of a function $z = f(x, y)$, taken along a curve given by $x = x(t)$, $y = y(t)$, at $t = t_0$. Denote $x(t_0) = a$ and $y(t_0) = b$ and assume x, y are differentiable functions at $t = t_0$. The formula (2.13), which gave us the expression for the total differential, yields

$$\frac{\Delta f}{\Delta t} = f_x(a, b)\frac{\Delta x}{\Delta t} + f_y(a, b)\frac{\Delta y}{\Delta t} + \varepsilon_1 \frac{\Delta x}{\Delta t} + \varepsilon_2 \frac{\Delta y}{\Delta t}$$

with ε_k tending to zero along with Δx and Δy. As $\Delta t \to 0$ we obtain

$$\frac{df(t_0)}{dt} = \frac{df}{dt}\bigg|_{t=t_0} = f_x(x(t_0), y(t_0))x'(t_0) + f_y(x(t_0), y(t_0))y'(t_0), \quad (2.31)$$

where the terms with ε_k vanished because $\varepsilon_k \to 0$ as $\Delta t \to 0$. Formula (2.31) is the chain rule for a function in two variables $z = f(x, y)$ if its variables depend on a single parameter t: $x = x(t)$, $y = y(t)$.

Example 2.24. Let $f(x, y) = x \cos y + e^x \sin y$ where $x(t) = t^2 + 1$ and $y(t) = t^3 + t$. We find the derivative of f at the point $(a, b) = (1, 0)$ corresponding to $t = 0$:

$$\left. \frac{df}{dt} \right|_{t=0} = f_x(a, b) x'(0) + f_y(a, b) y'(0) \quad \text{where} \quad \begin{cases} a = x(0) = 1, \\ b = y(0) = 0. \end{cases}$$

Here

$$f_x(x, y) = \cos y + e^x \sin y, \qquad x'(t) = 2t, \qquad x'(0) = 0,$$
$$f_y(x, y) = -x \sin y + e^x \cos y, \qquad y'(t) = 3t^2 + 1, \qquad y'(0) = 1,$$

and we obtain

$$\left. \frac{df}{dt} \right|_{t=0} = 1. \qquad \qquad \square$$

We can rewrite (2.31) in the form

$$\frac{df}{dt} = \frac{\partial f}{\partial x} \frac{dx}{dt} + \frac{\partial f}{\partial y} \frac{dy}{dt} \tag{2.32}$$

but we must regard this as a mere shorthand expression.

If we have a function $f = f(x, y, z)$ in three variables and seek the derivative along a curve $x = x(t)$, $y = y(t)$, $z = z(t)$ at point (a, b, c), where $a = x(t_0)$, $b = y(t_0)$, $c = z(t_0)$, we get a similar result:

$$\left. \frac{df}{dt} \right|_{t=t_0} = f_x(a, b, c) x'(t_0) + f_y(a, b, c) y'(t_0) + f_z(a, b, c) z'(t_0) \tag{2.33}$$

or in vector form

$$\left. \frac{df}{dt} \right|_{t=t_0} = \nabla f(a, b, c) \cdot \mathbf{r}'(t_0). \tag{2.34}$$

Next we seek the partial derivatives under the change of variables $x = x(s, t)$, $y = y(s, t)$ in the function $z = f(x, y)$. We assume that $x = x(s, t)$, $y = y(s, t)$, and f are differentiable. Substitution gives

$$z = f(x(s, t), y(s, t)).$$

Let us take the partial derivative of f with respect to s. In this case the variable t is held fixed, and we are led back to formula (2.32) for the chain rule with the evident change d/ds to $\partial/\partial s$. So

$$\frac{\partial f}{\partial s} = f_x(x(s,t), y(s,t)) \frac{\partial x(s,t)}{\partial s} + f_y(x(s,t), y(s,t)) \frac{\partial y(s,t)}{\partial s}.$$

Similarly

$$\frac{\partial f}{\partial t} = f_x(x(s,t), y(s,t)) \frac{\partial x(s,t)}{\partial t} + f_y(x(s,t), y(s,t)) \frac{\partial y(s,t)}{\partial t}.$$

We can write these more concisely as

$$\frac{\partial f}{\partial s} = \frac{\partial f}{\partial x}\frac{\partial x}{\partial s} + \frac{\partial f}{\partial y}\frac{\partial y}{\partial s}, \qquad \frac{\partial f}{\partial t} = \frac{\partial f}{\partial x}\frac{\partial x}{\partial t} + \frac{\partial f}{\partial y}\frac{\partial y}{\partial t}, \qquad (2.35)$$

or in the vector forms

$$\frac{\partial f}{\partial s} = \nabla f \cdot \mathbf{r}_s, \qquad \frac{\partial f}{\partial t} = \nabla f \cdot \mathbf{r}_t, \qquad (2.36)$$

where

$$\nabla = \mathbf{i}\frac{\partial}{\partial x} + \mathbf{j}\frac{\partial}{\partial y}, \qquad \mathbf{r} = x(s,t)\,\mathbf{i} + y(s,t)\,\mathbf{j}.$$

Example 2.25. Suppose $f(x,y) = \sin(xy)$ where $x = s - t$ and $y = s + t$. Then

$$\frac{\partial f}{\partial s} = (s+t)\cos(s^2 - t^2) + (s-t)\cos(s^2 - t^2),$$

$$\frac{\partial f}{\partial t} = -(s+t)\cos(s^2 - t^2) + (s-t)\cos(s^2 - t^2). \qquad \square$$

For a function $f = f(x, y, z)$ in three variables under the change $x = x(s,t,r)$, $y = y(s,t,r)$, $z = z(s,t,r)$, we obtain by extension

$$\frac{\partial f(\mathbf{x})}{\partial s} = \nabla f(\mathbf{x}) \cdot \mathbf{x}_s, \qquad \frac{\partial f(\mathbf{x})}{\partial t} = \nabla f(\mathbf{x}) \cdot \mathbf{x}_t, \qquad \frac{\partial f(\mathbf{x})}{\partial r} = \nabla f(\mathbf{x}) \cdot \mathbf{x}_r.$$

For a function $f(\mathbf{x})$ with $\mathbf{x} = (x_1, \ldots, x_n)$ and $x_k = x_k(t_1, \ldots, t_m)$,

$$\frac{\partial f(\mathbf{x}(t_1, \ldots, t_m))}{\partial t_i} = \nabla f(\mathbf{x}(t_1, \ldots, t_m)) \cdot \frac{\partial \mathbf{x}(t_1, \ldots, t_m)}{\partial t_i}.$$

To find second partial derivatives of f, we take the expression for the appropriate first derivative and apply the same differentiation process. For $z = f(x,y)$ where $x = x(s,t)$ and $y = y(s,t)$, we have by the product rule

$$f_{ss} = (f_x x_s + f_y y_s)_s$$

$$= (f_x x_s)_s + (f_y y_s)_s$$
$$= (f_x)_s x_s + f_x x_{ss} + (f_y)_s y_s + f_y y_{ss}$$
$$= (f_{xx} x_s + f_{xy} y_s) x_s + f_x x_{ss} + (f_{yx} x_s + f_{yy} y_s) y_s + f_y y_{ss}.$$

Analogously

$$f_{tt} = (f_{xx} x_t + f_{xy} y_t) x_t + f_x x_{tt} + (f_{yx} x_t + f_{yy} y_t) y_t + f_y y_{tt},$$
$$f_{st} = (f_{xx} x_t + f_{xy} y_t) x_s + f_x x_{st} + (f_{yx} x_t + f_{yy} y_t) y_s + f_y y_{st},$$
$$f_{ts} = (f_{xx} x_s + f_{xy} y_s) x_t + f_x x_{ts} + (f_{yx} x_s + f_{yy} y_s) y_t + f_y y_{ts}.$$

Similar formulas can be written for any number of variables and for derivatives of any order. See Problem 2.50 for an application.

We may encounter the derivatives of an expression $f(g(x, y))$ such as $f(x - cy)$. The external function f is a function in one variable and

$$\frac{\partial f(g(x, y))}{\partial x} = f'(g(x, y)) \frac{\partial g(x, y)}{\partial x},$$
$$\frac{\partial f(g(x, y))}{\partial y} = f'(g(x, y)) \frac{\partial g(x, y)}{\partial y}.$$

2.10 Change of Variables in Equations

Suppose we have a general partial differential equation of second order

$$F(x, y, f_x, f_y, f_{xx}, f_{xy}, f_{yy}) = 0$$

with respect to an unknown function f and we wish to express the equation x, y through variables u, v given by the formulas

$$x = g(u, v), \qquad y = h(u, v) \tag{2.37}$$

with smooth functions g, h such that the coordinates (x, y) are in one-to-one correspondence with the new coordinates (u, v).

Using the chain rule we have

$$f_x = f_u u_x + f_v v_x, \qquad f_y = f_u u_y + f_v v_y. \tag{2.38}$$

Let us start with the first derivatives u_x, u_y, v_x, v_y. For this we invert the relations (2.37) with respect to u, v, which is possible because (u, v) and (x, y) are in one-to-one correspondence:

$$u = u(x, y), \qquad v = v(x, y). \tag{2.39}$$

Substituting these into (2.37) we get the identities

$$x = g(u(x,y), v(x,y)), \qquad y = h(u(x,y), v(x,y)).$$

We can differentiate these with respect to x and y, obtaining

$$1 = g_u u_x + g_v v_x, \qquad\qquad 0 = h_u u_x + h_v v_x,$$

$$0 = g_u u_y + g_v v_y, \qquad\qquad 1 = h_u u_y + h_v v_y.$$

This system of four equations with respect to the four derivatives u_x, u_y, v_x, v_y has a solution expressed in terms of u, v. So $f_x = f_u u_x + f_v v_x$ and $f_y = f_u u_y + f_v v_y$ are also expressed in terms of u, v. Repeating the above process for their derivatives we get

$$f_{xx} = \frac{\partial(f_u u_x + f_v v_x)}{\partial u} u_x + \frac{\partial(f_u u_x + f_v v_x)}{\partial v} v_x,$$

$$f_{xy} = \frac{\partial(f_u u_x + f_v v_x)}{\partial u} u_y + \frac{\partial(f_u u_x + f_v v_x)}{\partial v} v_y,$$

$$f_{yy} = \frac{\partial(f_u u_y + f_v v_y)}{\partial u} u_y + \frac{\partial(f_u u_y + f_v v_y)}{\partial v} v_y.$$

Substituting these and relations (2.38) into the equation, we express it in terms of u, v. Note that the derivatives $u_x = u(x,y)_x$ and u_y, v_x, v_y should be expressed as functions of the variables u, v, so to find $\partial u_x / \partial u$ and other similar derivatives we do not require the derivatives u_{xx}, u_{xy}, u_{yy}. Similar transformations are available for equations and systems of partial differential equations if the functions depend on x, y, z or more independent variables.

Example 2.26. Let us transform Laplace's equation

$$u_{xx} + u_{yy} = 0$$

to the polar coordinate system. The relations between (x, y) and (r, θ) are

$$x = r\cos\theta, \qquad y = r\sin\theta, \qquad r^2 = x^2 + y^2, \qquad \tan\theta = y/x.$$

Note that in the last equality we cannot apply arctan to both sides as it is defined only in the first and fourth quadrants of the xy-plane (although the formulas will coincide at the end). Differentiation gives

$$2r r_x = 2x, \qquad 2r r_y = 2y, \qquad \frac{\theta_x}{\cos^2\theta} = -\frac{y}{x^2}, \qquad \frac{\theta_y}{\cos^2\theta} = \frac{1}{x},$$

hence

$$r_x = \frac{x}{r} = \cos\theta, \qquad \theta_x = -\frac{y}{x^2}\cos^2\theta = -\frac{r\sin\theta}{(r\cos\theta)^2}\cos^2\theta = -\frac{\sin\theta}{r},$$

$$r_y = \frac{y}{r} = \sin\theta, \qquad \theta_y = \frac{1}{r\cos\theta}\cos^2\theta = \frac{\cos\theta}{r}.$$

By the chain rule, the first partial derivatives of u are

$$u_x = u_r r_x + u_\theta \theta_x = u_r \cos\theta - u_\theta \frac{\sin\theta}{r}$$

$$u_y = u_r r_y + u_\theta \theta_y = u_r \sin\theta + u_\theta \frac{\cos\theta}{r}.$$

A second use of the chain rule (see Problem 2.51) gives

$$u_{xx} = u_{rr}\cos^2\theta - u_{r\theta}\frac{2\sin\theta\cos\theta}{r} + u_\theta\frac{2\sin\theta\cos\theta}{r^2} + u_r\frac{\sin^2\theta}{r} + u_{\theta\theta}\frac{\sin^2\theta}{r^2},$$

$$u_{yy} = u_{rr}\sin^2\theta + u_{r\theta}\frac{2\sin\theta\cos\theta}{r} - u_\theta\frac{2\sin\theta\cos\theta}{r^2} + u_r\frac{\cos^2\theta}{r} + u_{\theta\theta}\frac{\cos^2\theta}{r^2},$$

and by adding these equations we get

$$u_{rr} + \frac{1}{r}u_r + \frac{1}{r^2}u_{\theta\theta} = 0. \qquad\qquad \square$$

2.11   Taylor's Expansion

Let us extend the Taylor formula for a function in one variable to a function in n variables. It is easy to see that the directional derivative of $f(\mathbf{x})$ along unit vector $\mathbf{v} = (v_1, \ldots, v_n)$ is

$$D_\mathbf{v} f(\mathbf{x}) = \frac{d}{dt} f(x_1 + tv_1, \ldots, x_n + tv_n)\Big|_{t=0}. \qquad (2.40)$$

In the same way

$$D_\mathbf{v}^2 f(\mathbf{x}) = \frac{d^2}{dt^2} f(x_1 + tv_1, \ldots, x_n + tv_n)\Big|_{t=0} \qquad (2.41)$$

and for higher order directional derivatives

$$D_\mathbf{v}^k f(\mathbf{x}) = \frac{d^k}{dt^k} f(x_1 + tv_1, \ldots, x_n + tv_n)\Big|_{t=0}. \qquad (2.42)$$

Now for fixed $\mathbf{x}$ and a vector $\mathbf{v}$, let us consider the function $f(\mathbf{x}+t\mathbf{v}) = g(t)$ as a function of one variable t. Suppose f has all continuous derivatives up to the order $k+1$ in some vicinity of the fixed value $\mathbf{x}$. Then the function in the single variable t has continuous derivatives with respect to t up to order $k+1$ in some interval $(-\varepsilon, \varepsilon)$, hence g can be represented as a Taylor series in $(-\varepsilon, \varepsilon)$:

$$g(t) = g(0) + g'(0)t + \cdots + \frac{g^k(0)}{k!}t^k + R_k(t),$$

where the remainder term satisfies $R_k(t)/t^k \to 0$ as $t \to 0$. This yields Taylor's expansion for the function f in many variables along $\mathbf{v}$:

$$f(\mathbf{x}+t\mathbf{v}) = f(\mathbf{x}) + \left.\frac{df(\mathbf{x}+t\mathbf{v})}{dt}\right|_{t=0}\frac{t}{1!} + \cdots + \left.\frac{d^k f(\mathbf{x}+t\mathbf{v})}{dt^k}\right|_{t=0}\frac{t^k}{k!} + R_k(t)$$

$$= f(\mathbf{x}) + D_\mathbf{v} f(\mathbf{x})\frac{t}{1!} + \cdots + D_\mathbf{v}^k f(\mathbf{x})\frac{t^k}{k!} + R_k(t) \tag{2.43}$$

with $R_k(t)/t^k \to 0$ as $t \to 0$ independent of the direction of $\mathbf{v}$. The main part of this formula (without R_k) can be used to approximate f to the order k in a small vicinity of the point $\mathbf{x}$.

Example 2.27. Consider $f(x,y) = x^y$ near the point $(1,1)$. Here

$$f(\mathbf{x}) = f(x_1, x_2) = x_1^{x_2}, \qquad f(\mathbf{x}+t\mathbf{v}) = (x_1 + tv_1)^{x_2+tv_2}.$$

Differentiation gives

$$\left.\frac{df(\mathbf{x}+t\mathbf{v})}{dt}\right|_{t=0} = x_1^{x_2}\left(\frac{x_2}{x_1}v_1 + v_2 \ln x_1\right),$$

$$\left.\frac{d^2 f(\mathbf{x}+t\mathbf{v})}{dt^2}\right|_{t=0} = x_1^{x_2}\left[\left(\frac{v_1 x_2}{x_1} + v_2 \ln x_1\right)^2 + \frac{2v_1 v_2}{x_1} - \frac{v_1^2 x_2}{x_1^2}\right].$$

Evaluating at $x_1 = x_2 = 1$, we get the expansion $1 + v_1 t + v_1 v_2 t^2 + R_3(t)$. □

2.12   Derivatives of an Implicit Function

Let us start with the old problem of finding the derivative of an implicit function $y = y(x)$ defined by the equation $\phi(x,y) = 0$. We suppose that such a function exists and is differentiable. (In more complete books they state conditions for this.) Substitution gives

$$\phi(x, y(x)) = 0$$

which is an identity in some vicinity of the point $(x, y(x))$. Therefore we can differentiate both sides, obtaining another identity

$$\frac{d}{dx}\,\phi(x, y(x)) = 0.$$

The chain rule gives

$$\phi_x(x, y(x)) + \phi_y(x, y(x))y'(x) = 0 \tag{2.44}$$

hence

$$y'(x) = -\frac{\phi_x(x, y(x))}{\phi_y(x, y(x))}. \tag{2.45}$$

Note that in this relation $\phi_y(x, y(x))$ should not be zero. Besides the condition that there exist partial derivatives of ϕ at (x, y), the condition $\phi_y(x, y(x)) \neq 0$ is needed for existence of the function $y = y(x)$. Frequently (2.45) is written omitting the arguments but the reader should remember that the arguments are x and $y(x)$.

Example 2.28. Let us find the derivative of y defined by the equation $x^2 + y^2 = 1$. Of course we could express y explicitly as a function of x:

$$y(x) = \pm(1 - x^2)^{1/2}, \qquad y'(x) = \pm\frac{x}{(1 - x^2)^{1/2}}, \qquad |x| < 1. \tag{2.46}$$

But now we use (2.45):

$$\phi(x, y) = x^2 + y^2 - 1, \qquad \phi_x = 2x, \qquad \phi_y(x, y) = 2y,$$

hence

$$y'(x) = -x/y. \qquad \square$$

Remark 2.4. The last expression $-x/y$ of the example is defined for all $y \neq 0$. However, this is the value of the derivative of the implicit function y only if the point (x, y) lies on the circle, i.e., only if $x^2 + y^2 = 1$.

Note that if we start with the equation $x^2 + y^2 = -1$ and proceed formally, we obtain $-x/y$ for the derivative of the non-existent "implicit" function. So one should remember that points (x, y) in the derivative equation are not arbitrary but lie over the curve defined by the equation. $\square$

Example 2.29. If $x^y + y^x - \alpha = 0$ for $x, y > 0$, where α is a constant, then for the implicit function $y = y(x)$ we obtain

$$\frac{dy}{dx} = -\frac{yx^{y-1} + y^x \ln y}{xy^{x-1} + x^y \ln x}.$$

But again we should remember that (x, y) is not just any point on the plane; x and y must be related by the equation $x^y + y^x - \alpha = 0$. $\square$

Higher Derivatives of an Implicit Function

To find the second derivative of an implicit function $y = y(x)$ given by the equation $\phi(x, y(x)) = 0$, we differentiate both sides of (2.44) using the chain rule:

$$\phi_{xx}(x, y(x)) + 2\phi_{xy}(x, y(x))y'(x)$$
$$+ \phi_{yy}(x, y(x))[y'(x)]^2 + \phi_y(x, y(x))y''(x) = 0,$$

from which $y''(x)$ can be obtained. We have assumed that Clairaut's theorem applies. Higher derivatives of y can be calculated, provided the higher derivatives of ϕ exist.

Partial Derivatives of an Implicit Function

In a similar way we can find any partial derivatives for an implicit function $y = y(x_1, \ldots, x_n)$ defined by the equation

$$\phi(x_1, \ldots, x_n, y) = c.$$

Fixing all the variables x_i except x_k and y in the last equation and using the chain rule, we differentiate both sides of $\phi(x_1, \ldots, x_n, y(x_1, \ldots, x_n)) = c$ (which is an identity!), where $y = y(x_1, \ldots, x_n)$, with respect to x_k. The result is

$$\phi_{x_k} + \phi_y y_{x_k} = 0$$

and therefore

$$\frac{\partial y(x_1, \ldots, x_n)}{\partial x_k} = -\frac{\phi_{x_k}(x_1, \ldots, x_n, y(x_1, \ldots, x_n))}{\phi_y(x_1, \ldots, x_n, y(x_1, \ldots, x_n))}. \qquad (2.47)$$

Higher-order partial derivatives of y can be obtained in a similar fashion.

Example 2.30. The equation $2x + 3y + z = xz + 5$ defines z as an implicit function of x and y. We compute z_x and z_y. Setting $\phi(x, y, z) = 2x + 3y + z - xz - 5 = 0$ we obtain by (2.47)

$$z_x = -\frac{\phi_x}{\phi_z} = \frac{2 - z}{1 - x}, \qquad z_y = -\frac{\phi_y}{\phi_z} = -\frac{3}{1 - x}.$$

These can be evaluated at a point satisfying $\phi(x, y, z) = 0$, for instance at $P = (2, 1, 2)$ where we obtain $z_x = 0$ and $z_y = 3$. Of course for this simple example it is possible to determine z as an explicit function of x and y and then differentiate directly:

$$z = \frac{2x + 3y - 5}{x - 1}, \qquad z_x = \frac{3(1 - y)}{(x - 1)^2}, \qquad z_y = \frac{3}{x - 1}.$$

When $(x, y) = (2, 1)$, these replicate our previous results.   □

2.13   Tangent Plane to a Level Surface

Now we wish to draw a plane tangent to the *level surface* S given by the equation

$$f(x, y, z) = c = \text{constant}$$

at point (x_0, y_0, z_0). The point should be on S, therefore $f(x_0, y_0, z_0) = c$.

We start with the general equation for all planes passing through (x_0, y_0, z_0):

$$a(x - x_0) + b(y - y_0) + c(z - z_0) = 0 \qquad (2.48)$$

with arbitrary (a, b, c), remembering that the vector $\mathbf{N} = (a, b, c)$ is a normal to the plane. Now we should find $\mathbf{N}$. Recall that for a curve $\mathbf{r} = \mathbf{r}(t)$ the vector $\mathbf{r}'(t)$ is tangent. Let us draw a curve $\mathbf{r} = \mathbf{r}(t)$ on S through the point (x_0, y_0, z_0) such that $x_0 = x(t_0)$, $y_0 = y(t_0)$, $z_0 = z(t_0)$. So

$$f(x(t), y(t), z(t)) = c.$$

This equation is an identity; provided f and $\mathbf{r}(t)$ have derivatives in a vicinity of (x_0, y_0, z_0), we can differentiate it with respect to t at $t = t_0$. By the chain rule

$$f_x(x_0, y_0, z_0)x'(t_0) + f_y(x_0, y_0, z_0)y'(t_0) + f_z(x_0, y_0, z_0)z'(t_0) = 0$$

or, in vector notation,

$$\nabla f(x_0, y_0, z_0) \cdot \mathbf{r}'(t_0) = 0.$$

As $\mathbf{r} = \mathbf{r}(t)$ is an arbitrary curve on S through (x_0, y_0, z_0) we conclude that $\nabla f(x_0, y_0, z_0)$ is orthogonal to any tangent vector to a curve through (x_0, y_0, z_0) on S, hence that $\nabla f(x_0, y_0, z_0)$ is a perpendicular $\mathbf{N}$ to the tangent plane. From (2.48) we deduce the needed equation:

$$f_x(x_0, y_0, z_0)(x - x_0) + f_y(x_0, y_0, z_0)(y - y_0) + f_z(x_0, y_0, z_0)(z - z_0) = 0.$$
$$(2.49)$$

We can write the equation for the straight line through (x_0, y_0, z_0) containing **N**:

$$\frac{x - x_0}{f_x(x_0, y_0, z_0)} = \frac{y - y_0}{f_y(x_0, y_0, z_0)} = \frac{z - z_0}{f_z(x_0, y_0, z_0)}. \tag{2.50}$$

The corresponding parametric forms are

$$x = x_0 + t f_x(x_0, y_0, z_0),$$
$$y = y_0 + t f_y(x_0, y_0, z_0),$$
$$z = z_0 + t f_z(x_0, y_0, z_0).$$

As an illustration, Fig. 2.3 depicts portions of the tangent plane and normal line at a selected point on a sphere.

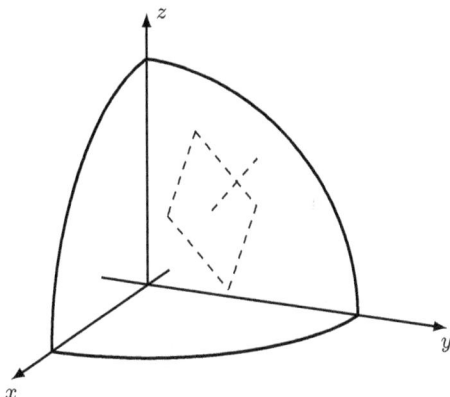

Fig. 2.3   Portions of the tangent plane and normal line at the point $x_0 = y_0 = z_0 = 1/\sqrt{3}$ on the sphere $x^2 + y^2 + z^2 = 1$.

Example 2.31. Let us write down the equation of the plane tangent to the ellipsoid $x^2 + 2y^2 + 3z^2 = 6$ at point $(1, 1, 1)$. After verifying that $(1, 1, 1)$ does lie on the ellipsoid, we calculate ∇f there:

$$f_x(1, 1, 1) = 2x|_{(1,1,1)} = 2,$$
$$f_y(1, 1, 1) = 4y|_{(1,1,1)} = 4,$$
$$f_z(1, 1, 1) = 6z|_{(1,1,1)} = 6.$$

Finally, (2.49) yields the required equation:

$$2(x - 1) + 4(y - 1) + 6(z - 1) = 0. \qquad \square$$

Tangent to a Level Curve

Similarly deduced is the equation for the line tangent to the level curve $f(x, y) = c$ at point (x_0, y_0):

$$f_x(x_0, y_0)(x - x_0) + f_y(x_0, y_0)(y - y_0) = 0. \tag{2.51}$$

Its derivation is left to the reader.

Equation (2.49) can be extended to any dimension as

$$\nabla f(x_1, \ldots, x_n) \cdot (\mathbf{r} - \mathbf{r}_0) = 0, \tag{2.52}$$

where f is a function in n variables, $\mathbf{r}$ is the position vector in $\mathbb{R}^n$, and $\mathbf{r}_0$ is the point on the level surface $f(x_1, \ldots, x_n) = c$ at which we describe the tangent $n - 1$ dimensional surface.

2.14   Extrema of a Function

Introduction and Review

A chief application of calculus is finding the extrema of a function. As the theory of extrema of a multivariable function is a natural extension of that for a function in one variable, it makes sense to briefly reconsider the structure of the latter theory.

The notion of maximum or minimum of a function arises from the graph of the function on some interval. The graph can have a few local peaks or valleys we call local extrema. More precisely, a function $y = f(x)$ attains a *local maximum (local minimum)* value at point $x = a$ if there is an ε-neighborhood, $\varepsilon > 0$, of a such that $f(a) \geq f(x)$ $(f(a) \leq f(x))$ for all $x \in [a - \varepsilon, a + \varepsilon]$. The definition of local extremum for a multivariable function differs only in how a neighborhood is specified.

A practical way to find possible extrema of $y = f(x)$ is given by Fermat's theorem, which states that if f is differentiable at a and has a local extremum at a then $f'(a) = 0$.[1] For a multivariable function F an analogue

---

[1]Pierre de Fermat (1601–1665) was a French lawyer, better known as a mathematician, who left us many nice theorems. He is most renowned for his Last Theorem in number theory, written in the margin of his copy of Diophantus' *Arithmetics* along with a note that the proof would not fit in such a small space. In the 19th century a German student was challenged to prove the simple-looking theorem: the equation $x^n + y^n = z^n$ has no integer solutions x, y, z for any integer $n \geq 3$. That student bequeathed a large monetary reward to anyone who could prove the theorem. Upon seeing this in the newspapers, many people became "mathematicians." A correct proof was published only in the mid 1990s.

of Fermat's condition is

$$\nabla F = \mathbf{0}. \tag{2.53}$$

A solution a to the equation $f'(x) = 0$ is called a critical point of f. For a multivariable function F the critical points are solutions of the vector equation (2.53), which is a system of scalar equations.

There are sufficient conditions for extrema. For a local maximum of f in one variable it is $f''(a) < 0$, and for a minimum it is $f''(a) > 0$. We find that a sufficient condition for a maximum value of F at $(a_1, \ldots, a_n)$ is that for every direction $\mathbf{v}$ the second directional derivative is negative:

$$D_{\mathbf{v}}^2 F(a_1, \ldots, a_n) < 0.$$

For a local minimum the inequality sign is simply reversed.

Finally, for a continuously differentiable function f on a closed segment $[a, b]$, the absolute (or global) maximum and minimum values (the greatest and least values) can be found by a simple algorithm. First we find all the critical points x_k on $[a, b]$. Next we calculate all the values $f(x_k)$, and supplement them with $f(a)$ and $f(b)$. We get the needed values by selecting the largest and smallest values from this list.

Let us proceed to discuss the extrema of multivariable functions in more detail.

Definition 2.7. We say that $(a_1, \ldots, a_n)$ is a *point of local maximum* (respectively, *minimum*) of a function $f = f(x_1, \ldots, x_n)$ if there exists $\varepsilon > 0$ such that

$$f(a_1, \ldots, a_n) \geq f(x_1, \ldots, x_n)$$

(respectively, $f(a_1, \ldots, a_n) \leq f(x_1, \ldots, x_n)$) for all $(x_1, \ldots, x_n)$ from the ε-neighborhood of the point $(a_1, \ldots, a_n)$, i.e., for all $(x_1, \ldots, x_n)$ from the ball

$$B_{\varepsilon}(a_1, \ldots, a_n) = \{(x_1, \ldots, x_n) \colon [(x_1 - a_1)^2 + \cdots + (x_n - a_n)^2]^{1/2} \leq \varepsilon\}.$$

Local Extrema of Functions in Two Variables

We start with Fermat's theorem for a two-variable function.

Theorem 2.7. *Let $f \colon D \subset \mathbb{R}^2 \to \mathbb{R}$ be differentiable at (a, b) in an open set D, and let (a, b) be a local extremum point for f. Then*

$$f_x(a, b) = 0, \qquad f_y(a, b) = 0. \tag{2.54}$$

Proof. Fix $y = b$. Then $f(x, b) = g(x)$ is a function in one variable x, for which $x = a$ is a local extremum point. By Fermat's theorem $g'(a) = 0$, and consequently $f_x(a, b) = 0$. It follows similarly that $f_y(a, b) = 0$. □

Since $D_{\mathbf{u}} f(a, b) = \nabla f(a, b) \cdot \mathbf{u}$, at a local extremum point the directional derivative must be zero at any direction $\mathbf{u}$.

As in the one-dimensional case, a solution (a, b) to equations (2.54) may be a non-extremum point for f. A point (a, b) at which (2.54) holds is called a *critical* or *stationary point*.

Example 2.32. To find critical points of $f(x, y) = x^2 + 2y^2 - 4x - 8y + 10$, we start with equations (2.54):

$$f_x(x, y) = 2x - 4 = 0 \implies x = 2, \qquad f_y(x, y) = 4y - 8 = 0 \implies y = 2.$$

So $(a, b) = (2, 2)$ is a critical point. The fact that it is a minimum point follows at once from another representation for f:

$$f(x, y) = (x - 2)^2 + 2(y - 2)^2 + 22.$$ □

Example 2.33. Let us find critical points for $f(x, y) = x^2 - y^2$. We get $f_x(x, y) = 2x = 0$ and $f_y(x, y) = -2y = 0$. So $(a, b) = (0, 0)$. Fix $y = 0$; then the function $f(x, 0) = x^2 > 0$ for $x \neq 0$. Now fix $x = 0$; the function $f(0, y) = -y^2 < 0$ for $y \neq 0$. Hence $(0, 0)$ cannot be an extremum point. This type of critical point, where the function has a maximum value along one direction and a minimum value along another, is called a *saddle point*. □

The vector form of (2.54), which is (2.53), extends directly to the case of n variables. For a function $z = f(x_1, \ldots, x_n)$, Fermat's theorem reads as follows.

Theorem 2.8. *Let $f \colon D \subset \mathbb{R}^n \to \mathbb{R}$ be differentiable at $(a_1, \ldots, a_n)$ in an open set D, and let $(a_1, \ldots, a_n)$ be a local extremum point for f. Then*

$$\nabla f(a_1, \ldots, a_n) = \mathbf{0}. \tag{2.55}$$

The proof mimics that of Theorem 2.7.

Sufficient Conditions for Extremum of a Function in Two Variables

We know that for a twice continuously differentiable function $y = f(x)$, a critical point $x = a$ is a local maximum point if $f''(a) < 0$ or a local minimum point if $f''(a) > 0$. A similar theorem holds for a function $f = f(x_1, \ldots, x_n)$, provided it is three times continuously differentiable in some neighborhood of a critical point.

Theorem 2.9. *If at a critical point $(a_1, \ldots, a_n)$ of $f = f(x_1, \ldots, x_n)$ the inequality*

$$D_{\mathbf{v}}^2 f(a_1, \ldots, a_n) > 0$$

holds for all directions $\mathbf{v}$, then $(a_1, \ldots, a_n)$ is a point of local minimum. If the inequality sign is reversed, then $(a_1, \ldots, a_n)$ is a point of local maximum.

We merely sketch the proof. Suppose $D_{\mathbf{v}}^2 f(a_1, \ldots, a_n) > 0$. The function

$$g(t) = f(a_1 + tv_1, \ldots, a_n + tv_n)$$

is, for fixed $\mathbf{v}$, a function in one variable t whose second derivative satisfies

$$g''(0) = D_{\mathbf{v}}^2 f(a_1, \ldots, a_n) > 0.$$

As f is twice differentiable in a neighborhood of $(a_1, \ldots, a_n) > 0$, there is a neighborhood of $t = 0$, say $t \in [-\varepsilon_{\mathbf{v}}, \varepsilon_{\mathbf{v}}]$ for some $\varepsilon_{\mathbf{v}} > 0$, in which $g(t) \geq g(0)$; equivalently,

$$f(a_1 + tv_1, \ldots, a_n + tv_n) \geq f(a_1, \ldots, a_n).$$

The condition that f possesses continuous derivatives up to third order in some vicinity of the critical point can be used along with Taylor's formula (2.43) to prove that $\inf \varepsilon_{\mathbf{v}} > 0$. Hence there is a ball centered at $(a_1, \ldots, a_n)$ in which $f(x_1, \ldots, x_n) \geq f(a_1, \ldots, a_n)$.

Because $D_{\mathbf{v}}^2 f$ is a quadratic form in the components of $\mathbf{v}$, the extremum problem reduces to finding the condition under which that form is positive definite or negative definite. For $n = 2$ we can formulate sufficient conditions for local extremum values of $f = f(x, y)$. (Conditions can be stated for any n but we do not present them here.) First we introduce the determinant

$$H(x, y) = \begin{vmatrix} f_{xx}(x, y) & f_{xy}(x, y) \\ f_{xy}(x, y) & f_{yy}(x, y) \end{vmatrix} = f_{xx}(x, y) f_{yy}(x, y) - f_{xy}^2(x, y).$$

Theorem 2.10. *Let all the partial derivatives of f up to third order be continuous in a disk centered at a critical point (a, b) of f, and suppose $\nabla f(a, b) = \mathbf{0}$. The following statements hold. (1) If $H(a, b) > 0$ and*

$f_{xx}(a, b) > 0$, then (a, b) is a point of local minimum. (2) If $H(a, b) > 0$ and $f_{xx}(a, b) < 0$, then (a, b) is a point of local maximum. (3) If $H(a, b) < 0$, then (a, b) is not a local extremum point.

Proof. Take $\mathbf{v} = (v_1, v_2)$ so that $D_{\mathbf{v}} f = f_x v_1 + f_y v_2$. The second directional derivative is

$$D_{\mathbf{v}}^2 f = D_{\mathbf{v}}(D_{\mathbf{v}} f)$$
$$= (f_x v_1 + f_y v_2)_x \, v_1 + (f_x v_1 + f_y v_2)_y \, v_2$$
$$= f_{xx}(v_1)^2 + 2 f_{xy} v_1 v_2 + f_{yy}(v_2)^2$$
$$= f_{xx} \left(v_1 + v_2 \frac{f_{xy}}{f_{xx}} \right)^2 + \frac{v_2^2}{f_{xx}} (f_{xx} f_{yy} - f_{xy}^2).$$

Again let $g(t) = f(a + tv_1, b + tv_2)$. Then $g'(0) = 0$ and $g''(0) = D_{\mathbf{v}}^2 f(a, b)$. So in case (1), for any unit vector $\mathbf{v}$, the point $t = 0$ is a local minimum point of $g(t)$ since $g'(0) = 0$ and $g''(0) > 0$. That is, $g(0) \le g(t)$ on some interval $[-\varepsilon_{\mathbf{v}}, \varepsilon_{\mathbf{v}}]$. As above, we omit the proof that there exists $\varepsilon_0 > 0$ for which $\varepsilon_{\mathbf{v}} \ge \varepsilon_0$ for any unit vector $\mathbf{v}$. The latter proves that there is a disk centered at (a, b) in which $f(x, y) \ge f(a, b)$. ☐

Example 2.34. Consider $f(x, y) = x^4 + y^4 - 4xy + 1$. First we find the critical points. Differentiating to get

$$f_x(x, y) = 4x^3 - 4y = 0, \qquad f_y(x, y) = 4y^3 - 4x = 0,$$

then combining and factoring, we find that x must satisfy

$$x(x - 1)(x + 1)(x^2 + 1)(x^4 + 1) = 0.$$

Since the roots are $x = 0$, $x = 1$, and $x = -1$, the critical points are $(0, 0)$, $(1, 1)$, and $(-1, -1)$. Let us classify them. The relations

$$f_{xx} = 12x^2, \qquad f_{xy} = -4, \qquad f_{yy} = 12y^2$$

give $H(x, y) = 144x^2 y^2 - 16$.

For $(x, y) = (0, 0)$ we have $H(0, 0) = -16 < 0$, so $(0, 0)$ is not an extremum point of f.

For $(x, y) = (1, 1)$ we have $H(1, 1) = 128 > 0$ and $f_{xx}(1, 1) > 0$, so $(1, 1)$ is a minimum point of f.

For $(x, y) = (-1-, 1)$ we have $H(-1, -1) = 128 > 0$ and $f_{xx}(-1, -1) > 0$, so $(-1, -1)$ is a minimum point of f as well.

Note that the function has two local minima and one critical point that is not extremal. ☐

Example 2.35. We seek the distance between the point $(3, 4, 5)$ and the plane $2x + y - z = 3$. This is of course the shortest distance from $(3, 4, 5)$ to a point of the plane.

The coordinates of a point (x, y, z) on the plane are not independent as required by Theorem 2.10; rather, they must satisfy the equation of the plane. So we are seeking the minimum value of the function

$$d = [(x - 3)^2 + (y - 4)^2 + (z - 5)^2]^{1/2}$$

for (x, y, z) such that $2x + y - z = 3$. Let us substitute $z = 2x + y - 3$ into the expression for d, thereby obtaining a function in two independent variables x and y. As it is easier to deal with the squared distance d^2, we choose to minimize

$$f(x, y) = (x - 3)^2 + (y - 4)^2 + (2x + y - 8)^2$$
$$= 5x^2 + 2y^2 + 4xy - 38x - 24y + 89.$$

The equations $f_x = 0$ and $f_y = 0$, which are

$$10x + 4y = 38, \qquad 4x + 4y = 24,$$

yield a single critical point $(7/3, 11/3)$ of f. It is geometrically clear that the squared distance $f(x, y)$ cannot have a maximum value, hence $(7/3, 11/3)$ must be the required point of minimum. But let us verify this by Theorem 2.10. We have $f_{xx} = 10 > 0$ for all x, y and $H(x, y) = 10 \cdot 4 - 4^2 > 0$, so $(7/3, 11/3)$ is the minimum point and the minimum distance is

$$[f(7/3, 11/3)]^{1/2} = (2/3)^{1/2}. \qquad \square$$

2.15   Absolute Extrema of a Function on a Compact Domain

Definition 2.8. A set D in $\mathbb{R}^n$ is *compact* if it is closed and bounded.

A simple bounded domain like a ball or a parallelepiped (with piecewise smooth boundary) is compact if it contains its boundary.

A fundamental fact for a compact set is that from any sequence $\{\mathbf{x}_k\}$ of points taken from D we can select a Cauchy subsequence. Moreover, because D is closed, the subsequence converges to a limit in D.

We have discussed local extrema of a function $z = f(\mathbf{x})$. We call M the *absolute maximum* of f in D if there is a point $\mathbf{x}_0 \in \mathbb{R}^n$ such that

$f(\mathbf{x}_0) = M$ and $f(\mathbf{x}) \leq M$ for all $\mathbf{x} \in D$. By reversing the inequality sign we can define the *absolute minimum* m of f.

The following theorem will be helpful in seeking points of absolute maximum and minimum in a compact subset of $\mathbb{R}^n$.

Theorem 2.11 (Weierstrass). *A real-valued function f continuous on a compact set $D \subset \mathbb{R}^n$ takes its maximum and minimum values in D.*

Proof. A continuous function defined on D is bounded. Were it unbounded from above, there would exist a sequence $\{\mathbf{x}_k\}$ convergent to $\mathbf{x}_0$ and for which $f(\mathbf{x}_k) \to \infty$ as $k \to \infty$. This would contradict the assumption that f is continuous at $\mathbf{x}_0$.

Let M be the supremum of the set of values taken by f on D. This means that for any $\mathbf{x} \in D$ we get $f(\mathbf{x}) \leq M$ and for any $\varepsilon > 0$ there exists $\mathbf{x}_k \in D$ such that $f(\mathbf{x}_k) > M - \varepsilon$. So we can find a sequence $\{\mathbf{x}_k\}$ in D such that $f(\mathbf{x}_k) \to M$. From $\{\mathbf{x}_k\}$ we can select a subsequence $\{\mathbf{x}_{k_n}\}$ convergent to some $\mathbf{x}_0 \in D$ and thus, by continuity of f, get $M = f(\mathbf{x}_0)$. Similarly obtained is the proof for the absolute minimum of f; alternatively, we could consider the function $-f$ for which the absolute minimum of f becomes the absolute maximum. $\qquad\square$

Example 2.36. The function

$$f(x, y) = x + y + (1 - x^2 - y^2)^{1/2}$$

is continuous on its closed and bounded (i.e., compact) domain in $\mathbb{R}^2$, which is the disk $D = \{(x, y) \colon x^2 + y^2 \leq 1\}$. It therefore attains maximum and minimum values on D. The domain $D_1 = \{(x, y) \colon x^2 + y^2 < 1\}$ for the function $g(x, y) = x + y + (1 - x^2 - y^2)^{-1/2}$ is open and so we cannot apply Weierstrass theorem here. It is evident that the function is not bounded from above, so it does not have a maximum value. It does have an absolute minimum. $\qquad\square$

Now let us discuss *how* to find the absolute extrema of a smooth function on a compact set. Recall how this was done for a differentiable function f on a segment $[a, b]$. It is clear (Fig. 2.4) that the absolute maximum of f can occur at a local maximum of f in (a, b), or at an endpoint where f may simply reach its largest value. So the procedure for finding the absolute maximum and minimum of f on $[a, b]$ consists of the following steps.

(1) By solving the equation $f'(x) = 0$, we find all critical points $x_1, \ldots, x_m$ of f on $[a, b]$. This set contains all the points of local maximum and minimum.

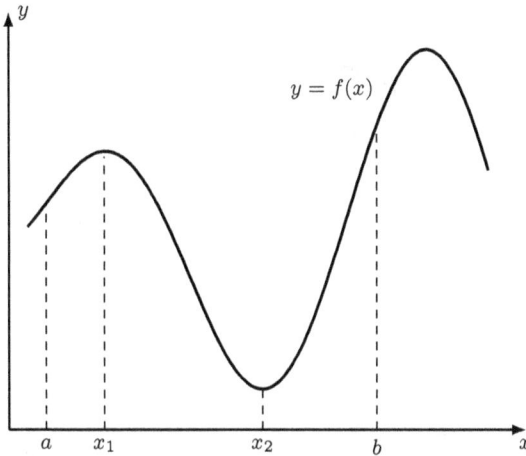

Fig. 2.4 The absolute maximum and minimum values of $y = f(x)$ on segment $[a, b]$ are max and min values of the numerical set $f(a)$, $f(x_1)$, $f(x_2)$, $f(b)$. Here x_1, x_2 are critical points of f on $[a, b]$, $f'(x_k) = 0$, and a, b are endpoints of $[a, b]$.

(2) Next we calculate the numerical set N of the values $f(x_1), \ldots, f(x_m)$ along with $f(a)$ and $f(b)$.
(3) Selecting the maximum and minimum values from N, we obtain the absolute maximum and minimum values of f on $[a, b]$.

Note that there is no need to determine whether a given critical point x_k is a point of maximum or minimum.

We can practically repeat this procedure for a continuously differentiable function $y = f(\mathbf{x})$ in n variables on a compact set $D \subset \mathbb{R}^n$, assuming the boundary of D is piecewise smooth. For simplicity, we take f to be continuously differentiable on an open set containing D. The procedure for finding the absolute maximum and minimum values is as follows.

(1) Find all critical points of f; that is, solve the simultaneous equations $f_{x_k}(\mathbf{x}) = 0$ for $k = 1, \ldots, n$.
(2) Select those critical points $\mathbf{x}_1, \ldots, \mathbf{x}_m$ that belong to D.
(3) Find any suspected extremum points of f on the boundary ∂D of D (see Fig. 2.5).
(4) For all the points $\mathbf{x}_k$ from steps 2 and 3, calculate $f(\mathbf{x}_k)$ and select the

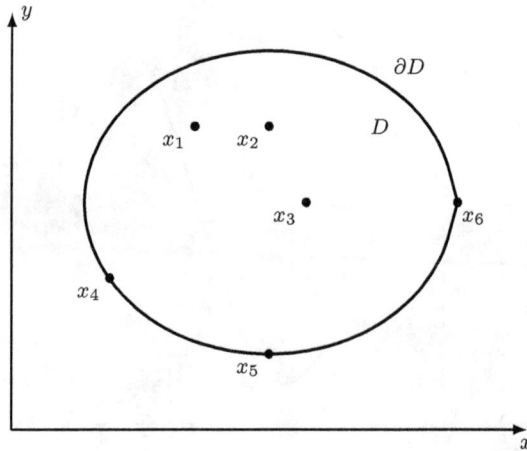

Fig. 2.5   The absolute max and min values of a smooth function $z = f(x, y)$ on a compact set D are the max and min values of the set $\{f(\mathbf{x}_k)\colon k = 1, 2, 3, 4, 5, 6\}$ where (a) $\mathbf{x}_1, \mathbf{x}_2, \mathbf{x}_3$ are all critical points of f inside D (i.e., solutions to $\nabla f(\mathbf{x}_k) = \mathbf{0}$), and (b) $\mathbf{x}_4, \mathbf{x}_5, \mathbf{x}_6$ are suspected max-min points on the boundary ∂D of D. The latter points include the critical points $\mathbf{x}_4, \mathbf{x}_5$ of f on ∂D and the angle point $\mathbf{x}_6$ of ∂D.

    max and min values from the resulting numerical set. These are the absolute maximum and minimum values of f on D.

    If the boundary ∂D of D is given by equations of the form $x = x(t)$ and $y = y(t)$ for $t \in [a, b]$, the suspected extremum points include critical points of the function in the variable t:

$$z = f(x(t), y(t)), \qquad t \in [a, b].$$

These are the points $\mathbf{x}_4$ and $\mathbf{x}_5$ in Fig. 2.5. We must also consider any angle point of ∂D, such as $\mathbf{x}_6$. The situation for $n > 2$ is even more complicated, as the boundary of D has a complicated structure even for a cube.

    Note that the parts of ∂D can be called surfaces (or *manifolds*) in $\mathbb{R}^n$. At smooth points locally ∂D can be placed in one-to-one correspondence with the points of some part of $\mathbb{R}^{n-1}$ and suspected extremum points of f on the smooth portion of f are critical points on ∂D. In step (3) we should study what happens on the set of non-smooth points. To understand the difficulties that can occur in this procedure, it is instructive to consider the boundary of a cube in three dimensions.

Again, in this procedure we need not study the nature of each of the suspected points.

Example 2.37. We seek the absolute extremum values of the function $f(x, y) = x^2 - 2x + y^2$ in the square $D = \{(x, y): |x| \leq 2, |y| \leq 2\}$. See Fig. 2.6. We solve the problem in a series of steps.

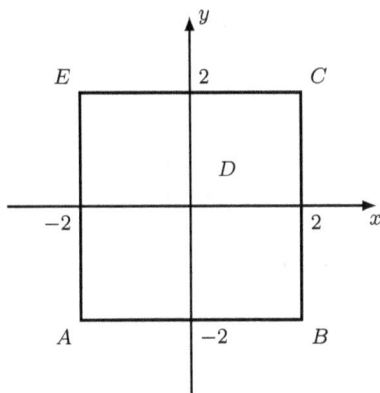

Fig. 2.6   Square D on which we seek the absolute extrema of $f(x, y) = x^2 - 2x + y^2$. Segments AB, BC, CE, and EA comprise ∂D. The point $(1, 0)$ is a critical point inside D.

(1) Solve $f_x = 2x - 2 = 0$ and $f_y = 2y = 0$. The only solution is $x = 1$ and $y = 0$.

(2) The critical point $(1, 0)$ lies inside D.

(3) Consider all suspected points on the boundary.

   (a) Denote by AB the section described by the relations $|x| \leq 2$ and $y = -2$. On AB the function f becomes $g(x) = f(x, -2) = x^2 - 2x + 4 = (x - 1)^2 + 3$. A critical point of g on $[-2, 2]$ is a solution of the equation $g'(x) = 2x - 2 = 0$, which gives $x = 1$. The point $(1, -2)$ lies inside $[-2, 2]$ and hence is a suspected point on AB. The endpoints of AB are $(-2, -2)$ and $(2, -2)$.

   (b) BC is the portion described by $x = 2$ and $|y| \leq 2$. Write $g(y) = f(2, y) = y^2$. The equation $g'(y) = 0$ gives $y = 0$. The collection of suspected points and endpoints is $(2, 0)$, $(2, -2)$, and $(2, 2)$.

   (c) CE is the portion described by $|x| \leq 2$ and $y = 2$. Write $g(x) = f(x, 2) = x^2 - 2x + 4$. The equation $g'(x) = 0$ coincides with the equation on AB, so we get the list of points $(1, 2)$, $(-2, 2)$, $(2, 2)$.

(d) EA is the portion with $x = -2$, $|y| \leq 2$ (i.e., $-2 \leq y \leq 2$; note that we do not need to keep the direction for y here). Set $g(y) = f(-2, y) = 8 + y^2$. The equation $g'(y) = 0$ implies $y = 0$, and the suspected points are $(-2, 0)$, $(-2, 2)$, $(-2, -2)$.

(4) The list of values of f at all the suspected points is

$$f(1, 0) = -1, \quad \begin{cases} f(-2, -2) = 12, & f(1, -2) = 3, \\ f(2, -2) = 4, & f(2, 0) = 0, \\ f(2, 2) = 4, & f(1, 2) = 3, \\ f(-2, 2) = 12, & f(-2, 0) = 8. \end{cases}$$

So the absolute maximum of f is 12, and the absolute minimum is -1. ☐

Example 2.38. Consider the absolute extrema of the same function $f(x, y) = x^2 - 2x + y^2$ inside the disk $(x - 1)^2 + y^2 \leq 1$ (Fig. 2.7).

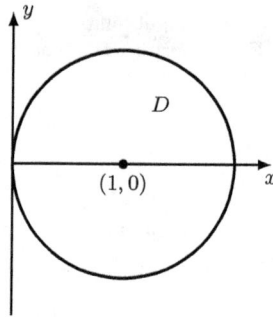

Fig. 2.7   Disk D on which we seek the absolute extrema of $f(x, y) = x^2 - 2x + y^2$. The boundary of D is given by the equations $x = 1 + \cos t$ and $y = \sin t$ for $t \in [0, 2\pi)$.

Since the critical point of f is $(1, 0)$, at least one of the absolute extrema is taken on the boundary ∂D, whose equations can be written in the parametric forms $x = 1 + \cos t$ and $y = \sin t$ for $t \in [0, 2\pi)$. The function f on ∂D is $g(t) = f(1 + \cos t, \sin t) = 0$ for all t. So the list of suspected values for the absolute extrema contains only two values: $f(1, 0) = -1$, and 0. The absolute maximum of f is zero; it is taken at any boundary point. The absolute minimum is -1, taken at $(1, 0)$. ☐

How should we go about finding the absolute extrema of a function $f(x, y, z)$ on the cube C shown in Fig. 2.8? First we should find all the

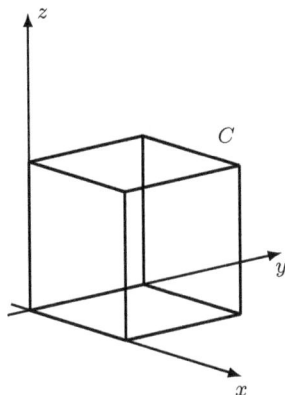

Fig. 2.8 Cube C with sides $S_1, \ldots, S_6$.

critical points — the solutions of the simultaneous equations $f_x(x, y, z) = 0$, $f_y(x, y, z) = 0$, and $f_z(x, y, z) = 0$ — and select those that lie within C. Next, C has six sides $S_1, \ldots, S_6$. We should describe each S_k in terms of x, y, z and restrict f to six different functions of two variables on the six square sides; this yields six problems of finding the absolute extrema for a function in two variables on a square (recall Example 2.37). Finally, we should write out the list of values for (a) the critical points of f in C and (b) the absolute extrema on all the sides, and select the maximum and minimum values from this list.

2.16 Constrained Extrema: Lagrange Multipliers

While seeking critical points for f on the boundary $\partial D \subset \mathbb{R}^2$ in Example 2.38, we were able to express the restriction of f to the boundary as a function in just one variable. But sometimes this substitution is difficult to realize, and we are left with an extremum problem involving a constraint (the equation for the boundary of D). This problem takes the general form

> *Find the extremum of the function* $z = f(x_1, \ldots, x_n)$ *subject to a constraint* $g(x_1, \ldots, x_n) = c = constant$.

More generally, we may face a list of constraints:

$$g_k(x_1, \ldots, x_n) = c_k \qquad (k = 1, \ldots, m \text{ with } m < n).$$

We saw constrained extremum problems in Section 2.13 when we had to restrict a function $z = f(x, y)$ to the domain boundary given by an equation $g(x, y) = 0$. But many other types of problems require us to find constrained extrema. These include seeking to maximize the volumes of various hollow shapes for a fixed surface area. Examples are given in Fig. 2.9. For each enclosure in Fig. 2.9 the problem can be formulated as

Find the maximum volume V of a figure if its area $A = c = constant$.

(a)   (b)   (c)

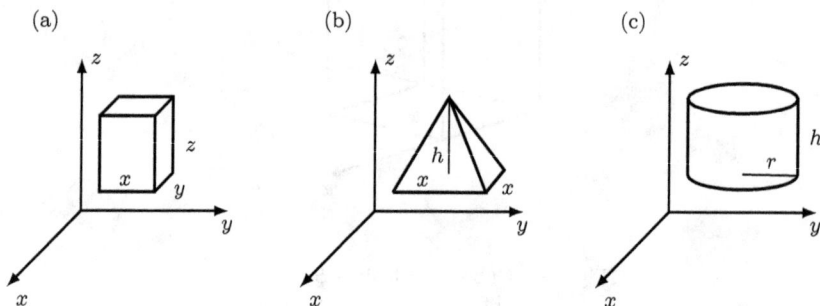

Fig. 2.9 (a) Rectangular parallelepiped with sides x, y, z. The volume is $V = xyz$. For a closed box the surface area is $A = 2(xy + yz + zx)$. For an open box without a lid, $A = xy + 2(yz + zx)$. (b) A right perfect pyramid with bottom sides x and altitude h. The volume is $V = x^2 h/(4\sqrt{3})$ and the surface area is $A = (3x/2)(x^2/12 + h^2)^{1/2}$. (c) A right circular cylinder. The volume is $V = 2\pi r^2 h$. The surface area is $A = 2\pi r^2 + 2\pi rh$. For the cylinder without a lid, $A = \pi r^2 + 2\pi rh$.

For the complete box of item (a), a problem could read

Find the maximum value of xyz if $2(xy + yz + zx) = 100$.

Note that in the equations we leave the measurement units implicit, but consistency is required in order for the problem to retain its physical meaning.

The reader knows that of all the plane figures with fixed perimeter, the circle has the greatest area. A similar result holds in three dimensions: of all the 3D figures with a given surface area, the sphere has the greatest volume. These facts are of course related to the volume extremum problems cited above, but there is an important difference. Problems such as those of Fig. 2.9 involve only finite numbers of variables $x_1, \ldots, x_n$. But in order to obtain "isoperimetric results" like the two just mentioned, one must express the volumes and areas as integrals with integrands dependent on functions describing the shapes of arbitrary figures. Such problems are studied in an area of mathematics called the calculus of variations, where a knowledge of multiple integrals is essential. They were first studied by Joseph-Louis Lagrange. His results for functions with a finite number of variables are presented below.

Now let us focus on the class of problems typified by Fig. 2.9.

Theorem 2.12 (Lagrange). *Let (x_0, y_0, z_0) be an extremum point of a function $f = f(x, y, z)$ subject to a constraint $g(x, y, z) = k = $ constant, and assume f, g are continuously differentiable in a vicinity of (x_0, y_0, z_0). Then the following relations hold:*

$$\nabla f(x_0, y_0, z_0) = \lambda \nabla g(x_0, y_0, z_0), \qquad g(x_0, y_0, z_0) = k \qquad (2.56)$$

or

$$\left.\begin{array}{l} f_x(x_0, y_0, z_0) = \lambda g_x(x_0, y_0, z_0) \\ f_y(x_0, y_0, z_0) = \lambda g_y(x_0, y_0, z_0) \\ f_z(x_0, y_0, z_0) = \lambda g_z(x_0, y_0, z_0) \end{array}\right\}, \qquad g(x_0, y_0, z_0) = k \qquad (2.57)$$

where λ is a constant called a Lagrange multiplier.

Proof. On the surface

$$g(x, y, z) = k, \qquad (2.58)$$

let us take a smooth curve

$$\mathbf{r} = \mathbf{r}(t) = x(t)\,\mathbf{i} + y(t)\,\mathbf{j} + z(t)\,\mathbf{k} \qquad (2.59)$$

through the point $\mathbf{r}(t_0) = (x_0, y_0, z_0)$. So $g(x(t), y(t), z(t)) = k$ in some vicinity of t_0. Since t_0 is an extreme point of the function

$$h(t) = f(x(t), y(t), z(t))$$

we have $h'(t_0) = 0$. The chain rule yields

$$\begin{aligned} & f_x(x(t_0), y(t_0), z(t_0))\, x'(t_0) \\ & + f_y(x(t_0), y(t_0), z(t_0))\, y'(t_0) \\ & + f_z(x(t_0), y(t_0), z(t_0))\, z'(t_0) = 0 \end{aligned}$$

or more concisely

$$\nabla f(x_0, y_0, z_0) \cdot \mathbf{r}'(t_0) = 0. \qquad (2.60)$$

Since $\mathbf{r}'(t_0)$ is tangent to the curve (2.59) at t_0, and (2.59) is an arbitrary curve on the surface (2.58) passing through (x_0, y_0, z_0), equation (2.60) states that $\nabla f(x_0, y_0, z_0)$ is normal to the surface (2.58) at (x_0, y_0, z_0) which as the level surface has also a normal $\nabla g(x_0, y_0, z_0)$ at this point. It follows that $\nabla f(x_0, y_0, z_0)$ and $\nabla g(x_0, y_0, z_0)$ are parallel. So there is a constant λ such that $\nabla f(x_0, y_0, z_0) = \lambda \nabla g(x_0, y_0, z_0)$. $\qquad\qquad\qquad\qquad$ □

Lagrange's equality (2.56) states also that at (x_0, y_0, z_0) the level surfaces $f(x, y, z) = f(x_0, y_0, z_0)$ and $g(x, y, z) = k$ are tangent. This tells us how to locate (x_0, y_0, z_0) as the point where two level surfaces touch. We should take the surface $f(x, y, z) = c$ with $c < f(x_0, y_0, z_0)$, so that the surfaces $f(x, y, z) = c$ and $g(x, y, z) = k$ do not intersect near (x_0, y_0, z_0), then increase $c \to f(x_0, y_0, z_0)$ to obtain the only point where the surfaces touch. At the limit position the level surfaces become tangent at (x_0, y_0, z_0).

A point (x_0, y_0, z_0) obtained by solving (2.57) is only a critical point for f under the constraint $g(x, y, z) = k$; i.e., Lagrange's equation is merely necessary for an extremum. (There are sufficient conditions similar to the sign condition for $f''(x_0)$ in the case of a function of one variable, but they are cumbersome in practice.) We should also stress that an extremum of f with a constraint is not, in general, an extremum of f without constraints.

In practice we typically seek the absolute extrema of a function f on a compact set with boundary $g(x, y, z) = k$. In this case on the boundary we should find all the critical points of f that can be found by Lagrange's multiplier procedure. So

(1) The critical points are solutions of equations (2.57).
(2) The boundary consists of the compact surface $g(x, y, z) = k$, and may involve compact curves that must be separately considered for suspected extremal points.
(3) Finally we calculate f at all suspected points inside the compact set and on its boundary and select the minimum and maximum values.

Clearly the Lagrange procedure applies to more general extremum problems for a function paired with finite numbers of constraints. We state without proof

Theorem 2.13. *Let* $\mathbf{r}_0 \in \mathbb{R}^n$ *be an extremum point of a smooth function* $f(\mathbf{r})$ *under smooth constraints* $g_k(\mathbf{r}) = c_k$ $(k = 1, \ldots, m < n)$. *There are*

Lagrange multipliers λ_k $(k = 1, \ldots, m)$ such that

$$\nabla f(\mathbf{r}_0) = \sum_{k=1}^{m} \lambda_k \nabla g_k(\mathbf{r}_0), \qquad g_k(\mathbf{r}_0) = c_k \quad (k = 1, \ldots, m). \qquad (2.61)$$

This is also just a necessary condition for extremum. Note that (2.61) constitutes $n+m$ simultaneous equations in $n+m$ unknowns, which are the n components of $\mathbf{r}_0$ and the m Lagrange multipliers λ_k. If the intersection of level sets $g(\mathbf{r})$ near $\mathbf{r}_0$ is a smooth manifold, the system (2.61) is typically well posed, although in certain cases a solution may not exist.

Example 2.39. Let us find the maximum volume of a rectangular closed box (Fig. 2.9(a)) if its surface area is 24 cm^2. We have $V = f(x, y, z) = xyz$, $A = g(x, y, z) = 2xy + 2yz + 2zx = 24$, and equations (2.57) become

$$(1) \quad yz = \lambda(2y + 2z),$$
$$(2) \quad xz = \lambda(2x + 2z),$$
$$(3) \quad xy = \lambda(2x + 2y),$$
$$(4) \quad xy + yz + zx = 12.$$

To find the unknown x, y, z, and λ, we multiply equation (1) by x, equation (2) by y, and equation (3) by z:

$$(a) \quad xyz = \lambda 2(xy + xz),$$
$$(b) \quad xyz = \lambda 2(xy + yz),$$
$$(c) \quad xyz = \lambda 2(xz + yz).$$

Subtracting (b) from (a) we get $2\lambda(xz - yz) = 0$ or

$$\lambda z(x - y) = 0$$

As λ, x, y, z must all be nonzero, we have $x = y$. Subtracting (c) from (b) we get $\lambda x(y - z) = 0$ so $y = z$. Thus (4) reduces to $3x^2 = 12$ or $x = \pm 2$ from which only $x = 2$ is valid. So the maximum volume is $(2 \text{ cm})^3 = 8$ cm^3. It is realized when the box is a cube. $\qquad \square$

Example 2.40. We find the maximum volume of a box without a cover, if the surface area is 18 cm^2. Now $V = f(x, y, z) = xyz$ and $A = g(x, y, z) = xy + 2yz + 2zx = 18$. Equations (2.57) are

$$(1) \quad yz = \lambda(y + 2z),$$
$$(2) \quad xz = \lambda(x + 2z),$$

$$(3) \quad xy = \lambda(2x + 2y),$$

$$(4) \quad xy + 2yz + 2zx = 18.$$

Repeating the manipulations of the previous example we obtain $x = y$ and $y = 2z$, with the fourth equation yielding $3y^2 = 18$ and $y = \sqrt{6}$. The maximum volume is $V = 3\sqrt{6}$ cm³. $\qquad\square$

As we see, in both examples we obtained unique solutions (except the second solution $V = 0$ that we removed) and these solutions give clearly maximum values.

Example 2.41. We seek to minimize the area S in Fig. 2.10, assuming (x_0, y_0) is a given, fixed point through which the line AB passes. To this

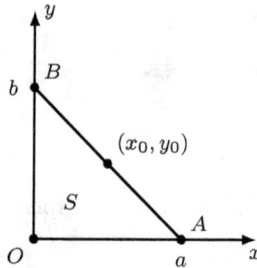

Fig. 2.10   Example 2.41.

end let us minimize $f(a, b) = ab$ subject to $y_0 = (-b/a)x_0 + b$. Defining $g(a, b) = ab - y_0 a - x_0 b$, we write Lagrange's equation $\nabla f = \lambda \nabla g$ in component form as

$$b = \lambda(b - y_0), \qquad a = \lambda(a - x_0).$$

These yield $b/a = y_0/x_0$, and use of the constraint equation gives $a = 2x_0$, $b = 2y_0$. Hence the extremum value for S is

$$S_m = \tfrac{1}{2}ab = 2x_0 y_0.$$

One way to verify that S_0 is a minimum is to use the inequality between the geometric and harmonic means:

$$(uv)^{1/2} \geq \frac{2}{\dfrac{1}{u} + \dfrac{1}{v}} \qquad (u > 0,\ v > 0)$$

where equality holds if and only if $u = v$. Setting $u = a/x_0$ and $v = b/y_0$, we have

$$\frac{a}{x_0} \cdot \frac{b}{y_0} \geq \left(\frac{2}{\dfrac{x_0}{a} + \dfrac{y_0}{b}} \right)^2 = 4$$

where we have used the intercept form $x/a + y/b = 1$ of the line AB. From this we obtain

$$\tfrac{1}{2}ab \geq 2x_0y_0$$

with equality holding if and only if $a/x_0 = b/y_0$. $\qquad\square$

Example 2.42. We seek the absolute maximum and minimum of $f(x,y) = 2x^2 + 3y^2 + 1$ on the elliptical domain $x^2 + 2y^2 \leq 4$. First we look for critical points inside the ellipse; the equations $f_x = 4x = 0$ and $f_y = 6y = 0$ yield a single critical point $(0,0)$. Now we seek suspected points on the boundary, solving (2.57) in the 2D version $f_x = \lambda g_x$, $f_y = \lambda g_y$, $x^2 + 2y^2 = 4$. These relations are

$$4x = \lambda 2x, \qquad 6y = \lambda 4y, \qquad x^2 + 2y^2 = 4.$$

The first, rewritten as $(4 - 2\lambda)x = 0$, gives $\lambda = 2$ or $x = 0$. For $x = 0$ the second equation gives an indefinite value for y, and the third equation gives $y^2 = 2$ or $y = \pm\sqrt{2}$. For $\lambda = 2$, x is indefinite and $y = 0$ so $x^2 = 4$ giving $x = \pm 2$. Hence $(0,0)$, $(\pm 2, 0)$, $(0, \pm\sqrt{2})$ are all suspected extremum points, and the suspected max-min values of f are

$$f(0,0) = 1, \qquad f(\pm 2, 0) = 9, \qquad f(0, \pm\sqrt{2}) = 7.$$

On the ellipse, the absolute minimum value of f is 1 and the absolute maximum value is 9. Of course, instead of using Lagrange's procedure, it would be easier to find the critical points on the ellipse by substituting $x = 2\cos t$ and $y = \sqrt{2}\sin t$ into f and finding its extrema on $[0, 2\pi]$. $\qquad\square$

2.17 Problems

Sets in the Euclidean Plane

2.1 Show that the following sets are open:

(a) $\{(x,y): (x+2)^2 + (y-1)^2 < 4\}$

(b) $\{(x,y): 0 < x < 1, \ 0 < y < 1 - x\}$

(c) the Euclidean plane $\mathbb{R}^2$

2.2 Show that the following sets are closed:

(a) $\{(x,y): (x+2)^2 + (y-1)^2 \leq 4\}$

(b) $[1,2]^2$

(c) the Euclidean plane $\mathbb{R}^2$

2.3 Construct a set in $\mathbb{R}^2$ that is neither open nor closed.

2.4 Give some examples of closed and bounded sets in $\mathbb{R}^2$ and $\mathbb{R}^3$.

Functions of Several Variables

2.5 Specify the domain of definition of each of the following functions, taking this to be the largest set for which the value of the function is real:

(a) $f(x,y) = (x-y)^{1/2}$ (g) $f(x,y) = \sin^{-1}(x-y)$

(b) $f(x,y) = \ln(x+y)$ (h) $f(x,y) = \cos^{-1}(x+y)$

(c) $f(x,y) = 1/(x^2+y^2)$ (i) $f(x,y) = \ln x + \ln y$

(d) $f(x,y) = \ln(x^2 - y^2 - 1)$ (j) $f(x,y) = \ln(xy)$

(e) $f(x,y) = (4 - x^2 - y^2)^{-1/2}$ (k) $f(x,y) = x + (xy)^{1/2}$

(f) $f(x,y) = e^{x+y}/(x-y)$ (l) $f(x,y) = (x + y^{1/2})^{1/2}$

2.6 For each of the following functions of three variables, specify the domain of definition:

(a) $f(x,y,z) = z/(x^2+y^2)^{1/2}$ (d) $f(x,y,z) = (x+1)(y-1)z$

(b) $f(x,y,z) = (xyz)^{1/2}$ (e) $f(x,y,z) = 1/(x+y+z)^{1/2}$

(c) $f(x,y,z) = y + (xz)^{1/2}$ (f) $f(x,y,z) = (\sin(x^2+y^2+z^2))^{1/2}$

2.7 Sketch each of the following functions:

(a) $f(x,y) = (x+y)^2/(xy)$ (c) $f(x,y) = xy + y/x$

(b) $f(x,y) = (x^2 - y^2)/(xy)$ (d) $f(x,y) = x^2(x^2+y^2)$

2.8 For each given function, find $f(y,x)$, $f(x,-y)$, $f(-x,-y)$, and $f(1/x, 1/y)$:

(a) $f(x,y) = xy^2$ (c) $f(x,y) = xy/(x^2+y^2)$

(b) $f(x,y) = (x+y)/(xy)$ (d) $f(x,y) = (x^2 - y^2)/(xy)$

2.9 Express the function $f(x,y,z) = 1 + xyz$ in (a) cylindrical coordinates, and (b) spherical coordinates.

Limits and Continuity

2.10 Prove that the following elementary inequalities hold for positive real numbers a, b:
$$ab \le \tfrac{1}{2}(a^2 + b^2), \qquad a + b \le [2(a^2 + b^2)]^{1/2}.$$
These can be useful in problems involving limits (cf., Example 2.3).

2.11 Show that any circular ε-neighborhood of a point $(a, b) \in \mathbb{R}^2$ contains a square neighborhood of (a, b).

2.12 For each function given below, find the limit of $f(x, y)$ as $(x, y) \to (1, 2)$:

(a) $f(x, y) = xy^{1/2}$

(b) $f(x, y) = x^2 + y^2 - 1$

(c) $f(x, y) = \cos(x^2 y)$

(d) $f(x, y) = e^{x+y}/(x - y)$

(e) $f(x, y) = \ln[(xy)^{1/2} - 1]$

(f) $f(x, y) = \exp[(xy)^{1/2}]$

(g) $f(x, y) = y^{-1} \sin xy$

(h) $f(x, y) = (x^2 + y^2)e^{-(x+y)}$

2.13 Show that the following functions do not have limits as $(x, y) \to (0, 0)$:

(a) $f(x, y) = x/(x + y)$

(b) $f(x, y) = (x - y)/(x + y)$

(c) $f(x, y) = (x^2 - y^2)/(x^2 + y^2)$

(d) $f(x, y) = |x|/(|x| + |y|)$

2.14 For each function $f(x, y)$ given below, calculate and compare the two iterated limits
$$\lim_{x \to 0}\left[\lim_{y \to 0} f(x, y)\right] \quad \text{and} \quad \lim_{y \to 0}\left[\lim_{x \to 0} f(x, y)\right].$$

(a) $f(x, y) = (x - y)/(x + y)$

(b) $f(x, y) = (x^2 y^2)/[x^2 y^2 + (x - y)^2]$

(c) $f(x, y) = [(1 + xy)^{1/2} - 1]/(x + y)$

2.15 Show that in each of the following cases, the double limit as $(x, y) \to (0, 0)$ exists, but at least one of the two possible iterated limits does not:

(a) $f(x, y) = x \sin(1/y)$

(b) $f(x, y) = (x + y) \cos(1/x) \cos(1/y)$

2.16 Show that the function $f(x, y) = xy/(x^2 + y^2)$ is continuous everywhere except at $(0, 0)$. Can it be defined at $(0, 0)$ in such a way that it becomes continuous there?

2.17 Find the limit
$$\lim_{(x,y) \to (1,1)} \frac{(x^2 - 2x + 1)(y - 1)}{(x^2 - 2x + 1) + 2(y - 1)^2}$$

or prove that it does not exist.

2.18 Find $\lim\limits_{\substack{x\to 1 \\ y\to 0}} \dfrac{\sin 3xy}{y}$.

2.19 Show that the function

$$f(x,y) = \begin{cases} \dfrac{y^3}{x^2+y^2}, & (x,y)\neq(0,0), \\ 0, & (x,y)=(0,0), \end{cases}$$

is continuous at $(0,0)$.

2.20 Prove that $\lim\limits_{(x,y)\to(x_0,y_0)} (ax+by) = ax_0 + by_0$.

2.21 For each function $f(x,y)$ given below, show that $f(x,y)\to 0$ as $(x,y)\to (0,0)$:

(a) $f(x,y) = x^2y/(x^2+y^2)$

(b) $f(x,y) = (x+y)/(x^2+y^2+1)$

(c) $f(x,y) = y\cos(y-x)^{-1}$

(d) $f(x,y) = (x^3+y^3)/(x^2+y^2)$

(e) $f(x,y) = |xy|$

(f) $f(x,y) = |x|+|y|$

(g) $f(x,y) = x\sin(1/y)$

(h) $f(x,y) = (x^2+y^2)\sin[(xy)^{-1}]$

(i) $f(x,y) = (x^3-y^3)/(x-y)$

(j) $f(x,y) = xy/(1+x^2y^2)$

(k) $f(x,y) = xy\cos x\cos y$

(l) $f(x,y) = xy\sin(1/x)\sin(1/y)$

(m) $f(x,y) = xy\cos(1/x)\cos(1/y)$

(n) $f(x,y) = xy(x^2-y^2)/(x^2+y^2)$

2.22 Find the set of discontinuity points for each of the following functions:

(a) $f(x,y,z) = \sin[1/(xyz)]$

(b) $f(x,y,z) = \cos[1/(xyz)]$

(c) $f(x,y,z) = 1/(\sin x\sin y\sin z)$

(d) $f(x,y,z) = [(x-a)^2+(y-b)^2+(z-c)^2]^{-1}$

(e) $f(x,y,z) = \exp[-1/(x^2+y^2+z^2)]$

2.23 Let $f(x,y)$ and $g(x,y)$ be defined in a neighborhood of (x_0,y_0) and continuous at this point. Show that the following functions are continuous at (x_0,y_0):

(a) $cf(x,y)$ for $c=$ constant

(b) $f(x,y)+g(x,y)$

(c) $f(x,y)-g(x,y)$

(d) $f(x,y)g(x,y)$

(e) $f(x,y)/g(x,y)$

Partial Differentiation

2.24 Calculate f_x and f_y for each of the following functions:

(a) $f(x,y) = x^y$

(b) $f(x,y) = \tan^{-1}(x/y)$

(c) $f(x,y) = x/y + y/x$

(d) $f(x,y) = \sin(ax + by)$

(e) $f(x,y) = x^{\sin y}$

(f) $f(x,y) = xy + x/y$

(g) $f(x,y) = e^{xy} \cos x \sin y$

(h) $f(x,y) = 2^{xy}$

(i) $f(x,y) = \ln \cos(x/y)$

(j) $f(x,y) = x^3 \cos y + y^6$

(k) $f(x,y) = (x - y)/(x + y)$

(l) $f(x,y) = \sin[(x + y)/(x - y)]$

2.25 For each function f given below, evaluate f_x and f_y at the point $(1, 1)$:

(a) $f(x,y) = 2^{x/y}$

(b) $f(x,y) = xy^{1/2}$

(c) $f(x,y) = x^3 \sin y$

(d) $f(x,y) = \sin[(x^2 + y^2)/(x + y)]$

(e) $f(x,y) = x^2 + y^{1/2}$

(f) $f(x,y) = xy \cos(xy)$

(g) $f(x,y) = x \sin(x + y)$

(h) $f(x,y) = \tan^{-1}(y/x)$

2.26 For each given function f, evaluate f_x, f_y, and f_z at the point $(1, 1, 1)$:

(a) $f(x,y,z) = xyz + y^{xz}$

(b) $f(x,y,z) = \tan^{-1}(1 + x + y/z)$

(c) $f(x,y,z) = x^3 y + y^3 z + z^3 x$

(d) $f(x,y,z) = \sin(xy/z)$

(e) $f(x,y,z) = e^{xy/z}$

(f) $f(x,y,z) = z^{x+y}$

(g) $f(x,y,z) = (xyz)^{1/3} + \ln(x + y + z)$

(h) $f(x,y,z) = x^2 + z^2 - xy - yz + 1$

(i) $f(x,y,z) = (x + y + z) \cos(xyz)$

(j) $f(x,y,z) = \cos^{-1}(xy/z)$

(k) $f(x,y,z) = x^{y/z}$

(l) $f(x,y) = \tan^{-1}(xy/z^2)$

2.27 Find all the first and second derivatives of each of the following functions:

(a) $f(x,y) = xe^y + ye^x$

(b) $f(x,y) = \sin(x^2 - y^2)$

(c) $f(x,y) = (x^2 + y^2)^{1/2}$

(d) $f(x,y) = \cos^{-1}(y/x)$

(e) $f(x,y) = (xy)^{1/2}$

(f) $f(x,y) = e^{x^2 - y} + 2x + 1$

2.28 Find all the first and second derivatives of each of the following functions:

(a) $f(x,y,z) = (xy)^z$

(b) $f(x,y,z) = z^{xy}$

(c) $f(x,y,z) = e^{x/y} + e^{y/z}$

(d) $f(x,y,z) = (x^2 + y^2) \cos z$

Concept of a Partial Differential Equation

2.29 Show that the following functions satisfy the indicated partial differential equations:

(a) $f(x,y) = x^2 + y^2$ and $y f_x - x f_y = 0$

(b) $f(x,y) = \ln(\sqrt{x} + \sqrt{y})$ and $x f_x + y f_y = 1/2$

(c) $f(x,y) = x^y$ and $(x/y) f_x + f_y / \ln x - 2f = 0$

(d) $f(x,y) = \ln(x^2 + y^2) + \tan^{-1}(y/x)$ and $f_{xx} + f_{yy} = 0$

(e) $f(x,y) = e^{\sin(ax+by)}$ and $b^2 f_{xx} - a^2 f_{yy} = 0$

Differentiability

2.30 Discuss the differentiability of each of the following functions at the point $(0,0)$:

(a) $f(\mathbf{x}) = x^2 + xy$

(b) $f(\mathbf{x}) = x^2 + y^2$

(c) $f(\mathbf{x}) = |xy|$

(d) $f(\mathbf{x}) = xy^2$

Total Differential

2.31 Find the total differentials of the following functions $z = f(x,y)$:

(a) $z = \ln(x^2 + y^2)$

(b) $z = x^y$

(c) $z = \tan^{-1}(y/x)$

(d) $z = y \ln x$

(e) $z = x e^{-xy}$

(f) $z = x/y$

(g) $z = \sinh(x+y)$

(h) $z = \cos(x/y)$

(i) $z = x^2 y^2$

(j) $z = \ln(x \cos y)$

(k) $z = \exp[\cos^2(x^2 + y^2)]$

(l) $z = xy/(x+y)$

2.32 Given $f(x,y,z) = z/(x^2 + y^2)$, evaluate the differential df at (a) point $(1,2,1)$ and (b) point $(1,1,1)$.

2.33 A function of three variables takes the form $u = f(x,y,z) = Ax + By + Cz + D$ where A, B, C, D are constants. What is its total differential?

2.34 The second-order differential of $f(x,y)$ is given by

$$d^2 f = \frac{\partial^2 f}{\partial x^2} dx^2 + \frac{\partial^2 f}{\partial x \partial y} dx\, dy + \frac{\partial^2 f}{\partial y^2} dy^2$$

provided f has continuous partial derivatives of second order. Compute $d^2 f$ for the following functions:

(a) $f(x,y) = xy/(x-y)$

(b) $f(x,y) = \ln(x^2 + y^2)$

2.35 Assuming $f(t)$ is continuous, find the total differential of

(a) $F(x,y) = \int_0^{x+y} f(t)\,dt,$ (b) $F(x,y) = \int_x^y f(t)\,dt.$

2.36 Approximate the error in the volume of each of the following solids, produced by small errors in the quantities that appear in its formula:

(a) right circular cone: $V = \pi r^2 h/3$, where r is the basal radius and h is the altitude

(b) spherical shell: $V = 4\pi(b^3 - a^3)/3$, where b is the outer radius and a is the inner radius

Nabla Operator, Directional Derivative

2.37 For each function f given below, find ∇f and the directional derivative of f in the direction of the vector $\mathbf{i} + \mathbf{j}$ at the point $(1, 2, 3)$:

(a) $f(x,y,z) = xyz$

(b) $f(x,y,z) = x^2 + y^2 + z^2$

(c) $f(x,y,z) = \ln(x^2 + y^2 + z^2)$

(d) $f(x,y,z) = x^{y/z}$

(e) $f(x,y,z) = (x/y)^z$

(f) $f(x,y,z) = xy + yz$

(g) $f(x,y,z) = x^2 + xy + yz$

(h) $f(x,y,z) = \sin(\pi xyz)$

(i) $f(x,y,z) = (x+y+z)/(x+y)$

(j) $f(x,y,z) = \ln(e^x + e^y)$

2.38 For each of the following functions f, find the directional derivative of f at point P in the direction of the vector from P to Q:

(a) $f(x,y,z) = x^2 - xy + y^2$, $P = (1,1,1)$, $Q = (2,2,2)$

(b) $f(x,y,z) = x^2 y^2 z^2$, $P = (1,2,2)$, $Q = (-1,-1,-1)$

(c) $f(x,y,z) = xy + yz + zx$, $P = (0,0,0)$, $Q = (1,2,3)$

(d) $f(x,y,z) = x^3 + y^3 + z^3$, $P = (1,2,-1)$, $Q = (3,2,1)$

2.39 For each given function f, find $\nabla^2 f$ and evaluate it at the point $(2,1,2)$:

(a) $f(x,y,z) = x^2 + y^2 + z^2$

(b) $f(x,y,z) = xy + yz$

(c) $f(x,y,z) = x^2 + xy + yz$

(d) $f(x,y,z) = \cos(x - y + z)$

(e) $f(x,y,z) = \sin(\pi xyz)$

(f) $f(x,y,z) = (x+y+z)/(x+y)$

(g) $f(x,y,z) = \ln(e^x + e^y)$

(h) $f(x,y,z) = xe^{xy}$

2.40 For each given vector field $\mathbf{A}$, find $\nabla \cdot \mathbf{A}$ and $\nabla \times \mathbf{A}$:

(a) $\mathbf{A} = x^2 y\,\mathbf{i} + y^2 z\,\mathbf{j} + z^2 x\,\mathbf{k}$

(b) $\mathbf{A} = x\,\mathbf{i} + y\,\mathbf{j} + z\,\mathbf{k}$

(c) $\mathbf{A} = x^2\,\mathbf{i} + y^2\,\mathbf{j} + z^2\,\mathbf{k}$

(d) $\mathbf{A} = \mathbf{c}$, a constant

(e) $\mathbf{A} = x\,\mathbf{i} + y\,\mathbf{j} + z\,\mathbf{k}$

(f) $\mathbf{A} = -y^2\,\mathbf{i} + x\,\mathbf{j} + z^2\,\mathbf{k}$

2.41 In cylindrical coordinates the Laplacian operator is given by

$$\nabla^2 u = u_{rr} + \frac{1}{r}u_r + \frac{1}{r^2}u_{\theta\theta} + u_{zz}.$$

Calculate $\nabla^2 u$ for the following field expressions:

(a) $u(r,\theta) = z + r\sin\theta$ \qquad\qquad (b) $u(r,\theta) = r + z\cos\theta$

2.42 Letting a, b, c be constants, show that the vector field

$$\mathbf{A} = (bz - cy)\,\mathbf{i} + (cx - az)\,\mathbf{j} + (ay - bx)\,\mathbf{k}$$

has zero divergence and constant rotor given by $\nabla \times \mathbf{A} = 2(a\,\mathbf{i} + b\,\mathbf{j} + c\,\mathbf{k})$.

2.43 Verify the following vector identities for sufficiently smooth fields:

(a) $\nabla(f + g) = \nabla f + \nabla g$ \qquad\qquad (e) $\nabla \times (\mathbf{F} + \mathbf{G}) = \nabla \times \mathbf{F} + \nabla \times \mathbf{G}$

(b) $\nabla(fg) = f\nabla g + g\nabla f$ \qquad\qquad (f) $\nabla \times (f\mathbf{F}) = f\nabla \times \mathbf{F} + \nabla f \times \mathbf{F}$

(c) $\nabla \cdot (\mathbf{F} + \mathbf{G}) = \nabla \cdot \mathbf{F} + \nabla \cdot \mathbf{G}$ \qquad\quad (g) $\nabla \times \nabla f = 0$

(d) $\nabla \cdot (f\mathbf{F}) = f\nabla \cdot \mathbf{F} + \nabla f \cdot \mathbf{F}$ \qquad\quad (h) $\nabla \cdot (\nabla \times \mathbf{F}) = 0$

2.44 Let $\mathbf{r}$ be the position vector, $r = |\mathbf{r}|$, and $\hat{\mathbf{r}} = \mathbf{r}/r$. Show that each of the following holds as an identity:

(a) $\nabla(\mathbf{r} \cdot \mathbf{r}) = 2\mathbf{r}$ \qquad\qquad\qquad (g) $\nabla^2(\mathbf{r} \cdot \mathbf{r}) = 6$

(b) $\nabla \cdot \mathbf{r} = 3$ \qquad\qquad\qquad\quad (h) $\nabla r = \hat{\mathbf{r}}$

(c) $\nabla \cdot \hat{\mathbf{r}} = 2/r$ \qquad\qquad\qquad (i) $\nabla(1/r^n) = -rn/r^{n+2}$

(d) $\nabla \cdot (r^n \mathbf{r}) = (n + 3)r^n$ \qquad\quad (j) $\nabla \times (\mathbf{r}/r^2) = 0$

(e) $\nabla \times \mathbf{r} = 0$ \qquad\qquad\qquad\quad (k) $\nabla \cdot (\mathbf{r}/r^2) = 1/r^2$

(f) $\nabla(1/r) = -\hat{\mathbf{r}}/r^2$

2.45 Let $\mathbf{r}$ be the position vector, $r = |\mathbf{r}|$, and $\hat{\mathbf{r}} = \mathbf{r}/r$. Assuming $\mathbf{a}, \mathbf{b}, \mathbf{c}$ are constant vector fields, show that each of the following holds:

(a) $\nabla(\mathbf{r} \cdot \mathbf{a}) = \mathbf{a}$ \qquad\qquad\qquad (g) $\nabla \cdot [(\mathbf{r} \times \mathbf{a}) \times \mathbf{r}] = -2\mathbf{a} \cdot \mathbf{r}$

(b) $\nabla \cdot (\mathbf{r} \times \mathbf{a}) = 0$ \qquad\qquad\quad (h) $\nabla \times (r\mathbf{a}) = \hat{\mathbf{r}} \times \mathbf{a}$

(c) $\nabla \times (\mathbf{r} \times \mathbf{a}) = -2\mathbf{a}$ \qquad\quad (i) $\nabla \times [(\mathbf{c} \times \mathbf{r}) \times \mathbf{a}] = \mathbf{c} \times \mathbf{a}$

(d) $\nabla \cdot (r\mathbf{a}) = \hat{\mathbf{r}} \cdot \mathbf{a}$ \qquad\qquad\quad (j) $\nabla \times [(\mathbf{c} \times \mathbf{r}) \times \mathbf{r}] = 3\mathbf{c} \times \mathbf{r}$

(e) $\nabla \cdot [\mathbf{a} \times (\mathbf{r} \times \mathbf{b})] = 2\mathbf{a} \cdot \mathbf{b}$ \qquad (k) $\nabla \cdot [(\mathbf{c} \cdot \mathbf{r})\mathbf{r}] = 4\mathbf{c} \cdot \mathbf{r}$

(f) $\nabla \cdot [(\mathbf{r} \times \mathbf{a}) \times \mathbf{c}] = -2\mathbf{a} \cdot \mathbf{c}$ \qquad (l) $\nabla \cdot [(\mathbf{c} \cdot \mathbf{r})\mathbf{a}] = \mathbf{a} \cdot \mathbf{c}$

(m) $\nabla \cdot (r^2 \mathbf{c}) = 2\mathbf{r} \cdot \mathbf{c}$

(n) $\nabla \times (r^2 \mathbf{c}) = 2\mathbf{r} \times \mathbf{c}$

(o) $\nabla \times [(\mathbf{c} \cdot \mathbf{r})\mathbf{r}] = \mathbf{c} \times \mathbf{r}$

(p) $\nabla \times [\mathbf{b}(\mathbf{r} \cdot \mathbf{a})] = \mathbf{a} \times \mathbf{b}$

(q) $\nabla[\mathbf{a} \cdot (\mathbf{c} \times \mathbf{r})] = \mathbf{a} \times \mathbf{c}$

(r) $\nabla \cdot [(\mathbf{a} \cdot \mathbf{r})(\mathbf{c} \times \mathbf{r})] = \mathbf{a} \cdot (\mathbf{c} \times \mathbf{r})$

(s) $\nabla(|\mathbf{c} \times \mathbf{r}|^2) = 2(\mathbf{c} \times \mathbf{r}) \times \mathbf{c}$

(t) $\nabla \cdot (\mathbf{c} \times \mathbf{r}) = 0$

Chain Rules

2.46 Find df/dt if

(a) $f(x,y) = x/y$ with $x = e^t$ and $y = \ln t$,

(b) $f(x,y) = e^{xy}$ with $x = \cos t$ and $y = \sin t$,

(c) $f(x,y,z) = x^2 y^3 z$ with $x = t$, $y = t^2$, $z = e^t$.

2.47 Find f_u and f_v if

(a) $f(x,y) = \sin(xy)$ with $x = uv$ and $y = u$,

(b) $f(x,y) = \ln(x^2 + y^2)$ with $x = uv$ and $y = u/v$,

(c) $f(x,y) = x^2 y^2$ with $x = u + v$ and $y = u/v$,

(d) $f(x,y) = x^2/y$ with $x = u - 3v$ and $y = 2u + v$,

(e) $f(x,y) = xy$ with $x = u \sin v$ and $y = v \sin u$.

2.48 Differentiate the following functions with respect to x:

(a) $f(x,y) = \ln(e^x + e^y)$ where $y = \frac{1}{3}x^3 + x$

(b) $f(x,y) = \tan^{-1}[(x+1)/y]$ where $y = (x+1)^2$

(c) $f(x,y,z) = (x^2 + y^2)^{1/2}$ where $y = e^x$

(d) $f(x,y) = x^y$ where $y = \ln x$

(e) $f(x,y) = \sin^{-1}(x/y)$ where $y = (x^2 + 1)^{1/2}$

2.49 Find df/dx if $f(x,y,z) = e^x(y - z)$ where $y = \sin x$ and $z = \cos x$.

2.50 Let $z = f(x,y)$ where $x = r \cos \theta$ and $y = r \sin \theta$. Find all the first and second partial derivatives of f with respect to r and θ.

2.51 Supply the details needed to complete Example 2.26.

2.52 Let f be a differentiable function of one variable. Express the differential dz in each of the following cases:

(a) $z = f(xy)$

(b) $z = f(x/y)$

(c) $z = f(x + y)$

(d) $z = f(x^2 - y^2)$

2.53 Consider a function of the form $u(t,x) = f(x - ct) + g(x + ct)$, where f and g are twice continuously differentiable functions and c is constant. Show that u satisfies the wave equation $u_{tt} = c^2 u_{xx}$.

Taylor Expansion

2.54 Compute the expansion (2.43) up to the t^2 term for the following functions $f(\mathbf{x}) = f(x_1, x_2)$, and evaluate at the points indicated:

(a) $f(\mathbf{x}) = x_1{}^{x_2}$ at $\mathbf{x} = (1, 0)$

(b) $f(\mathbf{x}) = e^{x_1 + x_2}$ at $\mathbf{x} = (0, 0)$

(c) $f(\mathbf{x}) = e^{x_1/x_2}$ at $\mathbf{x} = (0, 1)$

(d) $f(\mathbf{x}) = x_1 x_2^2$ at $\mathbf{x} = (1, 1)$

(e) $f(\mathbf{x}) = e^{x_1} \sin x_2$ at $\mathbf{x} = (0, 0)$

(f) $f(\mathbf{x}) = e^{-x_1} \cos x_2$ at $\mathbf{x} = (0, 0)$

Derivatives of Implicit Functions

2.55 In each case, state whether the equation $f(x, y) = 0$ determines y as an implicit function of x:

(a) $f(x, y) = x - y^2$

(b) $f(x, y) = x - \cos y$

(c) $f(x, y) = x - \tan^{-1} y$

(d) $f(x, y) = x - |y|$

2.56 Each of the following relations defines y as an implicit function of x. Find dy/dx:

(a) $x^y = y^x$

(b) $x^{1/2} + y^{1/2} = 1$

(c) $xy + \ln(xy) = 0$

(d) $y/x + \sin(y/x) = 1$

(e) $e^x + e^y = 2xy$

(f) $\cos(xy) + xy^2 - 5y = 0$

2.57 Find y' and y'':

(a) $x^2 + y^2 - a^2 = 0$

(b) $x^3 + 3axy + y^3 = 0$

(c) $y^2 - 2axy + x^2 - b^2 = 0$

(d) $ax^3 + x^3 y - ay^3 = 0$

2.58 Each of the following relations defines z as an implicit function of x and y. Find z_x and z_y:

(a) $xyz - e^z = 0$

(b) $z = x^2 + y^2 + z^2$

(c) $e^z + x^2 y + z = 0$

(d) $x^2 + y^2 - z^2 - xy + xz + 1 = 0$

(e) $e^z = \sin x \sin y$

(f) $z^3 - 3xyz - z = 0$

(g) $x(yz)^{1/2} + e^{y-z} = (x + y)/z$

Tangent Planes and Normal Lines

2.59 Show that the level surfaces of the function

$$f(x, y, z) = (ax^2 + by^2 + cz^2)^{-1} \qquad (a, b, c > 0)$$

are ellipsoids.

2.60 Discuss the level surfaces of the following functions:

(a) $f(x, y, z) = x - y + z$

(b) $f(x, y, z) = x^2 + y^2 + z^2$

(c) $f(x, y, z) = x^2 + y^2 - z^2$

(d) $f(x, y, z) = z/(x^2 + y^2)$

2.61 Find the plane tangent and the line normal to the sphere $x^2 + y^2 + z^2 = 1$ at the point $(1/\sqrt{3}, 1/\sqrt{3}, 1/\sqrt{3})$. That is, carry out the calculations required to produce Fig. 2.3.

2.62 Find the plane tangent to the given surface at the indicated point:

(a) the ellipsoid (1.33) at (x_0, y_0, z_0)

(b) $z = e^{xy}$ at (a, b, e^{ab})

(c) $z = \cos(xy)$ at $(a, b, \cos(ab))$

(d) $z = e^x \cos y$ at $(0, 0, 1)$

2.63 Find the line normal to the given surface through the indicated point.

(a) $z = e^x \sin y$ at $(0, 0, 0)$

(b) $z = e^x \cos y$ at $(0, 0, 1)$

(c) the ellipsoid (1.33) at (x_0, y_0, z_0)

(d) the paraboloid $x^2 + y^2 = az$ at $(1, 1, 2/a)$

2.64 Find the planes tangent to the surface $x^2 + y^2 + z^2 - 2y = 8$ whose normals are parallel to the vector $(1, 2, 2)$.

Local Extrema

2.65 For each given function, make a contour plot and find any critical points:

(a) $f(x, y) = x^3 + 4xy - y^2 + 2x + y - 2$

(b) $f(x, y) = (x^2 + y)e^{y/2}$

2.66 For each function given, investigate any local extrema:

(a) $f(x, y) = 2x^2 + (y - 2)^2$.

(b) $f(x, y) = (x^2 + y^2)[e^{-(x^2+y^2)} - 1]$

(c) $f(x, y) = -x^2 - xy - y^2 + x + y$

(d) $f(x, y) = x^2 + xy + y^2 - 3x - 2y$

(e) $f(x, y) = x^2 y^2 + (x^2 + y^2)/2 + xy + 1$

2.67 Investigate any local extrema, assuming $p, q > 0$:

(a) $f(x, y) = x^2/(2p) + y^2/(2q)$

(b) $f(x, y) = x^2/(2p) - y^2/(2q)$

Constrained Extrema

2.68 Find the extrema of $f(x,y)$ subject to the given condition:

(a) $f(x,y) = (x^2 + y^2)^{1/2}$ with $x + y = 1$

(b) $f(x,y) = x + y$ with $xy - 1 = 0$

(c) $f(x,y) = xy$ with $x + y = 1$

(d) $f(x,y) = x + y$ with $x^2 + y^2 = 1$

(e) $f(x,y) = 1/x + 1/y$ with $x + y = 2$

2.69 Extremize each given function on the circle $x^2 + y^2 = 2$:

(a) $f(x,y) = x^2 + xy + y^2$ (c) $f(x,y) = e^{xy}$

(b) $f(x,y) = e^{x+y}$ (d) $f(x,y) = \cos(xy)$

2.70 Rework Example 2.35 by the Lagrange multiplier approach.

2.71 More problems on constrained extrema:

(a) Find the rectangular parallelepiped having the largest volume for a given sum $12a$ of all of its edges.

(b) Find the maximum value of $f(x, y, z) = x^2 y^2 z^2$ over the sphere (1.32).

(c) Find the minimum distance from the point $(a, 0)$ to the parabola $y^2 = kx$, assuming $k > 0$ and $a < k/2$.

(d) Find the maximum area of a sector of a circle having fixed radius r, for a fixed perimeter P of the sector.

(e) Find the maximum volume of a rectangular parallelepiped with sides x, y, z if the sides satisfy the inequality $2x + 3y + 5z \le 2000$.

(f) Find the minimum surface area of an open-top rectangular box for a fixed volume V.

(g) Find the maximum area of a rectangle that can be inscribed in the ellipse (1.8).

(h) Find the maximum volume of a rectangular parallelepiped that can be inscribed in the ellipsoid (1.33).

Global Extrema

2.72 Find the global maximum and minimum values of each function on the region specified:

(a) $f(x,y) = x^2 + ay$ with $a > 0$; the region is specified by the inequalities $|x| \le 1$, $|y| \le 1$

(b) $f(x,y) = 1 - x + x^2 + 2y$; the region is bounded by the lines $x = 0$, $y = 0$, and $x + y = 1$

(c) $f(x,y) = xy(1-x-y)$; the region is bounded by the lines $x = 0$, $y = 0$, and $x + y = 1$

(d) $f(x,y) = xy$; the region is the disk $x^2 + y^2 \leq 1$

(e) $f(x,y) = 4xy - 2xy^2 + 2x$; the region is the triangle with vertices $(0,0)$, $(0,3)$, and $(3,0)$ in the xy-plane

2.73 Consider the function $f(x,y) = x^2 - y^2$. (a) Investigate its local extrema. (b) Find its global extrema on the disk $x^2 + y^2 \leq 4$. (c) Does $f(x,y)$ have extrema when subject to the constraint $x + y = 1$?

Some Inequalities of Applied Mathematics

2.74 (a) Prove the triangle inequality for two real numbers x and y:

$$|x+y| \leq |x| + |y|.$$

When does equality hold? (b) Show that

$$\left|\sum_{k=1}^{n} a_k\right| \leq \sum_{k=1}^{n} |a_k|$$

for real numbers $a_1, \ldots, a_n$.

2.75 Consider the function $f(x) = x^\alpha$ for $x > 0$ and $0 < \alpha < 1$. Applying the extended mean value theorem

$$f(x) = f(x_0) + f'(x_0)(x - x_0) + \tfrac{1}{2}f''(\xi)(x - x_0)^2$$

to f at $x_0 = 1$, we get the inequality

$$x^\alpha \leq \alpha x + (1 - \alpha) \qquad (x \geq 0,\ 0 < \alpha < 1).$$

Give a graphical interpretation of this result. Note that equality holds if and only if $x = 1$.

2.76 Let p and q be *conjugate indices*: that is, numbers exceeding unity and connected by the relation

$$\frac{1}{p} + \frac{1}{q} = 1.$$

(a) Show that this can be rewritten in any of the forms

$$(p-1)(q-1) = 1, \qquad q = \frac{p}{p-1}, \qquad p = \frac{q}{q-1},$$

$$p + q = pq, \qquad p = 1 + \frac{p}{q}, \qquad (p-1)q = p.$$

(b) Show that for any nonnegative real numbers A and B we have

$$AB \le \frac{A^p}{p} + \frac{B^q}{q}$$

with equality if and only if $B = A^{p-1}$. This is *Young's inequality*.

2.77 Let $a_1, \ldots, a_n$ and $b_1, \ldots, b_n$ be nonnegative numbers, and let p, q be as in the previous problem. Obtain *Hölder's inequality* for sums:

$$\sum_{i=1}^{n} a_i b_i \le \left(\sum_{i=1}^{n} a_i^p \right)^{1/p} \left(\sum_{i=1}^{n} b_i^q \right)^{1/q}.$$

This is a result of great significance in mathematical analysis. The case $p = q = 2$ is the *Cauchy–Schwarz inequality* for sums.

2.78 (e) Let $a_1, \ldots, a_n$ and $b_1, \ldots, b_n$ be nonnegative numbers, and let $p > 1$. Use Hölder's inequality to obtain *Minkowski's inequality* for sums:

$$\left[\sum_{i=1}^{n} (a_i + b_i)^p \right]^{1/p} \le \left[\sum_{i=1}^{n} a_i^p \right]^{1/p} + \left[\sum_{i=1}^{n} b_i^p \right]^{1/p}.$$

Remark 2.5. Versions of the triangle inequality, Hölder's inequality, the Cauchy–Schwarz inequality, and Minkowski's inequality for integrals will be given in the following chapters. $\qquad\square$

Remark 2.6. Also useful are the inequalities for the means of a discrete set of positive numbers $a_1, \ldots, a_n$:

$$\mathrm{HM} \le \mathrm{GM} \le \mathrm{AM} \le \mathrm{QM},$$

where

$$\mathrm{HM} = n \left(\sum_{k=1}^{n} a_k^{-1} \right)^{-1}, \qquad\qquad \mathrm{GM} = \left(\prod_{k=1}^{n} a_k \right)^{1/n},$$

$$\mathrm{AM} = \frac{1}{n} \sum_{k=1}^{n} a_k, \qquad\qquad \mathrm{QM} = \left(\frac{1}{n} \sum_{k=1}^{n} a_k^2 \right)^{1/2}.$$

The values HM, GM, AM, and QM are the harmonic mean, the geometric mean, the arithmetic mean, and the quadratic mean, respectively, of the a_k. Equality holds if and only if the a_k are all equal. $\qquad\square$

Chapter 3

Double Integrals

3.1 Some Optional Remarks on Integration

This section explains why the presentation of this book differs slightly from that of standard textbooks on integration. Without loss of continuity a reader can skip directly to Section 3.2.

On the Definite Integral

First we review the typical way of introducing the definite integral of a function $y = f(x)$ on a segment $[a, b]$.

(1) A partition P_n of $[a, b]$ defined by points x_k, where

$$x_1 = a < x_2 < x_3 < \cdots < x_{n+1} = b,$$

is introduced. From each segment $[x_k, x_{k+1}]$ of P_n, a point $\tilde{x}_k$ is selected. The length of the kth segment is

$$\Delta x_k = x_{k+1} - x_k.$$

The norm of the partition, $\|P_n\| = \max_k \Delta x_k$, is defined. See Fig. 3.1.

(2) For P_n a Riemann sum

$$R_n = \sum_{k=1}^{n} f(\tilde{x}_k)\,\Delta x_k, \qquad \tilde{x}_k \in [x_k, x_{k+1}]$$

is written down.

(3) It is announced that the definite integral

$$\int_a^b f(x)\,dx$$

is the limit of the Riemann sums R_n if this limit exists and is unique for any sequence of partitions and selection of $\tilde{x}_k$ when $\|P_n\| \to 0$.

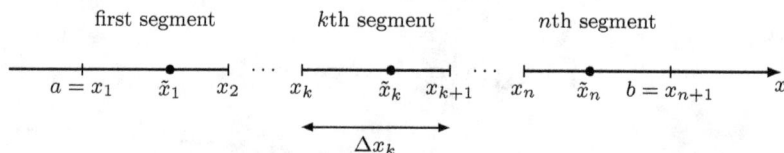

first segment kth segment nth segment

$a = x_1 \quad \tilde{x}_1 \quad x_2 \quad \cdots \quad x_k \quad \tilde{x}_k \quad x_{k+1} \quad \cdots \quad x_n \quad \tilde{x}_n \quad b = x_{n+1} \quad x$

Δx_k

Fig. 3.1 Notation for partitioning of an interval $[a, b]$. Numbering of segments corresponds to the numbering of their left endpoints.

This slightly cumbersome definition can present conceptual difficulties to a reader who has previously studied only limits of sequences and pointwise limits of functions. A student may have no idea what to do with multiple sequences of partitions and arbitrary selections of intermediate points $\tilde{x}_k$. Moreover it is unclear whether such a strange limit should actually exist. Intuition is aided by regarding the Riemann sum as an approximation to the area of a curvilinear "rectangle" bounded by the coordinate lines $x = a$ and $x = b$, the x-axis, and the curve $y = f(x) \geq 0$. This "convinces" students that a limit of Riemann sums should exist and equal the area of this "rectangle."

We propose to change the presentation of the definite integral, commencing with formulation of a real physical problem. Unfortunately the area problem for a curvilinear trapezoid is difficult to extend to three dimensions. Hence we shall consider the calculation of mass, in particular the mass of a straight rod with a nonuniform mass density $\gamma = \gamma(x)$. This is easily extended to the consideration of multiple integrals.

An approximate solution to this problem takes the form of a particular Riemann sum

$$R_n = \sum_{k=1}^{n} \gamma(x_k)\,\Delta x, \qquad \Delta x = \frac{b-a}{n}.$$

Figure 3.2 compares the terms of this sum with that of the more general Riemann sum

$$\sum_{k=1}^{n} \gamma(\tilde{x}_k)\,\Delta x_k.$$

Then we can announce that for a function continuous on $[a, b]$, the limit

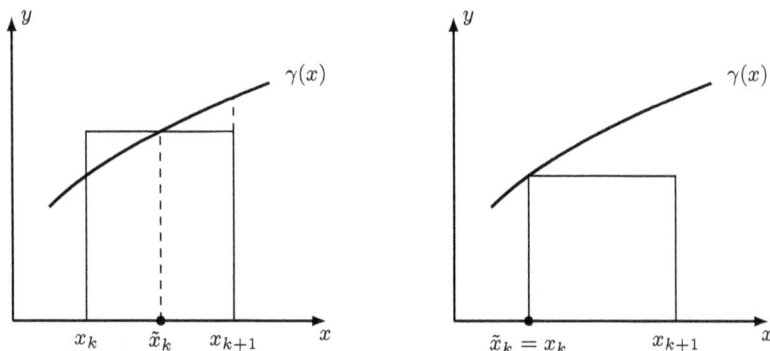

Fig. 3.2 *Left*, subinterval for a general Riemann sum; *right*, special case with intermediate point $\tilde{x}_k$ taken at the left endpoint x_k.

of the Riemann sums exists and is called the definite integral over $[a, b]$:

$$\int_a^b \gamma(x)\, dx. \tag{3.1}$$

In this way the student encounters the limit of an ordinary numerical sequence and the limit actually exists. A subsequent theorem can reveal that by using arbitrary partitions P_n and selection of intermediate points $\tilde{x}_k \in [x_k, x_{k+1}]$ when $\|P_n\| \to 0$, we get the unique limit denoted by (3.1); hence we can use the traditional definition of the definite integral, with $f = \gamma$, given in points (1)–(3) above (but now it is a theorem!). Moreover we can extend the theorem on existence of the integral to any piecewise continuous function f since, in the proof, nonnegativity of γ is not required.

Let us demonstrate how this program can be realized. Consider the problem of finding the mass m of a straight rod along the x-axis with a given linear mass density γ on the segment $[a, b]$. The density at point x can be defined as follows. Take the mass Δm of a segment $[x, x + h]$ of the rod. The average line density of this portion is $\Delta m/h$. Therefore, if it exists, the density at x is

$$\gamma(x) = \lim_{h \to 0+} \frac{\Delta m}{h}.$$

We assume this function γ is continuous on $[a, b]$. Next we uniformly partition $[a, b]$ into n equal segments by the points $x_k = a + (k-1)\,\Delta x$ $(k = 1, 2, \ldots, n+1)$ with $\Delta x = (b-a)/n$. Let m_k be the mass of the kth segment so that $m_k \approx \gamma(x_k)\,\Delta x$. Since mass is an additive quantity we have

$$m = \sum_{k=1}^{n} \Delta m_k \approx \sum_{k=1}^{n} \gamma(x_k)\,\Delta x.$$

On the right we see a term of an ordinary numerical sequence

$$\left\{ \sum_{k=1}^{n} \gamma(x_k)\,\Delta x \right\}$$

of Riemann sums, the limit as $n \to \infty$ of which can be announced as the mass m:

$$m = \lim_{n\to\infty} \sum_{k=1}^{n} \gamma(x_k)\,\Delta x.$$

Repeating this with an abstract continuous function f on $[a,b]$ (without the restriction that f is nonnegative), we get a similar limit to be called the definite integral of f over $[a,b]$ and denoted by

$$\int_a^b f(x)\,dx. \tag{3.2}$$

Of course we should prove that the limit exists, but this may seem "evident" to an engineering student. Hence we can formulate

Theorem 3.1. *(1) For a function f continuous on $[a,b]$, there exists the definite integral (3.2). (2) Let P_n be a partition of the interval $[a,b]$ with norm $\|P_n\| = \max_k \Delta x_k$. For any sequence of partitions P_n with $\|P_n\| \to 0$ as $n \to \infty$ and for any choice of intermediate points $\tilde{x}_k \in [x_k, x_{k+1}]$, the sequence of Riemann sums $\{R_n\}$ given by*

$$R_n = \sum_{k=1}^{n} f(\tilde{x}_k)\,\Delta x_k$$

has a limit equal to the definite integral:

$$\lim_{n\to\infty} R_n = \int_a^b f(x)\,dx.$$

Moreover these statements hold if f is merely piecewise continuous on $[a,b]$.

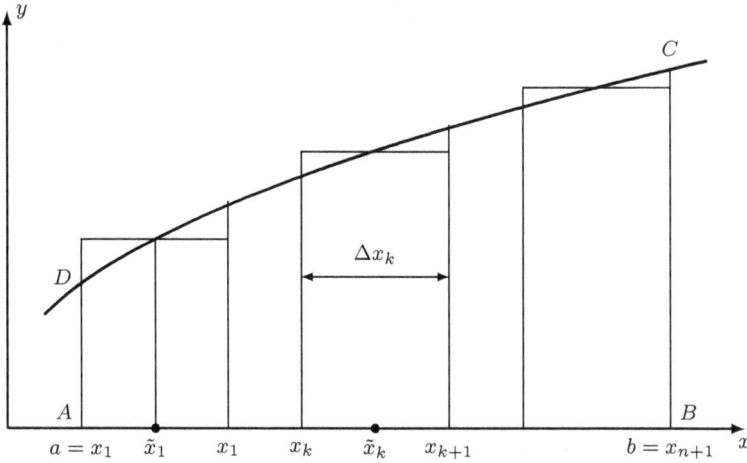

Fig. 3.3   Riemann-type approximation to the area under a curve given by $y = f(x)$ for $f(x) \geq 0$.

For an abstract function f, Fig. 3.3 presents the geometric meaning of a Riemann sum R_n as an approximation to the area of $ABCD$ by small rectangles.

But considerations based on the notion of mass can invoke a student's physical intuition. Indeed, suppose the linear mass density $\gamma(x)$ is continuous on an interval $[x_k, x_{k+1}]$. The mass Δm_k of the segment clearly satisfies

$$\min_{x \in [x_k, x_{k+1}]} \gamma(x) \cdot \Delta x_k \leq \Delta m_k \leq \max_{x \in [x_k, x_{k+1}]} \gamma(x) \cdot \Delta x_k.$$

By continuity $\gamma(x)$ takes on all values between its minimum and its maximum on $[x_k, x_{k+1}]$, hence there is a value γ_k^* such that $\gamma_k^* \Delta x_k = \Delta m_k$. Thus there is a point $x_k^* \in [x_k, x_{k+1}]$ such that

$$\Delta m_k = \gamma(x_k^*) \Delta x_k$$

and we really should have

$$m = \sum_{k=1}^{n} \gamma(x_k^*) \Delta x_k = \int_a^b \gamma(x) \, dx.$$

Although engineering oriented calculus textbooks normally omit the proof of Theorem 3.1, it is useful to note the ideas of the proof; they are practically the same for the case of multiple integration. It is important to realize that in mathematics there are no magical things that only mathematicians are able to understand.

The Main Ideas

Let us sketch a proof of Theorem 3.1. It consists of a few steps.

(1) Introduce a partition P_n for $[a, b]$ by points x_k where $x_1 = a < x_2 < \cdots < x_{n+1} = b$. Define a Riemann sum

$$R_n = \sum_{k=1}^{n} f(\tilde{x}_k)\,\Delta x_k$$

with $\Delta x_k = x_{k+1} - x_k$ and arbitrary $\tilde{x}_k \in [x_k, x_{k+1}]$.

(2) Because f is continuous on $[x_k, x_{k+1}]$, it attains minimum and maximum values $\underline{f}_k$ and $\overline{f}_k$ on $[x_k, x_{k+1}]$. That is, for any $\tilde{x}_k \in [x_k, x_{k+1}]$ the inequality

$$\underline{f}_k \le f(\tilde{x}_k) \le \overline{f}_k$$

holds. Hence for the partition there exist least and greatest Riemann sums $\underline{R}_n$ and $\overline{R}_n$ given by

$$\underline{R}_n = \sum_{k=1}^{n} \underline{f}_k\,\Delta x_k, \qquad \overline{R}_n = \sum_{k=1}^{n} \overline{f}_k\,\Delta x_k.$$

Clearly for any choice $\tilde{x}_k$ we have

$$\underline{R}_n \le R_n \le \overline{R}_n. \tag{3.3}$$

(3) Introduce a sequence of partitions P_n in such a way that partition P_n contains all the points x_k of all the previous partitions (P_n is said to be *finer* than P_m if $m < n$). Next take such a sequence of partitions with $\|P_n\| = \max_k \Delta x_k \to 0$ as $n \to \infty$. It is clear that for $n > m$,

$$\underline{R}_n \ge \underline{R}_m, \qquad \overline{R}_n \le \overline{R}_m.$$

So the numerical sequence $\{\underline{R}_n\}$ monotonically increases and $\underline{R}_n \le \overline{R}_1$. Thus there exists a limit

$$\lim_{n \to \infty} \underline{R}_n = \underline{R}.$$

The sequence $\{\overline{R}_n\}$ monotonically decreases and $\overline{R}_n \geq \underline{R}_1$, so it also has a limit,

$$\lim_{n \to \infty} \overline{R}_n = \overline{R},$$

for which (almost evidently)

$$\underline{R} \leq \overline{R}.$$

(4) It remains to prove that

$$\underline{R} = \overline{R} \tag{3.4}$$

as by a classic theorem for sequences, because of (3.3), we shall have

$$\lim_{n \to \infty} \underline{R}_n \leq \lim_{n \to \infty} R_n \leq \lim_{n \to \infty} \overline{R}_n$$

and hence by the definition there exists the definite integral

$$\int_a^b f(x)\,dx = \lim_{n \to \infty} R_n.$$

The decisive step is to prove (3.4) which requires some theoretical investigation of the nature of function continuity.

(5) It turns out that a function f that is continuous on a bounded segment $[a, b]$ is uniformly continuous on $[a, b]$. This means that for any $x \in [a, b]$ and any $\varepsilon > 0$ there exists $\delta > 0$ independent of x such that for any $y \in [a, b]$ satisfying $|x - y| \leq \delta$ we have $|f(x) - f(y)| \leq \varepsilon$. Hence for $\varepsilon > 0$ we can find $\delta > 0$ and a partition P_n such that $\|P_n\| < \delta$, and so for P_n

$$\overline{R}_n - \underline{R}_n = \sum_{k=1}^{n} |\overline{f}_k - \underline{f}_k|\,\Delta x_k \leq \sum_{k=1}^{n} \varepsilon\,\Delta x_k = \varepsilon(b - a).$$

Hence (3.4) holds and the proof sketch is complete. $\square$

A finite number of jump discontinuities in f does not affect the above proof, and hence for f piecewise continuous on $[a, b]$ the integral $\int_a^b f(x)\,dx$ exists. It turns out that it is sufficient if the set of all jump points has measure zero, meaning that it can be covered by countably many segments the sum of whose lengths is less than any assigned $\varepsilon > 0$.

To extend the existence proof for the definite integral to higher dimensions, it is necessary to show that a continuous function is uniformly continuous on a compact set and to study what is meant by sets of measure zero in higher dimensions. For our purposes it is enough to know that on the plane few smooth bounded curves have the zero measure, whereas few smooth bounded surfaces have zero measure in $\mathbb{R}^3$.

It is also instructive to compare the nature of convergence of an ordinary series $\sum_{i=1}^{\infty} a_i$ with that of a Riemann sum. For the series we construct a partial sum $S_n = \sum_{i=1}^{n} a_i$ and prove that the sequence $\{S_n\}$ converges to a limit as $n \to \infty$. Each term a_i is fixed and is included in S_{n+1} for $i = 1, \ldots, n$. For a Riemann sum $R_n = \sum_{i=1}^{n} f(\tilde{x}_i)\,\Delta x_i$, along with a change in n we get changeable summands $f(\tilde{x}_i)\,\Delta x_i$.

Specific Densities

Because we insist on introducing integrals via the mass problem, it makes sense to talk about the specific densities used for various types of solids (linear, planar, surface-like, three-dimensional). The specific density ρ that is tabulated for homogeneous materials in three-dimensions is the mass M of the material divided by its volume V:

$$\rho = M/V.$$

The SI units of this quantity are $[\rho] = \text{kg/m}^3$. If the density changes from point to point, a limit process is needed to define it. For point (x, y, z) we take a cube of side a centered at (x, y, z) with mass $M(a)$. The specific density at a point is

$$\rho(x, y, z) = \lim_{a \to 0+} \frac{M(a)}{V(a)}, \qquad V(a) = a^3. \tag{3.5}$$

In this way we have ρ as a function in three variables. For common engineering structures the density ρ is often piecewise constant (the structure consists of a few homogeneous materials). However we shall consider ρ as an arbitrary nonnegative function appropriate for the introduction of triple integrals.

For a homogeneous plate of constant thickness, we can bring in another type of specific density. Dividing the plate mass M by the plate area A, we get

$$\rho = M/A$$

with units $[\rho] = \text{kg/m}^2$. However the surface specific density ρ can change along with changes in plate thickness and in the 3D material density. We define the specific plate density through the limit

$$\rho(x,y) = \lim_{a \to 0+} \frac{M(a)}{A(a)}, \qquad A(a) = a^2, \tag{3.6}$$

where $M(a)$ is the plate mass inside the square with side a whose central axis is parallel to the z-axis and passes through the point (x,y). Now ρ is a function in two variables.

In the same way we can define the specific density at a point of a surface that represents a shell: a surface with thickness. The specific density is the limiting value of the mass inside some square cylinder, centered at a point with the axis normal to the surface at the point, divided by the area of the cylinder cut. The formula for the specific density looks like (3.6) with (x,y) replaced by the internal coordinates of the surface.

For homogeneous rod or bar along $x \in [a,b]$, we can introduce the linear specific density γ; we simply divide the total mass by the total length:

$$\gamma = M/L.$$

The units are $[\gamma] = $ kg/m. If the descriptive parameters change with x, a limit process is required in order to define the specific linear density:

$$\gamma(x) = \lim_{\Delta x \to 0+} \frac{M(\Delta x)}{\Delta x},$$

where $M(\Delta x)$ is the mass of the segment (a portion of the rod or bar) $[x - \Delta x/2, x + \Delta x/2]$. In a similar way we can work with the specific linear density of a cable that lies along a curve.

A General Remark

Engineering textbooks contain practically no rigorous proofs. Rather, like books on cooking, they present techniques and only some ideas about why these techniques work. It is often argued that this is appropriate with the present subject — that the aim of multivariable calculus is to teach how to integrate, that there is not enough class time to delve into the mathematics behind the recipes. But quite soon ordinary computers will calculate all needed integrals and solve other practical problems so easily that the main problem in mathematics instruction will have to shift from "How to do this?" to "Why does it work?" At that point it will have to contain more proofs than simple calculations in pencil.

However, if we continue to accept a lack of mathematical proof, it makes sense to treat multiple integrals from a physical viewpoint. At least this will simplify the concepts for the beginner.

3.2   Definition of Double Integral

Let us start by seeking the mass of a rectangular plate occupying the domain $R = [a, b] \times [c, d]$ with given plate density $\rho = \rho(x, y)$. We define

$$\rho(x, y) = \lim_{h \to 0+} \frac{m(h)}{4h^2},$$

where $m(h)$ is the mass of the square $[x - h, x + h] \times [y - h, y + h]$. So if ρ is a continuous function, the mass of a small portion ΔR of R with piecewise smooth boundary is

$$\Delta m \approx \rho(\tilde{x}, \tilde{y}) \, \Delta A, \tag{3.7}$$

where ΔA is the area of ΔR and $(\tilde{x}, \tilde{y})$ is an arbitrary point in ΔR. Note that if ρ is continuous then there is a point $(\tilde{x}_0, \tilde{y}_0) \in \Delta R$ such that $\Delta m = \rho(\tilde{x}_0, \tilde{y}_0) \, \Delta A$.

   We could start with arbitrary partitions of R restricted by the condition that the area of any partition cell can be found. But for the given R, most natural is the partition by straight lines parallel to the Cartesian axes. So we leave for now the problem of arbitrary partitions and create a uniform partition via straight lines parallel to the coordinate axes through points $(x_i = a + (i-1)\, \Delta x, 0)$ and $(0, y_j = c + (j-1)\, \Delta y)$ for $i, j = 1, 2, \ldots, n+1$, with $\Delta x = (b - a)/n$ and $\Delta y = (d - c)/n$. See Fig. 3.4. The quantities related to a rectangular cell with lower left point (x_i, y_j) will be labeled with indices ij; that is, the cell area is ΔA_{ij} and its mass is Δm_{ij}. We have

$$\Delta A_{ij} = \frac{(b - a)(d - c)}{n^2}.$$

   Taking ρ to be a continuous bounded function in R, we apply (3.7) to the ij-cell and obtain $\Delta m_{ij} \approx \rho(x_i, y_j) \, \Delta A_{ij}$. The sum of all cell masses is the plate mass

$$M \approx \sum_{i=1}^{n} \sum_{j=1}^{n} \rho(x_i, y_j) \, \Delta A_{ij} = \sum_{i=1}^{n} \sum_{j=1}^{n} \rho(x_i, y_j) \frac{(b - a)(d - c)}{n^2}. \tag{3.8}$$

It can be shown that as $n \to \infty$ the limit of these sums exists, so we announce that

$$M = \lim_{n \to \infty} \sum_{i=1}^{n} \sum_{j=1}^{n} \rho(x_i, y_j) \frac{(b - a)(d - c)}{n^2}.$$

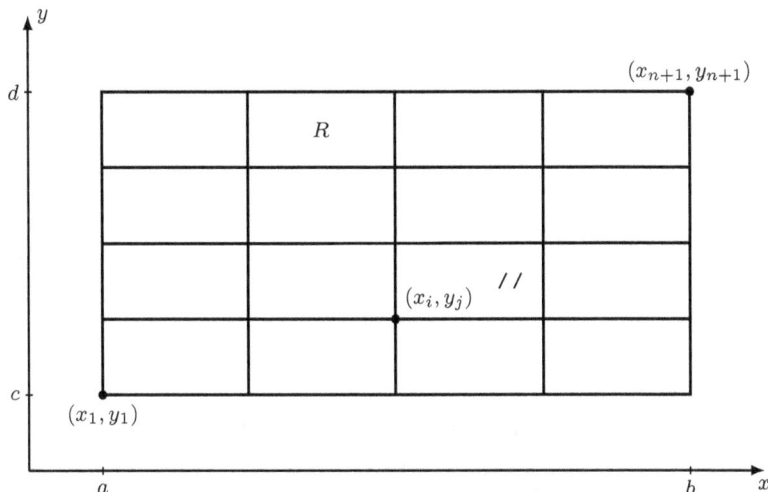

Fig. 3.4 Uniform partition of a rectangle R, with each side divided into n equal segments. The ij-cell lies to the upper right of the point (x_i, y_j).

Now we extend consideration to functions of class $C_b(R)$.

Definition 3.1. $C_b(R)$ is the set of functions defined on R, bounded and continuous except possibly for finite jump discontinuities across a finite set of smooth curves.

Definition 3.2. Let $f \in C_b(R)$. The *double integral* of f over R is

$$\iint_R f(x, y)\, dA = \lim_{n \to \infty} \sum_{i=1}^{n} \sum_{j=1}^{n} f(x_i, y_j)\, \Delta A_{mn}$$

$$= \lim_{n \to \infty} \sum_{i=1}^{n} \sum_{j=1}^{n} f(x_i, y_j)\, \frac{(b-a)(d-c)}{n^2}. \qquad (3.9)$$

It can be proved that the double integral exists. Clearly in this notation

$$M = \iint_R \rho(x, y)\, dA.$$

Example 3.1. Let us use Definition 3.2 to calculate

$$I = \iint_S xy\, dA \quad \text{where } S = [0, 1]^2.$$

Here $a = c = 0$, $b = d = 1$, $x_i = (i-1)/n$, and $y_j = (j-1)/n$. The Riemann sum is

$$\sum_{i=1}^{n}\sum_{j=1}^{n}\frac{i-1}{n}\frac{j-1}{n}\frac{1}{n^2} = \frac{1}{n^4}\sum_{i=1}^{n}(i-1)\sum_{j=1}^{n}(j-1) = \frac{1}{n^4}\left[\frac{n(n-1)}{2}\right]^2$$

and its limit as $n \to \infty$, which is $1/4$, is the value of I. $\qquad\square$

A General Rectangular Partition of P

In the definition of the one-dimensional definite integral, the partitions of the integration interval are made arbitrary. The same can be done for the double integral. Let us consider a general rectangular partition Π_{mn} of R. Namely, we shall use nonuniform partitions of the coordinate axes via ordered partition points x_i, y_j ($x_i < x_{i+1}$ and $y_j < y_{j+1}$ for all $1 \le i \le m$ and $1 \le j \le n$). The cells $\Delta R_{ij} = [x_i, x_{i+1}] \times [y_j, y_{j+1}]$ are nonuniform and an arbitrary point of ΔR_{ij} is denoted by $(\tilde{x}_i, \tilde{y}_j)$. We also define the norm of Π_{mn},

$$\|\Pi_{mn}\| = \max_{\substack{1 \le i \le m \\ 1 \le j \le n}} [(\Delta x_i)^2 + (\Delta y_j)^2]^{1/2},$$

which is the largest diagonal length of the rectangular partition cells. By (3.8) the mass of a plate occupying R would be

$$M \approx \sum_{i=1}^{m}\sum_{j=1}^{n} \rho(\tilde{x}_i, \tilde{y}_j)\, \Delta A_{ij}.$$

This prompts us to put forth an expression for an arbitrary function f,

$$\sum_{i=1}^{m}\sum_{j=1}^{n} f(\tilde{x}_i, \tilde{y}_j)\, \Delta A_{ij}, \tag{3.10}$$

which we shall term a *Riemann sum* for f over R.

Now we formulate a theorem which, without the condition that $f \in C_b(R)$, is usually given as the definition of the double integral.

Theorem 3.2. *Let $f \in C_b(R)$. The Riemann sequence $\{R_{mn}\}$ whose terms are*

$$R_{mn} = \sum_{i=1}^{m}\sum_{j=1}^{n} f(\tilde{x}_i, \tilde{y}_j)\, \Delta A_{ij}$$

has a limit on any sequence of partitions Π_{mn} *as* $\|\Pi_{nm}\| \to 0$. *This limit is independent of the selection of the partitions and of the points* $(\tilde{x}_i, \tilde{y}_j) \in \Delta R_{ij}$, *and it is equal to the double integral of* f *over* R *by Definition 3.2:*

$$\iint_R f(x,y)\, dA = \lim_{\|\Pi_{nm}\| \to 0} \sum_{i=1}^{m} \sum_{j=1}^{n} f(\tilde{x}_i, \tilde{y}_j)\, \Delta A_{ij}. \qquad (3.11)$$

The class $C_b(R)$ of functions for which we define the double integral could be extended, but this will be unnecessary for our purposes.

Fubini's Theorem

Let us consider a particular case of the Riemann sum:

$$\sum_{i=1}^{m} \sum_{j=1}^{n} f(x_i, y_j)\, \Delta A_{ij} = \sum_{i=1}^{m} \left(\sum_{j=1}^{n} f(x_i, y_j)\, \Delta y_j \right) \Delta x_i. \qquad (3.12)$$

The parenthetical term

$$\sum_{j=1}^{n} f(x_i, y_j)\, \Delta y_j$$

is a Riemann sum for the integral

$$\int_c^d f(x_i, y)\, dy = F(x_i)$$

and it looks like we could approximate the sum (3.12) by the expression

$$\sum_{i=1}^{m} F(x_i)\, \Delta x_i,$$

which is in turn a Riemann sum for the integral

$$\int_a^b F(x)\, dx.$$

So it seems that we could "approximate" the limit of the double Riemann sum by the *iterated integral*

$$\int_a^b \left(\int_c^d f(x,y)\, dy \right) dx. \qquad (3.13)$$

However, some difficulties lie in this approach. For some x_i the integral $F(x_i)$ may not exist if the line $x = x_i$ contains discontinuities of f.

Worse yet, in doing the inner-nested integration we must take a limit where Δx_i is constant while $\max_j \Delta y_j \to 0$. This will not satisfy the condition $\|\Pi_{mn}\| \to 0$, and Theorem 3.2 will not apply. However it can be proved that (3.13) does equal the corresponding double integral. Moreover, this integral can be calculated as the other possible iterated integral.

Theorem 3.3 (Fubini).   *Under the assumptions of Theorem 3.2 we have*

$$\iint_R f(x,y)\, dA = \int_a^b \left(\int_c^d f(x,y)\, dy \right) dx = \int_c^d \left(\int_a^b f(x,y)\, dx \right) dy.$$
$$(3.14)$$

Fubini's theorem allows us to calculate double integrals through simple iterated definite integrals. For simplicity we often omit the larger parentheses when writing iterated integrals; this is done starting in the next example.

Example 3.2. To calculate

$$I = \iint_R (x+y)\, dA, \qquad R = [0,1] \times [0,2],$$

we can use either expression in (3.14):

$$I = \int_0^1 \int_0^2 (x+y)\, dy\, dx = \int_0^1 2(x+1)\, dx = 3,$$

$$I = \int_0^2 \int_0^1 (x+y)\, dx\, dy = \int_0^2 \frac{2y+1}{2}\, dy = 3. \qquad \Box$$

Remark 3.1. The double integral need not be discussed in terms of rectangular partitions. The coordinate lines of polar coordinates or other general curvilinear coordinates can be used instead (Fig. 3.5). In such cases we can again start by approximating the mass using Riemann sums for the plane density function. One issue is how to represent ΔA_{ij}. If we seek only a theoretical result on the limit of the Riemann sums, we can even take partitions unrelated to the coordinate system. However for practical purposes the partitions should be defined by curvilinear coordinate lines. $\qquad \Box$

3.3   Double Integrals Over a General Domain

Let us use the physical interpretation of the value of a double integral when the integrand f denotes specific density. Suppose f is given over a bounded

(a)   (b)   (c)

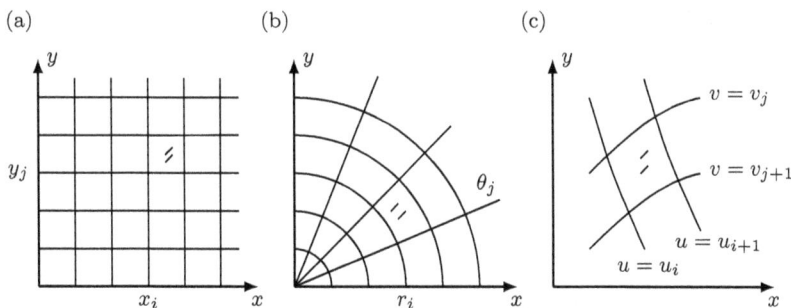

Fig. 3.5   The ij-cells of a planar domain defined by (a) the straight lines $x = x_i$, $y = y_j$, (b) the coordinate lines $r = r_i$, $\theta = \theta_j$ of the polar coordinate system, (c) the coordinate lines $u = u_i$, $v = v_j$ of a general curvilinear coordinate system in the xy-plane defined by the equations $x = x(u,v)$, $y = y(u,v)$.

domain $D \subset \mathbb{R}^2$ and is the density of a plate D. If we extend f outside of D by zero,

$$\tilde{f}(x,y) = \begin{cases} f(x,y), & (x,y) \in D, \\ 0, & (x,y) \notin D, \end{cases}$$

then clearly the mass M of a rectangular "plate" R including D, whose density is defined by $\tilde{f}$, is equal to the mass of D, and this prompts us to define the double integral over D by

$$\iint_D f(x,y)\, dA = \iint_R \tilde{f}(x,y)\, dA = M. \tag{3.15}$$

Next we apply (3.15) to an abstract function f having no physical meaning. The corresponding function $\tilde{f}$, defined on R, has jumps over the boundary of D in general. To apply Theorem 3.2 to the second integral in (3.15), we should state some conditions under which $\tilde{f} \in C_b(R)$ when $f \in C_b(\bar{D})$. It suffices for the boundary of D to consist of a finite number of smooth curves. This is enough for many engineering applications.

    Existence theorems are essential, but we also need practical methods by which integrals can be evaluated. Fubini's theorem is a helpful tool for domains having certain special, but still quite general, shapes.

Domains of Type 1

A *Type 1 domain* lies within a finite band $a \leq x \leq b$ and between the graphs of two bounded functions. In other words, the points $(x,y) \in D$

satisfy the inequalities $a \leq x \leq b$ and $c \leq h_1(x) \leq y \leq h_2(x) \leq d$ as shown in Fig. 3.6.

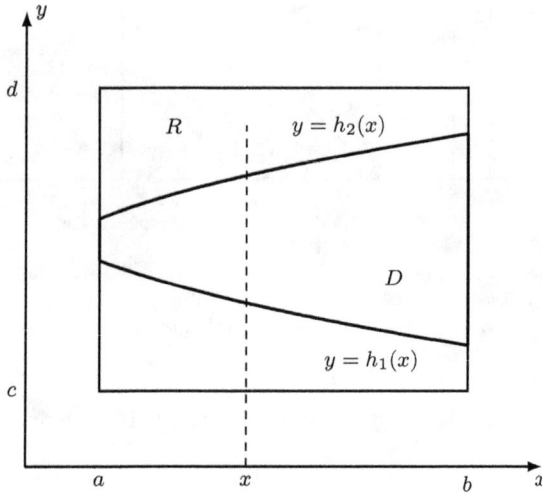

Fig. 3.6   Type 1 domain enclosed by a rectangle.

Note that a line $x = $ constant, $a < x < b$, intersects the boundary of D twice: at single "entrance" and "exit" points $(x, h_1(x))$ and $(x, h_2(x))$ respectively. See Fig. 3.7 for examples of Type 1 domains. Examples of domains that do not belong to Type 1 are presented in Fig. 3.8. But these

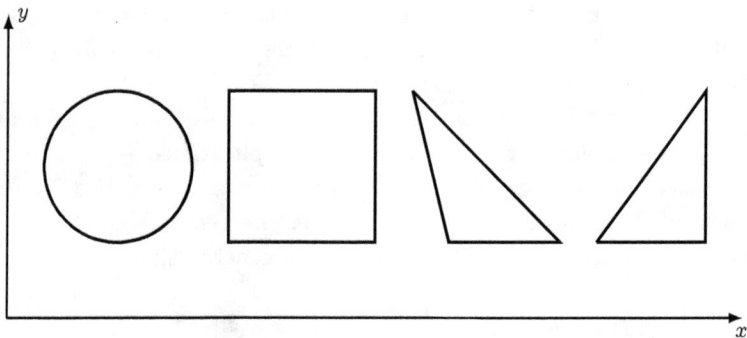

Fig. 3.7   Some Type 1 domains.

can be represented as unions of Type 1 domains, and this will extend the applicability Fubini's theorem.

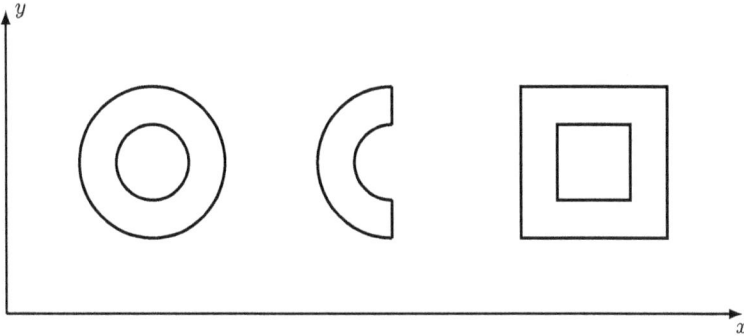

Fig. 3.8   Domains not of Type 1.

Let us discuss the evaluation of a double integral over a Type 1 domain. By (3.15) and Fubini's theorem,

$$\iint_D f(x,y)\,dA = \iint_R \tilde{f}(x,y)\,dA = \int_a^b \int_c^d \tilde{f}(x,y)\,dy\,dx.$$

But

$$\int_c^d \tilde{f}(x,y)\,dy = \int_c^{h_1(x)} \tilde{f}(x,y)\,dy + \int_{h_1(x)}^{h_2(x)} \tilde{f}(x,y)\,dy + \int_{h_2(x)}^d \tilde{f}(x,y)\,dy$$

$$= 0 + \int_{h_1(x)}^{h_2(x)} f(x,y)\,dy + 0$$

and hence for the domain of Type 1

$$\iint_D f(x,y)\,dA = \int_a^b \int_{h_1(x)}^{h_2(x)} f(x,y)\,dy\,dx. \tag{3.16}$$

Example 3.3. We find

$$I = \int_D xy\,dA$$

where D is the Type 1 domain shown in Fig. 3.9. Here $h_1(x) = 0$ and $h_2(x) = 1 - x$, so

$$I = \int_0^1 \int_0^{1-x} xy\,dy\,dx = \int_0^1 \frac{x(x-1)^2}{2}\,dx = \frac{1}{24}. \qquad \square$$

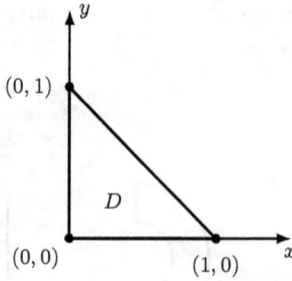

Fig. 3.9   Example 3.3.

Domains of Type 2

It is evident that the mass of a plate is invariant under shifts and rotations of the coordinate system. The same idea holds for an abstract integrand. Consider rotating the coordinate system about the origin by an angle $\pi/2$. Now the domain D lies inside a band $y \in [c, d]$. If it is such that any straight line parallel to a band edge, other than a band edge, intersects the boundary of D in two points, then it is a *Type 2 domain*.

A Type 2 domain lies in a band $c \leq y \leq d$, and any straight line $y = y_0$ intersects the boundary of D given by $x = g_1(y)$, $x = g_2(y)$ (we assume $g_1(y) < g_2(y)$), except at the band edges, in one entrance point and one exit point. See Figs. 3.10 and 3.11.

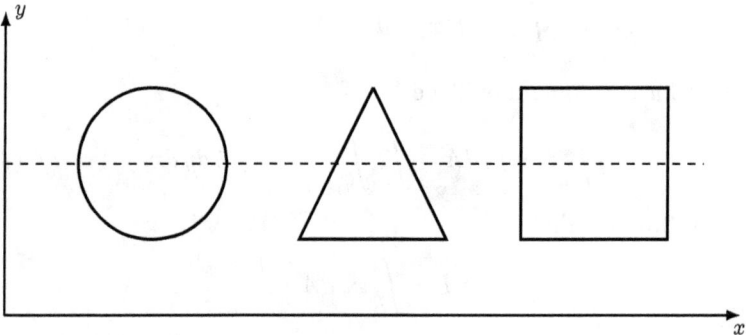

Fig. 3.10   Domains of Type 2.

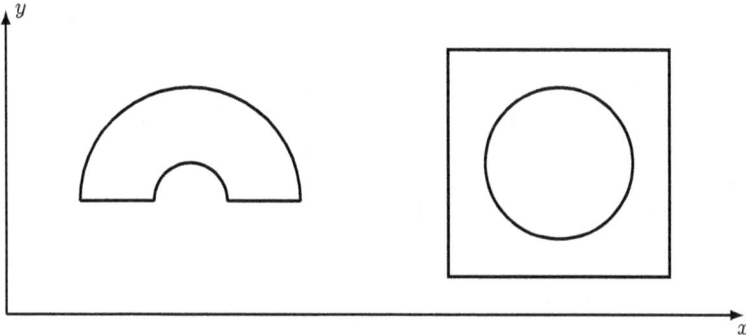

Fig. 3.11   Domains that do not belong to Type 2.

Practically repeating the steps for Type 1 domains, we apply Fubini's theorem and reduce a double integral for a Type 2 domain to an iterated integral:

$$\iint_D f(x,y)\,dA = \int_c^d \int_{g_1(y)}^{g_2(y)} f(x,y)\,dx\,dy. \qquad (3.17)$$

Figure 3.12 illustrates the roles played by the left and right portions of the boundary of D, i.e., by the functions $x = g_1(y)$ and $x = g_2(y)$.

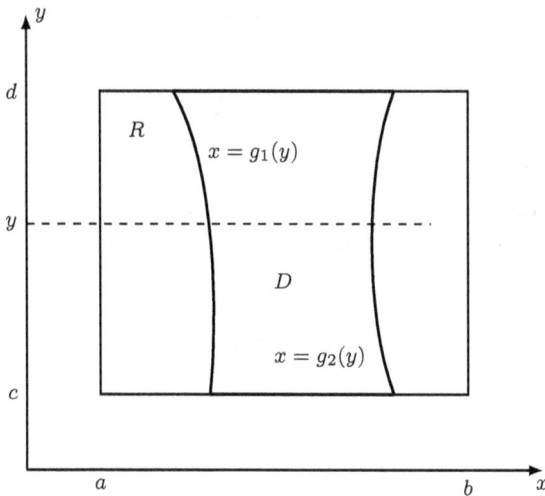

Fig. 3.12   General Type 2 domain.

Let us compare formulas (3.16) and (3.17). In each case the inner integral is such that its limits contain the argument that is not the integration variable. The outer integral has constant limits independent of either integration variable. Unless the integrand or the limits happen to involve some external parameter, the resulting double integral is a number.

Example 3.4. Let us return to Example 3.3. Since D is also of Type 2, we have

$$I = \int_0^1 \int_0^{1-y} xy \, dx \, dy = \int_0^1 \frac{y(y-1)^2}{2} \, dy = \frac{1}{24}. \qquad \square$$

For an integration domain of both types a judicious selection can make things easier, but this comes with experience.

Remark 3.2. By rotating the coordinate system through various angles, we can get other types of domains for which double integrals can be calculated similarly by Fubini's theorem. This involves lengthy formulas and is normally not done in elementary textbooks. $\qquad \square$

3.4   Some General Properties of Double Integrals

It is clear that if we have the union of two domains $D_1 \cup D_2$ that can share only boundary points, then the mass of the plate $D_1 \cup D_2$ is the sum of the masses of the plates D_1 and D_2. Correspondingly, for an abstract integrand we have

$$\iint_{D_1 \cup D_2} f(x,y) \, dA = \iint_{D_1} f(x,y) \, dA + \iint_{D_2} f(x,y) \, dA, \qquad (3.18)$$

which can also be seen formally. This property allows us to calculate double integrals for domains that are unions of Type 1 and Type 2 domains. For example, suppose D is as shown in Fig. 3.13.

We seek

$$\iint_{AEC} f(x,y) \, dA.$$

The domain is of both types. A Type 1 view would force us to split AEC into two triangles ABC and ECB, as the sides AC and CE are given by different equations. We should then use

$$\iint_{AEC} f(x,y) \, dA = \iint_{ABC} f(x,y) \, dA + \iint_{BEC} f(x,y) \, dA.$$

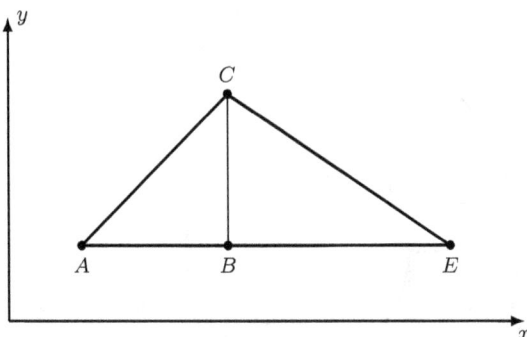

Fig. 3.13 Triangle decomposed into two triangles.

Note that for a simple function f, however, it is better to take AEC as a Type 2 domain because this leads to a single calculation.

Example 3.5. Let us calculate

$$I = \iint_D (2x - y) dA$$

where D is given in Fig. 3.13 with $A = (1,1)$, $E = (3,1)$, and $C = (2,2)$. Consider D as the union of triangles ABC and BEC. As line AC has equation $y = x$ and EC is given by $y = 4 - x$, we get

$$I = \int_1^2 \int_1^x (2x - y)\, dy\, dx + \int_2^3 \int_1^{4-x} (2x - y)\, dy\, dx = \frac{8}{3}.$$

Now we treat D as Type 2. The equations of AC and EC are $x = y$ and $x = 4 - y$ respectively. So

$$I = \int_1^2 \int_y^{4-y} (2x - y)\, dx\, dy = \frac{8}{3}. \qquad \square$$

Example 3.6. We calculate

$$\iint_S \frac{\partial f(x,y)}{\partial y} dA$$

where S is an annulus with inner radius 1 and outer radius 2. Referring to Fig. 3.14, let us split the surface with the straight line $y = 1$, getting

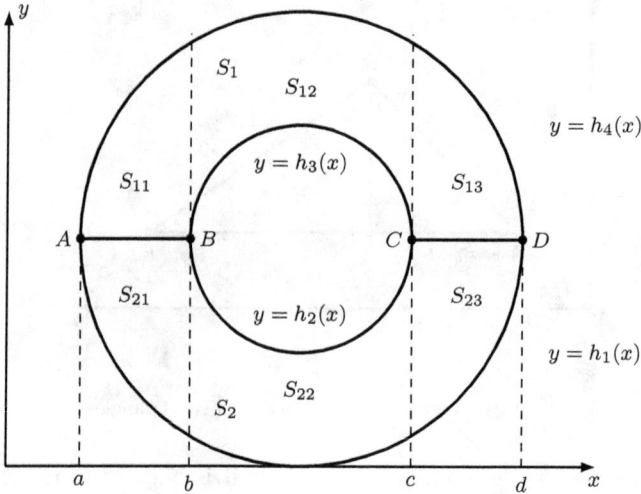

Fig. 3.14   Annulus, $S = S_1 \cup S_2$.

two subdomains S_1 and S_2 of Type 1. Integrals of this type will be met in applications of Green's theorem later. We have

$$\iint_S f_y(x,y)\, dA = \iint_{S_1} f_y(x,y)\, dA + \iint_{S_2} f_y(x,y)\, dA.$$

For each of the domains S_1 and S_2, we create three subdomains by splitting with the lines $x = b$ and $x = c$:

$$\iint_{S_i} f_y(x,y)\, dA = \iint_{S_{i1}} f_y(x,y)\, dA + \iint_{S_{i2}} f_y(x,y)\, dA + \iint_{S_{i3}} f_y(x,y)\, dA$$

for $i = 1, 2$. Calculating the six needed integrals, we get

$$\iint_{S_{11}} f_y(x,y)\, dA = \int_a^b \int_1^{h_4(x)} f_y(x,y)\, dy\, dx$$

$$= \int_a^b [f(x, h_4(x)) - f(x, 1)]\, dx,$$

$$\iint_{S_{21}} f_y(x,y) = \int_a^b \int_{h_1(x)}^1 f_y(x,y)\, dy\, dx$$

$$= \int_a^b [f(x, 1) - f(x, h_1(x))]\, dx,$$

$$\iint_{S_{12}} f_y(x,y)\, dA = \int_b^c \int_{h_3(x)}^{h_4(x)} f_y(x,y)\, dy\, dx$$

$$= \int_b^c [f(x, h_4(x)) - f(x, h_3(x))]\, dx,$$

$$\iint_{S_{22}} f_y(x,y)\, dA = \int_b^c \int_{h_1(x)}^{h_2(x)} f_y(x,y)\, dy\, dx$$

$$= \int_b^c [f(x, h_2(x)) - f(x, h_1(x))]\, dx,$$

$$\iint_{S_{13}} f_y(x,y)\, dA = \int_c^d \int_1^{h_4(x)} f_y(x,y)\, dy\, dx$$

$$= \int_c^d [f(x, h_4(x)) - f(x, 1)]\, dx,$$

$$\iint_{S_{23}} f_y(x,y)\, dA = \int_c^d \int_{h_1(x)}^1 f_y(x,y)\, dy\, dx$$

$$= \int_c^d [f(x, 1) - f(x, h_1(x))]\, dx.$$

Their sum is

$$\iint_S f_y(x,y)\, dA = \int_a^d f(x, h_4(x))\, dx - \int_a^d f(x, h_1(x))\, dx$$

$$+ \int_b^c f(x, h_2(x))\, dx - \int_b^c f(x, h_3(x))\, dx.$$

Note that the integrals over the internal segments AB and CD have disappeared. This is typical with various composite domains for double integrals. □

Other simple properties of the double integral follow from the properties of the limit for Riemann sums. First, the double integral is a *linear transformation*. If each $f_k \in C_b(D)$ and c is constant, then

$$\iint_D [cf_1(x,y) + f_2(x,y)]\, dA = c \iint_D f_1(x,y)\, dA + \iint_D f_2(x,y)\, dA.$$

This is used for calculations of double integrals.

Sometimes it is necessary to estimate (i.e., bound) the value of a double integral. There are many ways to do this. We present only simple estimates following immediately from the form of a Riemann sum and the limit properties:

$$\inf_{(x,y)\in D} f(x,y)\,A(D) \le \iint_D f(x,y)\,dA \le \sup_{(x,y)\in D} f(x,y)\,A(D), \qquad (3.19)$$

where $A(D)$ is the area of D. A similar inequality is

$$\left| \iint_D f(x,y)\,dA \right| \le \sup_{(x,y)\in D} |f(x,y)|\,A(D). \qquad (3.20)$$

We also have the *triangle* or *modulus inequality*

$$\left| \iint_D f(x,y)\,dA \right| \le \iint_D |f(x,y)|\,dA \qquad (3.21)$$

and the fact that if $f(x,y) \le g(x,y)$ for all $(x,y) \in D$ then

$$\iint_D f(x,y)\,dA \le \iint_D g(x,y)\,dA. \qquad (3.22)$$

If f is continuous in a closed domain D, then f takes on all the values between $\inf_{(x,y)\in D} f(x,y)$ and $\sup_{(x,y)\in D} f(x,y)$. Hence there is a point $(x_0, y_0) \in D$ such that

$$\iint_D f(x,y)\,dA = f(x_0, y_0)\,A(D).$$

This is a *mean value theorem for double integrals*.

3.5   Double Integrals in Polar Coordinates

Domains in Polar Coordinates

We have treated the double integral starting with partitions of the domain formed by straight lines parallel to the coordinate axes. This seems quite different from the construction of the ordinary definite integral, where the partitions should be arbitrary in order to announce that the integral exists.

So let us consider the double integral in non-Cartesian coordinates. We start with the polar coordinate system. Direct substitution of polar coordinates into the expression $dA = dx\,dy$ yields somewhat strange terms dr^2 and $r^2 d\theta^2$ in the area expression. We shall avoid the difficulty by introducing the double integral in polar coordinates, starting with construction of Riemann sums for the case where f is the plate density and the partitions are defined by the coordinate lines $r = r_i$ and $\theta = \theta_j$ of the polar coordinate system. This system is useful in engineering because many planar objects have the form of an annular sector, like the domain D in Fig. 3.15.

We begin by reviewing polar coordinates and noting some facts useful in problem solving. Any point A with coordinates (x, y) in the xy-plane can also be uniquely specified by two other numbers (r, θ) called the polar coordinates of A (Fig. 3.16). Here r is the distance OA and θ is the angle in radians between Ox and OA, taken positive when it is counterclockwise. The relations between Cartesian and polar coordinates are

$$r = (x^2 + y^2)^{1/2}, \qquad x = r\cos\theta, \qquad y = r\sin\theta. \qquad (3.23)$$

To keep r nonnegative and, to have the polar coordinates of A be unique, θ is traditionally taken from the interval $[0, 2\pi)$. However, for convenience we

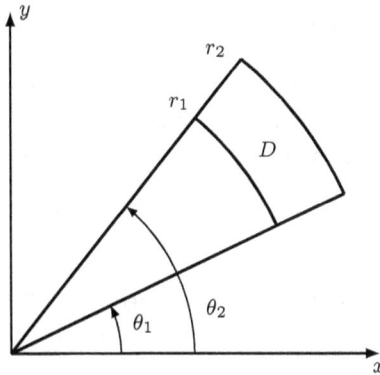

Fig. 3.15   Area sector in plane polar coordinates.

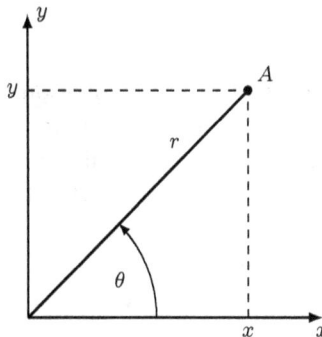

Fig. 3.16   Cartesian and polar coordinates of a point.

sometimes confine θ to certain other intervals of length 2π. For any such choice the pairing between Cartesian and polar coordinates is one-to-one for all points except the origin, which is in this sense a singular point.

We often describe a planar figure initially in Cartesian coordinates and subsequently use (3.23) to describe it in polar coordinates. Domain D of Fig. 3.15 could be described as falling between the two circles $x^2 + y^2 = r_1^2$ and $x^2 + y^2 = r_2^2$, and the two rays defined by the angles θ_1 and θ_2. The purely polar coordinate description would be that

$$(r, \theta) \in D \text{ if } (r, \theta) \in [r_1, r_2] \times [\theta_1, \theta_2]$$

or alternatively

$$D = \{(r, \theta): r_1 \leq r \leq r_2, \theta_1 \leq \theta \leq \theta_2\}.$$

Note that D is actually a *rectangle* in the plane with orthogonal $r\theta$ axes. In this plane negative r has no meaning.

Typically we employ polar coordinates when the object is a sector or a portion of one. If the sector is part of a disc whose center is some point (a, b) not at the origin, it makes sense to use translated polar coordinates:

$$x = a + r\cos\theta, \qquad y = b + r\sin\theta.$$

However, in some cases involving shifted circles we can still use traditional polar coordinates. For a circle $x^2 + (y - a)^2 = a^2$ centered at a point on the y-axis, we can use $x = r\cos\theta$ and $y = r\sin\theta$ to recast the circle as $r = 2a\sin\theta$ where θ runs from 0 to π.

The Double Integral

Calculation of double integrals over disks or circular sectors is typically more convenient in polar coordinates. The double integral must be defined in such a way that its value is independent of whether polar or Cartesian coordinates are used. We start with a partition of D by circles $r = r_i$ and rays $\theta = \theta_j$; these constitute the coordinate lines in polar coordinates (Fig. 3.17). The partition produces "curvilinear rectangle" type ij-cells. They are really rectangles in the $r\theta$-plane with orthogonal axes for r and θ. All the quantities related with cell D_{ij} have indices ij: the mass is Δm_{ij}, the area is ΔA_{ij}, the radial extent is $\Delta r_i = r_{i+1} - r_i$, and the angular extent is $\Delta\theta_j = \theta_{j+1} - \theta_j$. The mass of the plate is

$$M = \sum_{i,j} \Delta m_{ij} \approx \sum_{i,j} \rho(\tilde{r}_i, \tilde{\theta}_j)\, \Delta A_{ij}. \qquad (3.24)$$

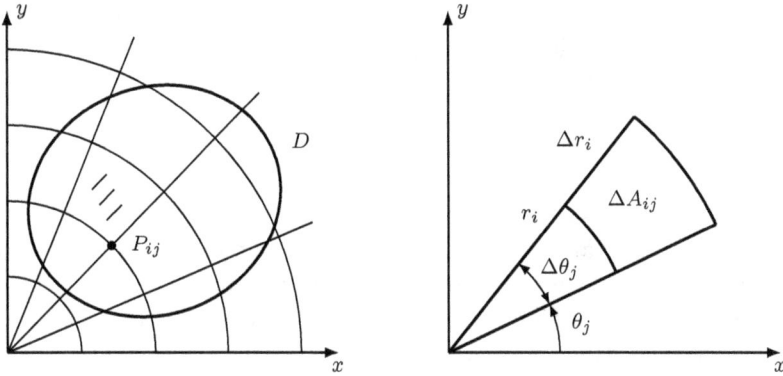

Fig. 3.17 *Left*, partition of a domain in the polar coordinate system. Here $P_{ij} = (r_i, \theta_j)$ and the (i, j)-cell D_{ij} is hatched. *Right*, area element.

We could start with a rectangular plate in the $r\theta$-plane. But rather than repeating the approach with rectangular partitions, we take an arbitrary domain D. Let us understand that, for the subsequent limiting procedure to introduce the integral, we can place zero value for the cells that cover the boundary, as the limit of this part of the sum is zero. The area of D_{ij} is

$$\Delta A_{ij} = \tfrac{1}{2}(r_i + \Delta r_i)^2 \, \Delta \theta_j - \tfrac{1}{2}r_i^2 \, \Delta \theta_j$$
$$= \tfrac{1}{2}[2r_i \, \Delta r_i + (\Delta r_i)^2] \, \Delta \theta_j$$
$$= r_i \, \Delta r_i \, \Delta \theta_j + \varepsilon_{ij} \tag{3.25}$$

where $\varepsilon_{ij} = \tfrac{1}{2}(\Delta r_i)^2 \, \Delta \theta_j$. For $r_i \neq 0$ we get

$$\lim_{(\Delta r_i, \Delta \theta_j) \to (0,0)} \frac{\varepsilon_{ij}}{r_i \, \Delta r_i \, \Delta \theta_j} = 0 \tag{3.26}$$

and in the expression for M we can replace ΔA_{ij} by its approximation

$$M \approx \sum_{i,j} \rho(\tilde{r}_i, \tilde{\theta}_j) \, r_i \, \Delta r_i \, \Delta \theta_j.$$

This expression for $\rho \in C_b(D)$ is a Riemann sum for the double integral

$$\iint_{D_{r\theta}} \rho(r, \theta) \, r \, dr \, d\theta$$

to which the sequence of the Riemann sums converges for any polar partition sequence such that $\max_{ij}[(\Delta r_i)^2 + (\Delta \theta_j)^2]^{1/2} \to 0$. Note that the notation $D_{r\theta}$ is for D, but in the $r\theta$-plane. The mass of D is

$$\iint_D \rho \, dA = \iint_{D_{r\theta}} \rho(r, \theta) \, r \, dr \, d\theta.$$

Remark 3.3. Let us show that the ε_{ij} term in (3.25) can be ignored in the integral for M. For any $\varepsilon > 0$ we can select the radius r_0 such that for the circle $r < r_0$ the absolute value of the terms of the sum $\sum_{ij} \rho(\tilde{r}_i, \tilde{\theta}_j) \, \varepsilon_{ij}$ for the cells intersecting with $r < r_0$ is less than $\varepsilon/2$. We estimate the absolute value of the sum (denoted below with a prime) of the remaining terms. For this we denote $\max_D \rho = \rho_0$ and take $\Delta r_i < \delta$ which can be as small as we wish. Now

$$\left| \sum_{ij}' \rho(\tilde{r}_i, \tilde{\theta}_j) \varepsilon_{ij} \right| < \frac{1}{2} \sum_{ij}' \rho_0 \frac{\delta}{r_0} r_i \, \Delta r_i \, \Delta \theta_j < \frac{1}{2} \rho_0 \frac{\delta}{r_0} A(D)$$

which can be made less than $\varepsilon/2$ since δ can be taken sufficiently small. $\qquad \square$

Clearly we can repeat all of this for an abstract function f. For f given in Cartesian coordinates we shall get

$$\iint_D f(x, y) \, dA = \iint_{D_{r\theta}} f(r \cos \theta, r \sin \theta) \, r \, dr \, d\theta. \qquad (3.27)$$

Note that for polar coordinates the infinitesimal area element

$$dA = r \, dr \, d\theta \qquad (3.28)$$

carries a factor r. Such a factor will appear in dA for any curvilinear coordinate system.

It is worth noting that in the $r\theta$-plane with orthogonal coordinate axes for r and θ, a domain like the intersection of a sector with a ring will be represented as a rectangle. Such occurrences invite applications of Fubini's theorem. However, in polar coordinates only Type 1 domains are usually employed (Fig. 3.18, right). Fubini's theorem gives

$$\iint_D f(x, y) \, dA = \int_{\theta_1}^{\theta_2} \int_{h_1(\theta)}^{h_2(\theta)} f(r \cos \theta, r \sin \theta) \, r \, dr \, d\theta. \qquad (3.29)$$

Observe that the notation for the double integral on the left side of (3.29) is universal and serves in polar coordinates as well. Omission of the factor r from dA is a common mistake.

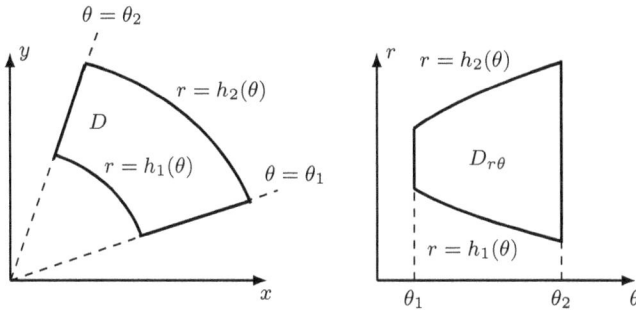

Fig. 3.18   Usual type of domain D for applying Fubini's theorem in polar coordinates r, θ. *Left, xy-plane; right, θr-plane.*

In mechanics, expressions like (3.28) are often written almost intuitively. Of course the tangent to a circle is orthogonal to the radius at each point on the circle. When dealing with very small cells, we can approximate a "curvilinear rectangle" such as the one shown in Fig. 3.19 by an actual rectangle. Simple multiplication of the side lengths $AB = r\,\Delta\theta$ and $AD = \Delta r$ gives $r\,\Delta r\,\Delta\theta$, which seems to generate (3.28).

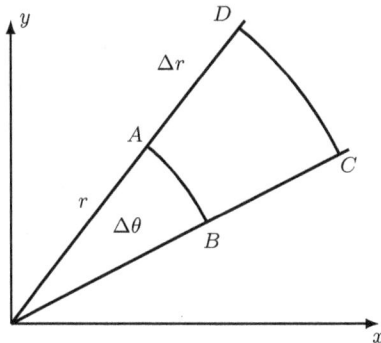

Fig. 3.19   Intuitive generation of the dA expression through multiplication of the two lengths Δr and $r\Delta\theta$.

Example 3.7. Let us evaluate the integral

$$I = \iint_D (x+y)\, dA$$

where D is the right half of the disk $x^2 + y^2 \leq 1$. We have

$$D_{r\theta} = \{(r, \theta): 0 \leq r \leq 1, -\pi/2 \leq \theta \leq \pi/2\}.$$

So

$$I = \int_{-\pi/2}^{\pi/2} \int_0^1 (r\cos\theta + r\sin\theta)\, r\, dr\, d\theta.$$

We can simplify the integral using the relation $\sin(-\theta) = -\sin\theta$ and symmetry of D with respect to the x-axis so the integral with $\sin\theta$ vanishes:

$$I = \int_{-\pi/2}^{\pi/2} \int_0^1 r^2 \cos\theta\, dr\, d\theta = \frac{1}{3} \int_{-\pi/2}^{\pi/2} \cos\theta\, d\theta = \frac{2}{3}. \qquad \square$$

Example 3.8. We find the mass m of a portion of an annulus. The domain is bounded by the circles $x^2 + y^2 = 1$ and $x^2 + y^2 = 4$, the x-axis, and the straight line $y = x$ ($x > 0$). The mass density at a point is proportional to the distance of the point from the origin: $\rho(r, \theta) = kr$ with some coefficient of proportionality k. The domain of integration is

$$D_{r\theta} = \{(r, \theta): 1 \leq r \leq 2, 0 \leq \theta \leq \pi/4\}.$$

So

$$m = \iint_{D_{r\theta}} \rho(r, \theta)\, r\, dr\, d\theta = \int_0^{\pi/4} \int_1^2 kr^2\, dr\, d\theta = \frac{7k\pi}{12}. \qquad \square$$

Example 3.9. Consider the area bounded by the spiral $r = \theta$ and the positive y-axis. The integrand is $f(r, \theta) = 1$, so

$$A = \int_0^{\pi/2} \int_0^\theta 1\, r\, dr\, d\theta = \frac{1}{2} \int_0^{\pi/2} \theta^2\, d\theta = \frac{\pi^3}{48}. \qquad \square$$

Example 3.10. We find

$$I = \int_D (x^2 + y^2)^{1/2}\, dA$$

where D is the set of points (x, y) satisfying $x^2 - 4x + y^2 \leq 0$. The boundary equation $(x - 2)^2 + y^2 = 2^2$ specifies a circle of radius 2 with center $(2, 0)$. It makes sense to integrate in polar coordinates. The circle equation is $r^2 \cos^2\theta - 4r\cos\theta + r^2 \sin\theta = 0$ or $r = 4\cos\theta$. Clearly the y-axis is tangent to the circle at $r = 0$ so $\theta \in [-\pi/2, \pi/2]$. Thus

$$I = \int_{-\pi/2}^{\pi/2} \int_0^{4\cos\theta} r^2\, dr\, d\theta = \frac{64}{3} \int_{-\pi/2}^{\pi/2} \cos^3\theta\, d\theta = \frac{256}{9}. \qquad \square$$

3.6   Double Integral in General Curvilinear Coordinates

We could also call this section "Change of Variables in the Double Integral." Indeed it is similar to our previous change from x, y to r, θ. Changes of variable in double integration are aimed at easier calculation, just as for ordinary definite integration. But while in the latter case our aim is often to simplify the integrand, with double integrals it often is to reduce the domain of integration to a rectangle in the new coordinate system.

So suppose we change variables from x, y to new variables u, v:

$$x = g(u, v), \qquad y = h(u, v),$$

which we shall denote mostly by

$$x = x(u, v), \qquad y = y(u, v).$$

This change should be such that the points in the domain of integration D are in one-to-one correspondence with the transformed domain D_{uv} in the variables u, v, except perhaps at certain singular points like the origin O for polar coordinate transformation. See Fig. 3.20.

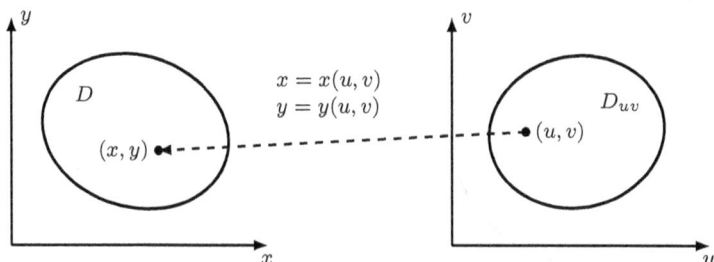

Fig. 3.20   Coordinate transformation.

We assume the functions g and h are continuously differentiable on the closure of D_{uv} (or that they are continuously differentiable in an open domain including D_{uv}), and that the mapping from D to D_{uv} is invertible at all points (except possibly for a few singular points) of D. This mapping evidently puts the boundaries of D and D_{uv} in one-to-one correspondence. Hence we can depict D_{uv} in the uv-plane with the u- and v-axes orthogonal.

Example 3.11. Let us consider the transformation to polar coordinates when D is the sector $0 \leq r \leq a$, $\theta_1 \leq \theta \leq \theta_2$ of Fig. 3.21. We see that in

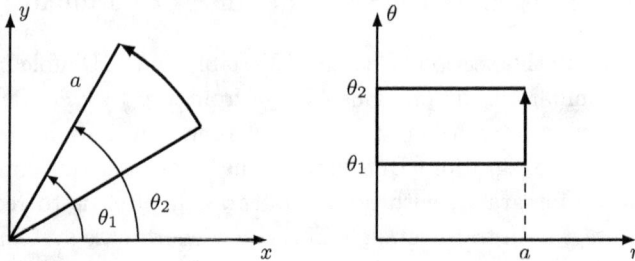

Fig. 3.21   Example 3.11. Sector $0 \le r \le a$, $\theta_1 \le \theta \le \theta_2$ in the xy and $r\theta$ coordinate planes.

the coordinates (r, θ) the image of the sector is a rectangle. Note that to the origin O there corresponds the straight line from θ_1 to θ_2 with $r = 0$. It is instructive to examine a circle of radius a under the transformation.

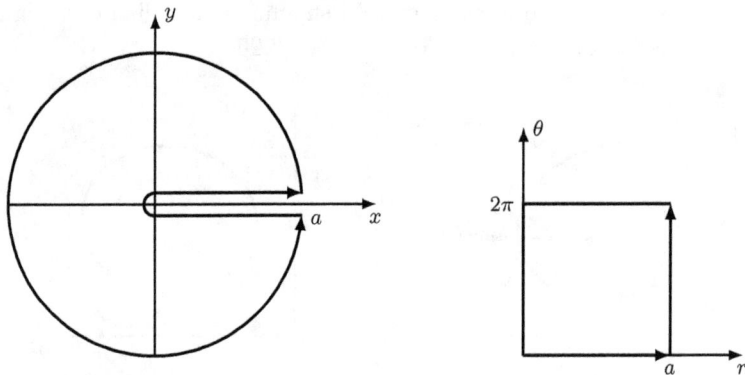

Fig. 3.22   Example 3.11. The circle of radius a in the xy and $r\theta$ planes.

In the xy-plane we should follow a cut around the positive x-axis as shown in Fig. 3.22. The upper side of the cut corresponds to the segment $[0, a]$ at $\theta = 0$. The lower side of the cut corresponds to the segment at $\theta = 2\pi$, running from $r = a$ to $r = 0$. This cut excludes the positive x-axis in order to preserve one-to-one correspondence.   □

Let us extend some ideas for polar coordinates to the general case. First, fixing the variables $u = u_0$, $v = v_0$ one at a time, we determine two coordinate lines in the plane:

$$x = x(u, v_0) \qquad \text{and} \qquad x = x(u_0, v)$$
$$y = y(u, v_0) \qquad \qquad\qquad y = y(u_0, v)$$

Taking a few values for u_0 and v_0, we get a coordinate grid which in the uv-plane (assuming the u- and v-axes are orthogonal) constitutes rectangular partitioning by straight lines parallel to the u or v axes. The corresponding curves in the xy-plane are shown in Fig. 3.23.

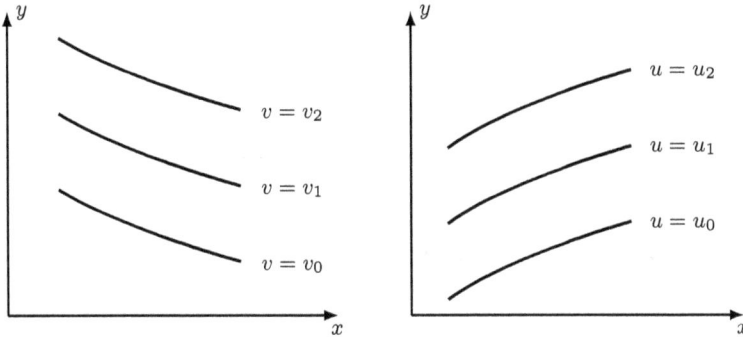

Fig. 3.23   Coordinate lines.

Using finite sets of curvilinear coordinate lines, we can create new types of partitions in D. See Fig. 3.24.

To construct the integral in the (u, v) coordinates, we return to the problem of finding the mass M of a plate D, understanding that the result should not depend on the type of partitioning. The quantities related to the ij-cell of the partition in Fig. 3.24 will be tagged with indices i, j. These include its mass Δm_{ij}, its area ΔA_{ij}, an interior point $(\tilde{u}_i, \tilde{v}_j)$, the density $\rho(\tilde{u}_i, \tilde{v}_j)$ at that point, and $\Delta u_i = u_{i+1} - u_i$, $\Delta v_j = v_{j+1} - v_j$. Again, assuming "small" partition cells we can write

$$M = \sum_{i,j} \Delta m_{ij} \approx \sum_{i,j} \rho(\tilde{u}_i, \tilde{v}_j) \, \Delta A_{ij}, \qquad (3.30)$$

where the sum is over all elements of the partition for D. As before, we can omit those elements that are not fully inside D as for a piecewise continuous function ρ and a piecewise smooth boundary this part of the sum will tend to zero when, introducing the integral, we produce the limit passage. Next

Fundamentals of Multivariable Calculus

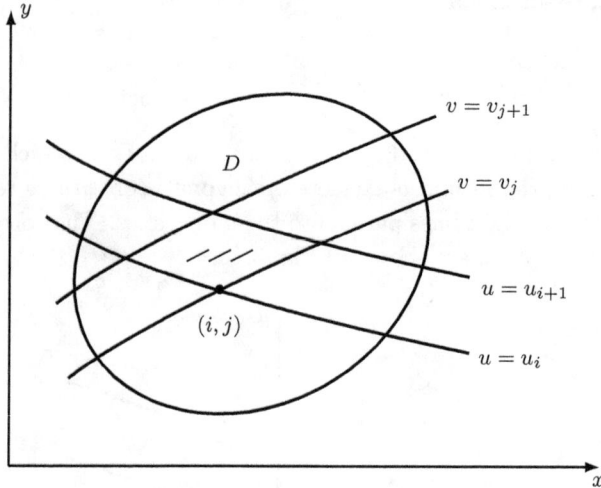

Fig. 3.24   General curvilinear partition of a domain D. Cell ij is shown hatched.

we approximate ΔA_{ij} as follows. For the sides of the ij-cell let us draw vectors

$$d_u \mathbf{r}(u_i, v_j) = \mathbf{r}_u(u_i, v_j) \, \Delta u_i,$$
$$d_v \mathbf{r}(u_i, v_j) = \mathbf{r}_v(u_i, v_j) \, \Delta v_j,$$

tangent to the corresponding sides and having magnitudes that approximate the side lengths (Fig. 3.25).

As the coordinate lines are smooth, it can be shown that ΔA_{ij} can be approximated by the area of the parallelogram constructed over the tangent vectors:

$$\Delta A_{ij} \approx |d_u \mathbf{r}(u_i, v_j) \times d_v \mathbf{r}(u_i, v_j)| = |\mathbf{r}_u(u_i, v_j) \times \mathbf{r}_v(u_i, v_j)| \, \Delta u_i \, \Delta v_j. \quad (3.31)$$

(We take Δu_i and Δv_j to be positive.) Here the tangent vectors are given in $\mathbb{R}^2$ and the cross product is done in $\mathbb{R}^3$ so we "add" the third dimension in a trivial manner:

$$\mathbf{r}_u = x_u \, \mathbf{i} + y_u \, \mathbf{j} + 0 \, \mathbf{k}, \qquad \mathbf{r}_v = x_v \, \mathbf{i} + y_v \, \mathbf{j} + 0 \, \mathbf{k}.$$

Now we can write

$$M \approx \sum_{i,j} \rho(\tilde{u}_i, \tilde{v}_j) \, |\mathbf{r}_u(u_i, v_j) \times \mathbf{r}_v(u_i, v_j)| \, \Delta u_i \, \Delta v_j.$$

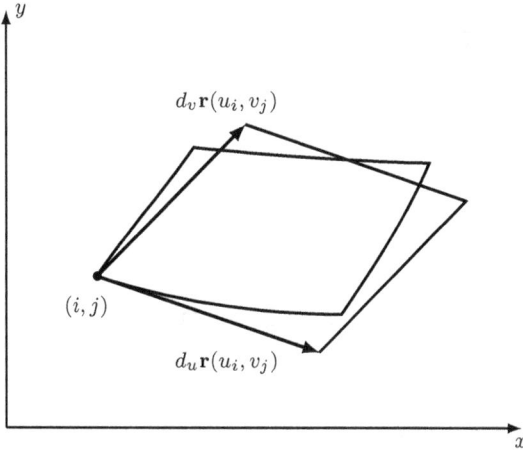

Fig. 3.25 Approximating the area of the ij-cell.

Before announcing the limit passage to the final result, let us change $\mathbf{r}_u(u,v) \times \mathbf{r}_v(u,v)$ to a more convenient form using determinants:

$$\mathbf{r}_u \times \mathbf{r}_v = \begin{vmatrix} \mathbf{i} & \mathbf{j} & \mathbf{k} \\ x_u & y_u & 0 \\ x_v & y_v & 0 \end{vmatrix} = \begin{vmatrix} x_u & y_u \\ x_v & y_v \end{vmatrix} \mathbf{k}.$$

The last determinant on the right has a special name — the *Jacobian*, in honor of Carl Jacobi, or *functional determinant* — and a special notation

$$\begin{vmatrix} x_u & y_u \\ x_v & y_v \end{vmatrix} = \frac{\partial(x,y)}{\partial(u,v)}. \tag{3.32}$$

Example 3.12. For polar coordinates we have

$$\frac{\partial(x,y)}{\partial(r,\theta)} = \begin{vmatrix} (r\cos\theta)_r & (r\sin\theta)_r \\ (r\cos\theta)_\theta & (r\sin\theta)_\theta \end{vmatrix} = r(\cos^2\theta + \sin^2\theta) = r. \qquad \Box$$

So

$$M \approx \sum_{i,j} \Delta m_{ij} \approx \sum_{i,j} \rho(\tilde{u}_i, \tilde{v}_j) \left| \frac{\partial(x,y)}{\partial(u,v)} \right|_{\substack{u=u_i \\ v=v_j}} \Delta u_i \, \Delta v_j.$$

This is a Riemann sum for the double integral over D_{uv} with the integrand

$$\rho(u,v) \left| \frac{\partial(x,y)}{\partial(u,v)} \right|.$$

As the integrand is bounded and piecewise continuous in D_{uv}, and in D_{uv} the corresponding partition is accomplished by straight lines parallel to the u- and v-axes, we conclude that supposing $\rho(u,v)|\mathbf{r}_u \times \mathbf{r}_v| \in C_b(D_{uv})$, under the usual assumptions of double integration, i.e., that the biggest of the diagonals $[(\Delta u_i)^2 + (\Delta v_j)^2]^{1/2} \to 0$ and the points $(\tilde{u}_i, \tilde{v}_j)$ are arbitrary in the ij-cells, we get the unique limit for the mass

$$M = \iint_{D_{uv}} \rho(u,v)) \left| \frac{\partial(x,y)}{\partial(u,v)} \right| dA_{uv}, \quad dA_{uv} = du\, dv.$$

Because this is the same mass M we calculated in Cartesian coordinates, we have

$$\iint_D \rho(x,y)\, dA = \iint_{D_{uv}} \rho(u,v)) \left| \frac{\partial(x,y)}{\partial(u,v)} \right| dA_{uv}, \quad dA_{uv} = du\, dv.$$

Now we may "forget" that $f = \rho$ was a mass density and use this construction for any bounded piecewise continuous function $f \in C_b(D_{uv})$. So we have introduced the double integral in curvilinear coordinates or, equivalently, how to change variables in double integrals:

$$\iint_D f(x,y)\, dA = \iint_{D_{uv}} f(x(u,v), y(u,v)) \left| \frac{\partial(x,y)}{\partial(u,v)} \right| dA_{uv}, \quad dA_{uv} = du\, dv.$$

We should add that the change of ΔA_{ij} to (3.31) can be done in the limits of Riemann sums. Indeed for smooth functions $x = x(u,v)$, $y = y(u,v)$ and coordinate lines that are not tangent at any point of intersection it can be shown that there is a function $\phi(t)$ such that

$$\left| \frac{\Delta A_{ij} - |\mathbf{r}_u(u_i, v_j) \times \mathbf{r}_v(u_i, v_j)|\, \Delta u_i\, \Delta v_j}{(\Delta u_i)^2 + (\Delta v_j)^2} \right|$$

$$< \phi([(\Delta u_i)^2 + (\Delta v_j)^2]^{1/2}) \quad \text{where } \phi(t) \to 0 \text{ as } t \to 0.$$

Example 3.13. We show how an iterated integral

$$\int_0^1 \int_x^1 f(x,y)\, dy\, dx$$

transforms under the relations $x = x(u,v) = u+v$ and $y = y(u,v) = u-v$. The Jacobian of the transformation is

$$\frac{\partial(x,y)}{\partial(u,v)} = \begin{vmatrix} (u+v)_u & (u-v)_u \\ (u+v)_v & (u-v)_v \end{vmatrix} = \begin{vmatrix} 1 & 1 \\ 1 & -1 \end{vmatrix} = -2$$

and the inverse transformation is given by

$$u = \frac{x+y}{2}, \qquad v = \frac{x-y}{2}.$$

The boundaries of the domain (a right triangle in the xy-plane) transform as follows (Fig. 3.26).

(1) The line segment $x = 0$, $0 \le y \le 1$ in the xy-plane maps into the line segment $u = -v$, $-\frac{1}{2} \le v \le 0$ in the uv-plane.
(2) The line segment $y = x$, $0 \le x \le 1$ in the xy-plane maps into the line segment $v = 0$, $0 \le u \le 1$ in the uv-plane.
(3) The line segment $y = 1$, $0 \le x \le 1$ in the xy-plane maps into the line segment $u = v + 1$, $-\frac{1}{2} \le v \le 0$ in the uv-plane.

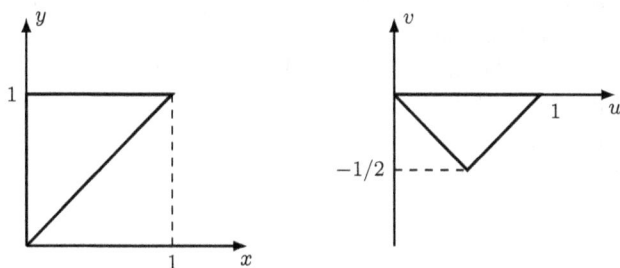

Fig. 3.26   Example 3.13.

We have

$$\int_0^1 \int_x^1 f(x,y)\, dy\, dx = \int_{-1/2}^0 \int_{-v}^{v+1} f(u+v, u-v) \, |-2| \, du\, dv. \qquad \square$$

Intuitively, $|\partial(x,y)/\partial(u,v)|$ is a quantity multiplied by the area of the elementary rectangle in the uv-plane to get the area of the corresponding elementary domain in the xy-plane constituted by u and v coordinate lines. It must be positive in order to preserve the orientation of the uv-coordinate system (and the orientation of the corresponding boundary contour of the image). Nearly all mathematics and physics textbooks employ a right-handed coordinate system. Left-handed systems are used on occasion, however, notably in areas such as the strength of materials where it is convenient to take the positive vertical direction downward (Fig. 3.27).

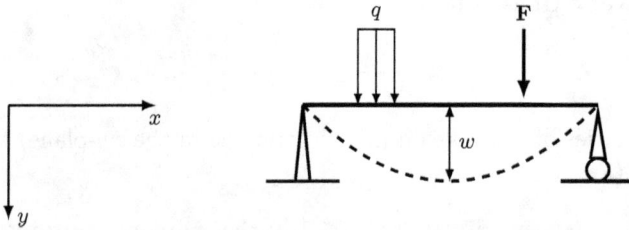

Fig. 3.27 Quantities of interest in a beam deflection problem. The z-axis is directed toward the reader.

Coordinate Rotation

Rotation of a domain D with respect to the coordinate system can sometimes produce a Type 1 or Type 2 domain. This is equivalent to a rotation of the coordinate system itself. Moreover, under coordinate rotation through an angle $\pi/2$, Type 2 domains become Type 1 domains. Clearly coordinate rotation could be useful in other instances as well, so let us discuss the issue for a moment. Suppose we rotate the Cartesian xy-system through an angle α to produce new Cartesian uv-system as in Fig. 3.28. Point A

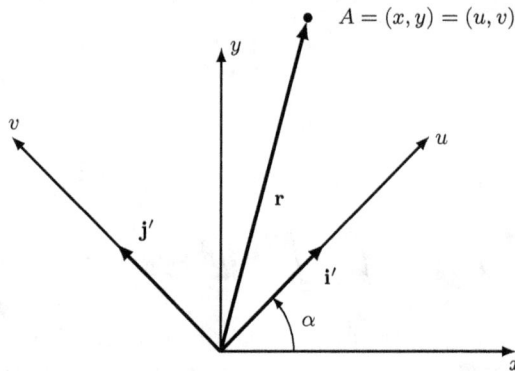

Fig. 3.28   Coordinate rotation.

has coordinates (x, y) in the xy-system and (u, v) in the uv-system. Its position vector is given by either of the expressions

$$\mathbf{r} = x\,\mathbf{i} + y\,\mathbf{j} = u\,\mathbf{i}' + v\,\mathbf{j}'$$

where $\mathbf{i}', \mathbf{j}'$ are the unit basis vectors in the uv-system. Dot multiplying the last equality by $\mathbf{i}$ we get

$$x = u\,\mathbf{i}' \cdot \mathbf{i} + v\,\mathbf{j}' \cdot \mathbf{i} = u\cos\alpha + v\cos(\tfrac{\pi}{2} + \alpha) = u\cos\alpha - v\sin\alpha.$$

Dot multiplying by $\mathbf{j}$ instead, we get

$$y = u\,\mathbf{i}' \cdot \mathbf{j} + v\,\mathbf{j}' \cdot \mathbf{j} = u\cos(\tfrac{\pi}{2} - \alpha) + v\cos\alpha = u\sin\alpha + v\cos\alpha.$$

These relations yield

$$\frac{\partial(x,y)}{\partial(u,v)} = \begin{vmatrix} \cos\alpha & \sin\alpha \\ -\sin\alpha & \cos\alpha \end{vmatrix} = \cos^2\alpha + \sin^2\alpha = 1,$$

which is clear even without calculation since under mere coordinate rotation the areas of elements cannot change. So we have the following formula for the change of a double integral under rotation through angle α:

$$\iint_D f(x,y)\,dA = \iint_{D_{uv}} f(u\cos\alpha - v\sin\alpha,\, u\sin\alpha + v\cos\alpha)\,dA_{uv},$$

where $dA_{uv} = du\,dv$.

Remark 3.4. In the definition of the integral $\int_a^b f(x)\,dx$, it is required that the limit of the Riemann sums $\sum_i f(\tilde{x}_i)\,\Delta x_i$ under $\max_i \Delta x_i \to 0$ should be the same for an arbitrary sequence of partitions of $[a,b]$ and arbitrary selection of points $\tilde{x}_i$. As the Riemann sums $\sum_{ij} f(\tilde{x}_i, \tilde{y}_j)\,\Delta A_{ij}$ for double integrals $\iint_D f(x,y)\,dA$ look quite similar to $\sum_i f(\tilde{x}_i)\,\Delta x_i$ it seems natural to treat the double integral for arbitrary partitions of D as well. However for some partitions of D it is impossible to find the areas ΔA_{ij} with reasonable formulas. So we should restrict the set of possible partitions as was done in this section. $\qquad\square$

Remark 3.5. Let us emphasize why, for various integrals, we choose to start with the problem of finding the mass of a body with given density. The main point is that engineering students believe that the notion of mass is simple and that they know its main properties (although possibly these are not so simple). First of all they "know" that the measure of a body mass is unique and that by partitioning a body we get the sum of the masses of the parts equal to the mass of the whole, which is the initial point to construct an integral. But more important is the following property of the Riemann sums which we explain, for simplicity, for the case of continuous density ρ for double integrals. A term of a Riemann sum is $\rho(\tilde{x}, \tilde{y})\,\Delta A$, where ΔA is the area of a partition cell C we assume has uncomplicated shape. Clearly we have the inequalities

$$\min_{(x,y)\in C} \rho(x,y)\,\Delta A \leq \rho(\tilde{x},\tilde{y})\,\Delta A \leq \max_{(x,y)\in C} \rho(x,y)\,\Delta A.$$

The average density $\bar{\rho}$ of the cell with mass Δm is $\Delta m/\Delta A$. Clearly

$$\rho_{\min} = \min_{(x,y)\in C} \rho(x,y) \leq \bar{\rho} \leq \max_{(x,y)\in C} \rho(x,y) = \rho_{\max}.$$

As ρ is a continuous function and on the closed cell takes all the values in the interval $[\rho_{\min}, \rho_{\max}]$, there is a point $(x^*, y^*) \in C$ such that $\rho(x^*, y^*) = \bar{\rho}$. By selecting this point in the Riemann term, we get the precise mass of cell C. Such a selection can be done in each term of the Riemann sum. This means that for any partition, with appropriate $(\tilde{x}_i, \tilde{y}_j)$ selected, the value of the corresponding Riemann sum is the mass of the body. Showing this for any sequence of partitions, and for any selection of points inside the cells, is a task for pure mathematics. □

3.7 Some Applications of Double Integrals

We have started to construct the double integral by calculating the mass M of a plate D having density $\rho = \rho(x,y)$. Setting $\rho = 1$ we obtain a formula for the calculation of area. Moreover, 2D density functions occur in fields such as electromagnetics, population biology, and so on.

Some of the applications in mechanics are aimed at characterizing how mass is distributed in space. An example is center of mass. We begin with a system of n particles. Let the kth particle have mass m_k and position vector $\mathbf{r}_k$ in $\mathbb{R}^3$ under the action of force $\mathbf{F}_k$. Newton's second law

$$m_k \frac{d^2 \mathbf{r}_k}{dt^2} = \mathbf{F}_k$$

can be summed over k:

$$\sum_{k=1}^{n} m_k \frac{d^2 \mathbf{r}_k}{dt^2} = \sum_{k=1}^{n} \mathbf{F}_k. \tag{3.33}$$

On the right we have the resultant force

$$\mathbf{F} = \sum_{k=1}^{n} \mathbf{F}_k$$

acting on the system of particles. If we denote the total mass of the system by

$$M = \sum_{k=1}^{n} m_k$$

and introduce the position vector

$$\mathbf{r}_c = \frac{1}{M} \sum_{k=1}^{n} m_k \mathbf{r}_k \qquad (3.34)$$

we can rewrite the above equation as

$$M \frac{d^2 \mathbf{r}_c}{dt^2} = \mathbf{F}.$$

Hence we have found a special point in space with position $\mathbf{r}_c$ such that if we collect all the masses at $\mathbf{r}_c$, this mass under the resultant force $\mathbf{F}$ moves precisely as this point moves. The point c defined by $\mathbf{r}_c$ is called the *center of mass*.

More generally, the center of mass is introduced by supposing that the particles can interact through forces $\mathbf{f}_{ki}$ (the force from the ith mass acting on the kth mass). By Newton's third law $\mathbf{f}_{ki} = -\mathbf{f}_{ik}$, so the equation of motion for the kth particle is

$$m_k \frac{d^2 \mathbf{r}_k}{dt^2} = \mathbf{F}_k + \sum_{i=1,\, i \neq k}^{n} \mathbf{f}_{ki}.$$

Summing this over k, we again get (3.33) and hence the same center of mass location. Moreover, we can assume that the particles are connected with rigid massless bars. In this case we shall get the same location for the center of mass, prompting us to apply (3.34) to any rigid body with a distributed mass.

It is worth noting that we can get the same position vector $\mathbf{r}_c$ if we ask where to position an immovable support to balance the system in a uniform gravitational field.

Let us see how to introduce the center of mass for planar mass distribution. For example the center of mass of n point masses m_{ij} with coordinates (x_i, y_j) on the xy-plane is

$$x_c = \frac{1}{M} \sum_{i,j} m_{ij} x_i, \qquad y_c = \frac{1}{M} \sum_{i,j} m_{ij} y_j, \qquad (3.35)$$

where $M = \sum_{ij} m_{ij}$ is the total mass of the points. The moments with respect to the y and x axes are respectively

$$M_y = \sum_{i,j} m_{ij} x_i, \qquad M_x = \sum_{i,j} m_{ij} y_j.$$

Replacing m_{ij} with an approximation to the mass of an ij-cell, $\rho(x_i, y_j) \, \Delta A_{ij}$, we get approximations to the moments for a plate:

$$M_y \approx \sum_{i,j} x_i \rho(x_i, y_j) \, \Delta A_{ij}, \qquad M_x \approx \sum_{i,j} y_j \rho(x_i, y_j) \, \Delta A_{ij}.$$

Taking limits in these Riemann sums, we obtain

$$M_x = \iint_D y \rho(x, y) \, dA, \qquad M_y = \iint_D x \rho(x, y) \, dA.$$

So the plate center of mass has coordinates

$$x_c = M_y / M, \qquad y_c = M_x / M.$$

Example 3.14. A plate having mass density $\rho(x, y) = y$ occupies the region between the curves $y = x^2$ and $x = y^2$. To find its center of mass, we calculate

$$M_y = \int_0^1 \int_{x^2}^{\sqrt{x}} xy \, dy \, dx = \frac{1}{12}, \qquad M_x = \int_0^1 \int_{x^2}^{\sqrt{x}} y^2 \, dy \, dx = \frac{3}{35},$$

and

$$M = \int_0^1 \int_{x^2}^{\sqrt{x}} y \, dy \, dx = \frac{3}{20}.$$

Therefore $(x_c, y_c) = (5/9, 4/7)$. $\qquad\qquad\qquad\qquad\qquad\qquad\qquad\square$

By taking $\rho(x, y) \equiv 1$ we come to the idea of an *area centroid* of D, located at coordinates

$$x_c = \frac{1}{A} \iint_D x \, dA, \qquad y_c = \frac{1}{A} \iint_D y \, dA.$$

Example 3.15. Consider the quarter disk described by $x^2 + y^2 \le a^2$ for $0 \le x \le a$. Here $A = \pi a^2 / 4$ and we have by symmetry

$$y_c = x_c = \frac{4}{\pi a^2} \int_0^a \int_0^{\sqrt{a^2 - x^2}} x \, dy \, dx = \frac{4a}{3\pi}. \qquad\qquad\square$$

There is also a useful formula for the central moment of inertia:

$$I = \iint_D (x^2 + y^2)\rho \, dA. \tag{3.36}$$

This appears in the law governing rotation of the plate about an axis perpendicular to D through the origin:

$$I \frac{d^2\alpha}{dt^2} = m,$$

where α is the angle of rotation and m describes a torque applied to the plate. Note the similarity to Newton's second law.

We also can mention the inertia tensor of a solid body. For a plate we present two of its components participating in the problems of plate rotation about the Ox and Oy axes:

$$I_{xx} = \iint_D y^2 \rho(x, y) \, dA, \qquad I_{yy} = \iint_D x^2 \rho(x, y) \, dA.$$

Example 3.16. For the quarter disk of Example 3.15, assuming a uniform unit mass density, we have (again by symmetry)

$$I_{xx} = I_{yy} = \int_0^a \int_0^{\sqrt{a^2 - x^2}} x^2 \, dy \, dx = \frac{\pi a^4}{16}. \qquad \square$$

3.8 Cauchy–Schwarz Inequality

We have mentioned some useful inequalities allowing us to estimate the values of double integrals. They are based on the corresponding elementary inequalities applied to Riemann sum approximations to the integrals.

In a similar way we can extend Cauchy's inequality

$$\sum_{i=1}^n a_i b_i \leq \left(\sum_{i=1}^n a_i^2 \right)^{1/2} \left(\sum_{i=1}^n b_i^2 \right)^{1/2}$$

to integral form:

$$\left| \int_D f(x, y) g(x, y) \, dA \right| \leq \left(\int_D f^2(x, y) \, dA \right)^{1/2} \left(\int_D g^2(x, y) \, dA \right)^{1/2}.$$

The extension of Hölder's inequality is

$$\left| \int_D f(x, y) g(x, y) \, dA \right| \leq \left(\int_D |f(x, y)|^p \, dA \right)^{1/p} \left(\int_D |g(x, y)|^q \, dA \right)^{1/q}$$

with

$$1/p + 1/q = 1, \qquad p > 1.$$

The limit procedure used here will yield many other inequalities for both double and triple integrals.

3.9   Problems

Review of the Definite Integral and Its Applications

3.1   Use a limit of Riemann sums to calculate the integral

$$\int_0^2 (x + 1)\, dx.$$

3.2   Argue that the Riemann integral

$$\int_0^1 f(x)\, dx, \qquad f(x) = \begin{cases} 1, & x \text{ irrational,} \\ 0, & x \text{ rational,} \end{cases}$$

does not exist.

3.3   The mass density on the segment $[0, L]$ of the x-axis is given by $\gamma(x) = x(L - x)$. What is the total mass of the segment?

Double Integration

3.4   Use the definition of the double integral to evaluate

(a)

$$I = \iint_D x\, dy\, dx \quad \text{where} \quad D = [0, 1]^2,$$

(b)

$$I = \iint_D xy\, dy\, dx \quad \text{where} \quad D = [-1, 1] \times [0, 2].$$

3.5   Verify the following results:

(a)

$$\int_0^\pi \int_0^{\sqrt{\cos\theta}} r^3\, dr\, d\theta = \frac{\pi}{8}$$

(b)

$$\int_0^\pi \int_1^{\cos\theta + 1} r^2 \sin\theta\, dr\, d\theta = \frac{2}{3}$$

(c)

$$\int_0^{2\pi} \int_1^{\cos\theta} r^3 \sin\theta \, dr \, d\theta = 0$$

3.6 Evaluate

$$I_1 = \int_0^3 \int_0^2 yx^2 \, dy \, dx, \qquad I_2 = \int_0^1 \int_{\sqrt{x}}^1 e^{-y^3} \, dy \, dx, \qquad I_3 = \int_0^1 \int_{x^2}^x xy \, dy \, dx.$$

3.7 Integrate the following functions over the square $[0,1]^2$:

(a) $f(x,y) = e^{x+y}$

(b) $f(x,y) = y^2/(1+x^2)$

(c) $f(x,y) = 1/(x+y+1)^2$

(d) $f(x,y) = xy$

(e) $f(x,y) = x^2 + y^2$

3.8 Let A be the triangular region in the xy-plane given by $0 \le x \le 1, 0 \le y \le x$. Calculate the double integral of $f(x,y) = xy$ over A two ways as an iterated integral.

3.9 In each instance, change the order of integration:

(a)

$$I = \int_0^1 \int_y^{\sqrt{y}} f(x,y) \, dx \, dy$$

(b)

$$I = \int_{-1}^1 \int_0^{\sqrt{1-x^2}} f(x,y) \, dy \, dx$$

(c)

$$I = \int_0^1 \int_0^x f(x,y) \, dy \, dx$$

(d)

$$I = \int_0^1 \int_0^{x^2} f(x,y) \, dy \, dx$$

3.10 Express each sum of double integrals as just one double integral:

(a)

$$\int_0^{1/2} \int_y^{2y} f(x,y) \, dx \, dy + \int_{1/2}^1 \int_y^1 f(x,y) \, dx \, dy$$

(b)

$$\int_0^1 \int_0^x f(x,y) \, dy \, dx + \int_1^2 \int_{x-1}^1 f(x,y) \, dy \, dx$$

3.11 Integrate the following functions over the given regions A, performing each computation in two different ways:

(a) $f(x,y) = xy$, region A bounded by the lines $y = 3x$, $y = \frac{1}{3}x$, and $x = 1$

(b) $f(x,y) = x + y$, region A bounded by the curves $y = x^2$ and $y = x^{1/2}$ for $0 \le x \le 1$

(c) $f(x,y) = y^2$, region $A = \{(x,y): |x| + |y| \le 1\}$

(d) $f(x,y) = x^2 y$, region A bounded by the curves $y = x^2$ and $y = x$ for $0 \le x \le 1$

3.12 Change each given integral to polar coordinates:

(a)
$$I = \int_A e^{x^2 + y^2} \, dx \, dy$$

where $A = \{(x,y): x^2 + y^2 \le 1\}$ is the unit disk in the xy-plane

(b)
$$I = \int_A (x^2 + y^2)^2 \, dx \, dy$$

where $A = \{(x,y): x^2 + y^2 \le 1\}$ is the unit disk in the xy-plane

(c)
$$I = \int_A y^2 \, dx \, dy$$

where A is the region in the $z = 0$ plane bounded by the curves $y = 1/x$, $y = 2/x$, $y = x/2$, and $y = 2x$

3.13 Integrate the following functions over the quarter disk $x^2 + y^2 \le 1$, $x \ge 0$, $y \ge 0$:

(a) $f(x,y) = x^2 + y^2$ (c) $f(x,y) = (1 - x^2 - y^2)^{1/2}$

(b) $f(x,y) = \exp(x^2 + y^2)$ (d) $f(x,y) = \ln(1 + x^2 + y^2)$

3.14 Let $a > 0$ and $0 \le \alpha < \pi/2$ be given constants. Sketch the set of points (r, θ) for which $0 \le r \le a \cos \theta$ and $0 \le \theta \le \alpha$. Integrate the function $f(r, \theta) = r \sin^2 \theta$ over this region.

Areas, Mean Values, Centroids, and Moments of Inertia

3.15 Find the area enclosed by the graph of $r = \cos \theta$ between the positive x-axis and the ray $\theta = \pi/4$.

3.16 Sketch each pair of curves and find the enclosed area:

(a) $x^2 - y^2 = 1$ and $x = 2$

(b) $x^2 = 2py$ and $y^2 = 2px$ where $p > 0$ is a constant

(c) $y = 1 - x^2$ and $y = 1 - x$ for $0 \le x \le 1$, $0 \le y \le 1$

(d) $x^2 + y^2 = 1$ and $y = (x - 1)^2$

3.17 Find the area enclosed by the ellipse (1.8).

3.18 Let A be the plane region lying between the curves $y = x^2$ and $y = 1$ for $|x| \le 1$. Calculate the area of A as a Type I integral and as a Type II integral. Then evaluate $I = \iint_A x^2 \, dA$.

3.19 Assuming a is a positive constant, find the area enclosed by each of the following curves:

(a) $r = a(1 + \cos \theta)$ 　　　　　　(c) $r = a \cos \theta$

(b) the spiral $r = \theta$ for $0 \le \theta \le \pi$ 　　(d) $r = 1 - \cos \theta$

3.20 Find the areas of the regions bounded by the following sets of curves in the plane:

(a) the circle $x^2 + (y - 3)^2 = 9$, the line $y = x$, and the line $y = x/2$

(b) the circle $r = 1$ and the circle $r = 2\sin \theta$ for $0 \le \theta \le \pi$

(c) the two circles $r = a \cos \theta$ and $r = b \cos \theta$ where $b > a$

3.21 Find the mean value of the function over the specified plane region:

(a) $f(x, y) = ax + y$ over the triangle bounded by the lines $x = 0$, $y = 0$, and $x + y = b$

(b) $f(x, y) = (1 - x^2 - y^2)^{1/2}$ over the disk $x^2 + y^2 \le 1$

3.22 Find the centroid of

(a) the region between the line $y = 0$ and the curve $y = \sin x$ for $0 \le x \le \pi$,

(b) the region above the line $y = -x$ and below the parabola $y = 2 - x^2$,

(c) the region inside the ellipse (1.8) for $x \ge 0$, and

(d) the cardioid described in Problem 3.19(a).

3.23 Find the center of mass and the moments of inertia I_{xx} and I_{yy} for a homogeneous plate occupying

(a) the rectangular region $|x| \le b/2$, $|y| \le h/2$,

(b) the annular region $a \le r \le b$, $0 \le \theta < 2\pi$, $z = 0$.

3.24 A homogeneous plate occupies the region bounded by the ellipse (1.8). Find the values of (a) I_{xx}, (b) I_{yy}, and (c) I from (3.36).

More General Coordinate Transformations

3.25 Sketch the region R in the xy-plane described by the inequality $x - 1 \leq y \leq x + 1$. What is the image of R under the transformation equations $u = x - y$ and $v = x$? State the equations of the inverse transformation.

3.26 Find the Jacobian of each of following transformations, and evaluate it at the point $(1, 1)$:

(a) $x = ue^v$ and $y = u - e^v$ (c) $x = 2u + v$ and $y = u - 2v$

(b) $x = e^u$ and $y = e^v$ (d) $x = u + v$ and $y = uv$

3.27 Let $x = au + bv$ and $y = cu + dv$. Find $\partial(x, y)/\partial(u, v)$ and $\partial(u, v)/\partial(x, y)$.

3.28 Consider the transformation $x + y = u$, $y = vx$. Calculate its Jacobian. Find the inverse transformation and its Jacobian. How does the square $[1, 2] \times [0, 1]$ map from the uv-plane to the xy-plane?

3.29 Let $f(x, y) = xy$ and calculate the integral of Example 3.13 in both the xy and uv systems.

3.30 Let R be the square region having corner points $(1, 0)$, $(0, 1)$, $(-1, 0)$, and $(0, -1)$ in the xy-plane. Find a transformation that maps R into a region S having sides parallel to the coordinate axes in the uv-plane. What is the inverse transformation, and what are the Jacobians of the two transformations?

3.31 Consider the integral

$$I = \iint_R x^2 y \, dA$$

where the region R is enclosed by the triangle having vertices $(-1, 0)$, $(0, 1)$, and $(1, 0)$. (a) Compute I in two different ways as an iterated integral. (b) Find the image S of R under the transformation relations $u = y - x$ and $v = y + x$. (c) Compute I by integration over S; compare with the result from part (a).

Other Applications of the Double Integral

3.32 Find the total mass of

(a) the disk $x^2 + y^2 \leq a^2$ when the mass density is given by $\rho(x, y) = 2a + x$;

(b) the square plate occupying the region $[0, 1]^2$ when the mass density is given by $\rho(x, y) = (x^2 + y^2)^{1/2}$.

3.33 A plate occupying the region $[-1, 1]^2$ carries an electric charge described by density function $|x|$ millicoulombs per square meter. Find the total electric charge on the plate.

3.34 A plate of uniform unit density occupies the region of Example 3.14. Find its center of mass.

Chapter 4

Triple Integrals

Now we construct the triple integral for which

$$\iiint_E dV \qquad \text{and} \qquad \iiint_E \rho(x, y, z) \, dV$$

are respectively the volume of a 3D figure E in $\mathbb{R}^3$ and the mass of a solid body E with density $\rho = \rho(x, y, z)$. Introduction of the triple integral will mimic the scheme used for the double integral. Namely, we commence with the problem of finding the mass of E with given ρ, starting with the evident fact that any partition of a solid of mass M into small parts with masses m_k is such that $M = \sum_k m_k$. Next we extend construction of the integral by replacing ρ with an arbitrary function. Finally, a 3D version of Fubini's theorem will allow us to calculate triple integrals for domains of three types.

4.1 Triple Integral Over a Parallelepiped

To avoid a few difficulties we start with the problem of a solid rectangular parallelepiped

$$P = [a, b] \times [c, d] \times [s, t].$$

Let $\rho = \rho(x, y, z)$ be the mass density function, assumed continuous on P. Since P is closed, ρ is also bounded on P. To approximate the mass M of P, we create a uniform partition of P by subdividing each of the segments $[a, b]$, $[c, d]$, and $[s, t]$ into n equal parts by points x_i, y_j, and z_k respectively. The partition of P is produced by all the planes through the points (x_i, y_j, z_k) parallel to one of the three coordinate planes xy, xz, or yz (Fig. 4.1).

The side lengths of the small cell

$$\Delta P_{ijk} = [x_i, x_{i+1}] \times [y_j, y_{j+1}] \times [z_k, z_{k+1}]$$

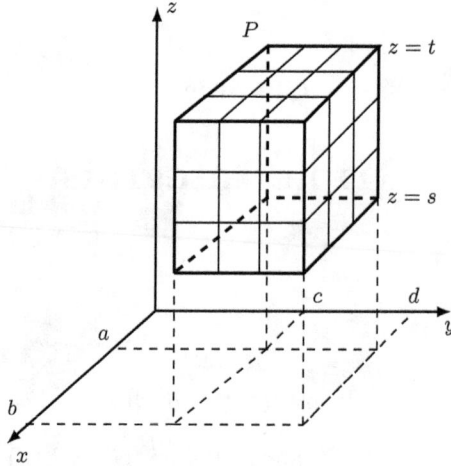

Fig. 4.1   Uniform partition of rectangular parallelepiped $P = [a,b] \times [c,d] \times [s,t]$.

are

$$\Delta x_i = (b-a)/n, \qquad \Delta y_j = (d-c)/n, \qquad \Delta z_k = (t-s)/n,$$

and its volume ΔV_{ijk} is

$$\Delta V_{ijk} = (b-a)(d-c)(t-s)/n^3 = \Delta V_n.$$

The values associated with ΔP_{ijk} are tagged with the indices ijk. With Δm_{ijk} the mass of ΔP_{ijk}, the mass of P becomes

$$M = \sum_{i=1}^{n}\sum_{j=1}^{n}\sum_{k=1}^{n} \Delta m_{ijk}.$$

Clearly the value Δm_{ijk} lies between the masses of hypothetical cells ΔP_{ijk} having homogeneous densities equal to the minimum and maximum densities of the cell:

$$\min_{\Delta P_{ijk}} \rho(x,y,z)\,\Delta V_n \leq \Delta m_{ijk} \leq \max_{\Delta P_{ijk}} \rho(x,y,z)\,\Delta V_n.$$

As ρ is a continuous function, there is a point $(\tilde{x}_i, \tilde{y}_j, \tilde{z}_k) \in \Delta P_{ijk}$ such that $\Delta m_{ijk} = \rho(\tilde{x}_i, \tilde{y}_j, \tilde{z}_k)\,\Delta V_n$. Hence

$$M = \sum_{i=1}^{n}\sum_{j=1}^{n}\sum_{k=1}^{n} \rho(\tilde{x}_i, \tilde{y}_j, \tilde{z}_k)\,\Delta V_n.$$

Unfortunately we don't know the values $(\tilde{x}_i, \tilde{y}_j, \tilde{z}_k)$. So, using continuity of ρ, let us write down the approximations

$$\Delta m_{ijk} \approx \rho(x_i, y_j, z_k)\,\Delta V_n$$

and

$$M \approx \sum_{i=1}^{n}\sum_{j=1}^{n}\sum_{k=1}^{n} \rho(x_i, y_j, z_k)\,\Delta V_n. \tag{4.1}$$

It can be proved that for ρ continuous the limit of the sum on the right as $n \to \infty$ exists, and we know the limit of

$$\sum_{i=1}^{n}\sum_{j=1}^{n}\sum_{k=1}^{n} \rho(\tilde{x}_i, \tilde{y}_j, \tilde{z}_k)\,\Delta V_n$$

as $n \to \infty$ is M. So it is natural expect that

$$M = \lim_{n\to\infty} \sum_{i=1}^{n}\sum_{j=1}^{n}\sum_{k=1}^{n} \rho(x_i, y_j, z_k)\,\Delta V_n.$$

This will follow from a theorem formulated below, which in particular states that the limit

$$\lim_{n\to\infty} \sum_{i=1}^{n}\sum_{j=1}^{n}\sum_{k=1}^{n} \rho(\tilde{x}_i, \tilde{y}_j, \tilde{z}_k)\,\Delta V_n$$

is the same for any choice of points $(\tilde{x}_i, \tilde{y}_j, \tilde{z}_k)$ in ΔP_{ijk}. We denote this limit

$$\iiint_{P} \rho(x, y, z)\,dV = M$$

and call it the triple integral of ρ over P.

Changing ρ to f, an arbitrary continuous function on P, we introduce *the triple integral* of f over the parallelepiped P:

$$\iiint_{P} f(x, y, z)\,dV = \lim_{n\to\infty} \sum_{i=1}^{n}\sum_{j=1}^{n}\sum_{k=1}^{n} f(x_i, y_j, z_k)\,\Delta V_n.$$

As with ρ, it can be proved that the limit on the right exists. This definition of the triple integral through uniform partitions of P is quite restrictive. Now we extend the permissible integrands to $f \in C_b(P)$. Let E be a bounded region with boundary consisting of a few smooth surfaces given by continuously differentiable vector functions $\mathbf{r}(u, v)$. By $C_b(E)$ we denote the class of functions $f: E \subset \mathbb{R}^3 \to \mathbb{R}$ that are bounded in E and also continuous in E except, possibly, over a finite number of smooth surfaces, lines, or points where they can have jump discontinuities. We introduce $C_b(E)$ for future consideration and now we shall use $C_b(P)$.

So we create a nonuniform partition Π_{mnp} of P by the planes parallel to the coordinate planes defined by the nonuniform partitions of the segments $[a,b]$, $[c,d]$, and $[s,t]$ with respective sets of points

$$x_1 = a, \ x_2, x_3, \ldots, x_m, \ x_{m+1} = b,$$

$$y_1 = c, \ y_2, y_3, \ldots, y_n, \ y_{n+1} = d,$$

$$z_1 = s, \ z_2, z_3, \ldots, z_p, \ z_{p+1} = t.$$

For this partition of P we get small ijk-cells, and if $f(x,y,z) = \rho(x,y,z)$ we can write

$$M \approx \sum_{i=1}^{m}\sum_{j=1}^{n}\sum_{k=1}^{p} f(\tilde{x}_i, \tilde{y}_j, \tilde{z}_k)\,\Delta V_{ijk},$$

where $\Delta V_{ijk} = \Delta x_i\,\Delta y_j\,\Delta z_k$ and $(\tilde{x}_i, \tilde{y}_j, \tilde{z}_k)$ lies in ΔP_{ijk}. We have defined the triple integral using uniform partitions. Now we formulate

Theorem 4.1. *Let $f \in C_b(P)$. For any sequence of partitions $\{\Pi_{mnp}\}$ and selection of points $(\tilde{x}_i, \tilde{y}_j, \tilde{z}_k)$, if*

$$\max_{i,j,k}\left[(\Delta x_i)^2 + (\Delta y_j)^2 + (\Delta z_k)^2\right]^{1/2} \to 0$$

then there follows the existence of the unique limit

$$\lim_{\substack{m\to\infty \\ n\to\infty \\ p\to\infty}} \sum_{i=1}^{m}\sum_{j=1}^{n}\sum_{k=1}^{p} f(\tilde{x}_i, \tilde{y}_j, \tilde{z}_k)\,\Delta V_{ijk}$$

which is denoted by

$$\iiint_P f(x,y,z)\,dV.$$

The expression

$$\sum_{i=1}^{m}\sum_{j=1}^{n}\sum_{k=1}^{p} f(\tilde{x}_i, \tilde{y}_j, \tilde{z}_k)\,\Delta V_{ijk}$$

will be called a Riemann sum for this triple integral.

As is common in engineering books, we leave this theorem without proof. In such books it is usual to define the triple integral though the limit formulated in the theorem, and to announce that it exists for some class of functions f. It is possible to introduce more general partitions of P for which Theorem 4.1 also holds. For example, they may be defined by curvilinear coordinate systems.

Fubini's Theorem

Now let us write out a particular form of the triple Riemann sum:

$$\sum_{i=1}^{m}\sum_{j=1}^{n}\sum_{k=1}^{p} f(x_i, y_j, z_k)\,\Delta V_{ijk} = \sum_{i=1}^{m}\sum_{j=1}^{n} \left(\sum_{k=1}^{p} f(x_i, y_j, z_k)\,\Delta z_k \right) \Delta y_j\,\Delta x_i.$$

For fixed i, j, the parenthetical expression on the right is a Riemann sum for the integral

$$\int_{s}^{t} f(x_i, y_j, z)\,dz = g(x_i, y_j).$$

It appears that we can "approximate" the triple Riemann sum by

$$\sum_{i=1}^{m}\sum_{j=1}^{n} g(x_i, y_j)\,\Delta y_j\,\Delta x_i$$

which is a Riemann sum for the double integral

$$\iint_{R_{xy}} g(x, y)\,dA_{xy} = \iint_{R_{xy}} \left(\int_{s}^{t} f(x, y, z)\,dz \right) dA_{xy},$$

where R_{xy} is the rectangle $[a, b] \times [c, d]$ and dA_{xy} is the area element $dA_{xy} = dx\,dy$.

Have we "established" that by taking the limit of the triple Riemann sums in this way, we can calculate the triple integral iteratively, first computing an ordinary definite integral with respect to z, then the double integral over R_{xy}? Unfortunately two difficulties occur. First, for some (x, y) the integral $\int_{s}^{t} f(x, y, z)\,dz$ may not exist; we have assumed f can have jumps in P. Second, the type of limit we have used does not obey the conditions of Theorem 4.1 where $\max_{i,j,k} [(\Delta x_i)^2 + (\Delta y_j)^2 + (\Delta z_k)^2]^{1/2} \to 0$. Indeed we take first the limit with fixed Δx_i and Δy_j. Nonetheless, it is possible to prove the following.

Theorem 4.2 (Fubini). *For $f \in C_b(P)$, the triple integral can be calculated in three forms:*

$$\iiint_P f(x, y, z)\, dV = \iint_{D_{xy}} \int_s^t f(x, y, z)\, dz\, dA$$

$$= \iint_{D_{yz}} \int_a^b f(x, y, z)\, dx\, dA$$

$$= \iint_{D_{xz}} \int_c^d f(x, y, z)\, dy\, dA, \qquad (4.2)$$

where $D_{yz} = [c, d] \times [s, t]$, and $D_{xz} = [s, t] \times [a, b]$, and dA denotes the area elements in the corresponding coordinate planes.

Note that

$$\iint_{D_{xy}} g(x, y)\, dA,$$

as an ordinary double integral of g over D_{xy}, can be calculated by Fubini's theorem in either of the iterated forms

$$\iint_{D_{xy}} g(x, y)\, dA = \int_a^b \int_c^d g(x, y)\, dA = \int_c^d \int_a^b g(x, y)\, dA$$

and in curvilinear coordinate systems including polar coordinates. So (4.2) can be extended to various forms of thrice-iterated integrals, the first of which for P is

$$\iiint_P f(x, y, z)\, dV = \int_a^b \int_c^d \int_s^t f(x, y, z)\, dz\, dy\, dx.$$

Example 4.1. Take $f(x, y, z) = xyz$ on $P = [0, a] \times [0, b] \times [0, c]$. Then

$$\iiint_P f(x, y, z)\, dV = \int_0^a \int_0^b \int_0^c xyz\, dz\, dy\, dx$$

$$= \int_0^a x\, dx \int_0^b y\, dy \int_0^c z\, dz = \frac{a^2 b^2 c^2}{8}. \qquad \square$$

4.2   Triple Integral Over a General Bounded Domain

By analogy with the double integral for general domains, let us introduce the triple integral of f given on a bounded general domain $E \subset \mathbb{R}^3$ lying inside some rectangular parallelepiped P. We again extend f to $\tilde{f}$:

$$\tilde{f}(x, y, z) = \begin{cases} f, & (x, y, z) \in E, \\ 0, & (x, y, z) \notin E. \end{cases}$$

If we consider f to be mass density in E, then clearly the mass of E equals that of P as the density of P outside E is zero. So it is natural to write

Definition 4.1.

$$\iiint_E f(x, y, z)\, dV = \iiint_P \tilde{f}(x, y, z)\, dV. \tag{4.3}$$

The following properties can be proved formally, but are evident if we take f as a mass density ρ.

If E consists of two parts E_1 and E_2 having only boundary points in common, then clearly, if $f(x, y, z)$ is the density of E, the mass of E is the sum of the masses of E_1 and E_2. This also holds for a general function f:

$$\iiint_{E_1 \cup E_2} f(x, y, z)\, dV = \iiint_{E_1} f(x, y, z)\, dV + \iiint_{E_2} f(x, y, z)\, dV.$$

The integrals over E_1 and E_2 could be computed in different coordinate systems.

Linearity of the triple integral follows from that of any Riemann sum with respect to f, and the limit properties. So

$$\iiint_E [c f_1(x, y, z) + f_2(x, y, z)]\, dV$$

$$= c \iiint_E f_1(x, y, z)\, dV + \iiint_E f_2(x, y, z)\, dV$$

where $f_1(x, y, z)$ and $f_2(x, y, z)$ are integrable functions over E and c is an arbitrary constant.

Now let us present some estimates for triple integrals. If we return to the idea that f is the density of a solid E, and increase it to $\sup_E f$ at each point, the mass of the solid cannot decrease so

$$\iiint_E f(x, y, z)\, dV \leq \iiint_E \sup_E f(x, y, z)\, dV$$

$$= \sup_E f \cdot \iiint_E dV = \sup_E f \cdot V(E),$$

where $V(E)$ is the volume of E. Clearly this inequality is valid for any part of E. A similar but "opposite" inequality holds if we decrease f to $\inf_E f$, and altogether we have

$$\inf_E f \cdot V(E) \le \iiint_E f(x, y, z) \, dV \le \sup_E f \cdot V(E).$$

In the theory of partial differential equations important roles are played by two inequalities we present without proof. They are Hölder's inequality

$$\iiint_E |f(x, y, z)g(x, y, z)| \, dV \le \left(\iiint_E |f(x, y, z)|^p \, dV \right)^{1/p}$$
$$\times \left(\iiint_E |g(x, y, z)|^q \, dV \right)^{1/q}$$

with f, g piecewise continuous on E and $1/p + 1/q = 1$ for $p > 1$, and Minkowski's inequality

$$\left(\iiint_E |f(x, y, z) + g(x, y, z)|^p \, dV \right)^{1/p} \le \left(\iiint_E |f(x, y, z)|^p \, dV \right)^{1/p}$$
$$+ \left(\iiint_E |g(x, y, z)|^p \, dV \right)^{1/p}$$

for $p > 1$. This inequality means that $\left(\iiint_E |f|^p \, dV \right)^{1/p}$ can serve as a norm on the set of functions that are integrable with degree p.

For f continuous on closed E there is a theorem analogous to the mean value theorem for functions of a real variable. We present this without proof.

Theorem 4.3. *Let f be continuous on a closed bounded domain E. There is a point $(\tilde{x}, \tilde{y}, \tilde{z}) \in E$ such that*

$$\iiint_E f(x, y, z) \, dV = f(\tilde{x}, \tilde{y}, \tilde{z}) \cdot V(E). \tag{4.4}$$

We also have the *triangle inequality*

$$\left| \iiint_E f(x, y, z) \, dV \right| \le \iiint_E |f(x, y, z)| \, dV \tag{4.5}$$

and the fact that if $f(x, y, z) \le g(x, y, z)$ for all $(x, y, z) \in E$ then

$$\iiint_E f(x, y, z) \, dV \le \iiint_E g(x, y, z) \, dV. \tag{4.6}$$

4.3 Triple Integration Over Domains of Special Types

Fubini's formulas (4.2) allow us to calculate triple integrals over special domains of three types, each time reducing a triple integral to an iterated integral.

The domain of Type 1 is a part of a cylinder with the generator parallel to the z-axis and bounded by surfaces $z = g_1(x, y)$ and $z = g_2(x, y)$ with $g_1(x, y) \leq g_2(x, y)$. So for any (x, y) in the interior of D_{xy}, the projection of E onto the xy-plane, the straight line through (x, y) and parallel to the z-axis intersects the boundary of E exactly twice if $g_1(x, y) < g_2(x, y)$. We can think of this line as entering E at point $(x, y, z_1 = g_1(x, y))$ and leaving E at point $(x, y, z_2 = g_2(x, y))$. See Fig. 4.2, left. Examples include vertical cylinders, cubes, balls, and pyramids.

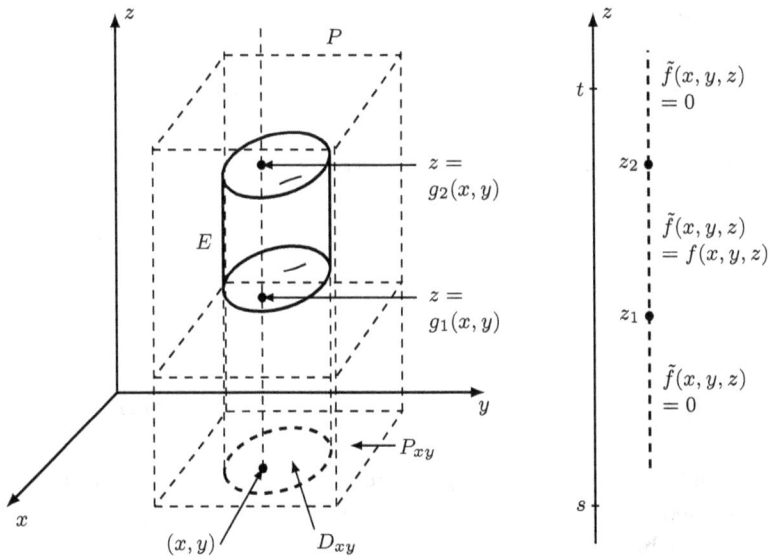

Fig. 4.2 *Left.* Solid E is inside parallelepiped P. D_{xy} is the orthogonal projection of E onto the xy-plane; P_{xy} is the orthogonal projection of P onto the xy-plane. $(x, y, g_1(x, y))$ is the point of entrance of the line through $(x, y, 0)$ and parallel to the z-axis, while $(x, y, g_2(x, y))$ is the point of exit. *Right.* Line of integration (dashed).

Let us discuss how to calculate triple integrals over a Type 1 domain using Fubini's formula

$$\iiint_P f(x,y,z)\, dV = \iint_{D_{xy}} \int_s^t f(x,y,z)\, dz\, dA_{xy}$$

for a straight parallelepiped P containing E, where D_{xy} is the orthogonal projection of E onto the xy-plane. First we calculate the inner-nested integral for a fixed $(x,y) \in D_{xy}$. The right half of Fig. 4.2 shows the line of integration. We have

$$\int_s^t \tilde{f}(x,y,z)\, dz = \int_s^{g_1(x,y)} \tilde{f}(x,y,z)\, dz$$
$$+ \int_{g_1(x,y)}^{g_2(x,y)} \tilde{f}(x,y,z)\, dz + \int_{g_2(x,y)}^t \tilde{f}(x,y,z)\, dz.$$

By definition of $\tilde{f}$ this is

$$\int_s^t \tilde{f}(x,y,z)\, dz = \int_{g_1(x,y)}^{g_2(x,y)} f(x,y,z)\, dz$$

since $\tilde{f}(x,y,z) = 0$ for $z \notin [g_1(x,y), g_2(x,y)]$. As $\tilde{f} = 0$ outside E, we also have

$$\int_s^t \tilde{f}(x,y,z)\, dz = 0, \qquad (x,y) \notin D_{xy}.$$

So for the domain E of Type 1,

$$\iiint_E f(x,y,z)\, dV = \iint_{D_{xy}} \int_{g_1(x,y)}^{g_2(x,y)} f(x,y,z)\, dz\, dA_{xy}.$$

To the double integral over D_{xy} we are free to apply the two-dimensional Fubini's theorem. It may also be advantageous to employ polar coordinates, which brings us to calculation of the triple integral in cylindrical coordinates.

Example 4.2. We find

$$I = \iiint_E z\, dV,$$

where E lies inside the bar bounded by the planes $x = 0$, $x = 1$, $y = 0$, and $y = 1$, assuming E is restricted from below and above by the planes

$z = x + y$ and $z = 10 - x - y$, respectively. Refer to Fig. 4.3. Clearly D_{xy} is the square $[0, 1] \times [0, 1]$. We have

$$I = \iint_{D_{xy}} \int_{x+y}^{10-x-y} z \, dz \, dA$$

$$= \int_0^1 \int_0^1 \frac{1}{2} \left[(10 - x - y)^2 - (x + y)^2 \right] dy \, dx$$

$$= \int_0^1 (45 - 10x) \, dx = 40. \qquad \square$$

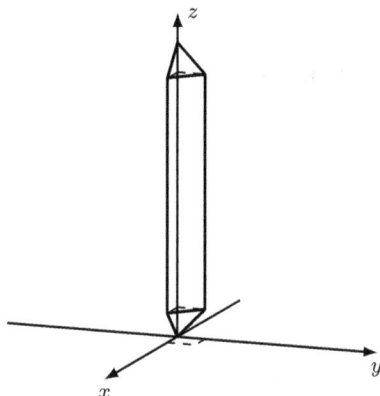

Fig. 4.3   Prism-shaped volume of Example 4.2.

Example 4.3. Figure 4.4 shows the plane $z = x$ cutting a wedge from the cylindrical region described by the inequalities $x^2 + y^2 \le 1$ and $z \ge 0$. The volume of the first-octant portion of the wedge is

$$V = \int_0^1 \int_0^{\sqrt{1-x^2}} \int_0^x dz \, dy \, dx = \frac{1}{3}. \qquad \square$$

Example 4.4. Suppose we wish to integrate a function $f(x, y, z)$ over the volume region E that lies between the paraboloid $z = (x^2 + y^2)/2$ and the sphere $x^2 + y^2 + z^2 = 3$. The intersection of these surfaces is determined by simultaneous solution of their two equations. Using the first equation to eliminate x and y from the second equation, we get $z^2 + 2z = 3$. Rejecting

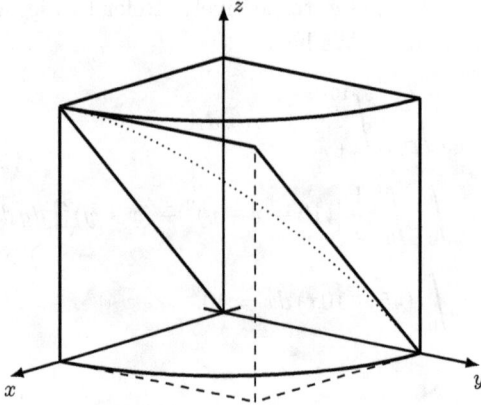

Fig. 4.4   Wedge cut from a cylinder (Example 4.3).

the solution $z = -3$, we accept $z = 1$ and find that the boundary of D_{xy} is
the circle $x^2 + y^2 = 2$. See Fig. 4.5. By Fubini's formula,

$$\iint_{D_{xy}} \int_{(x^2+y^2)/2}^{\sqrt{3-x^2-y^2}} f(x,y,z)\, dz\, dA_{xy}$$

$$= \int_{-\sqrt{2}}^{\sqrt{2}} \int_{-\sqrt{2-x^2}}^{\sqrt{2-x^2}} \int_{(x^2+y^2)/2}^{\sqrt{3-x^2-y^2}} f(x,y,z)\, dz\, dy\, dx. \qquad \square$$

More on Cylindrical Coordinates

In the last example both boundary surfaces were surfaces of revolution
with respect to the z-axis. This should prompt us to think of cylindrical
coordinates. Using polar coordinates in D_{xy} (which in this case should
perhaps be redenoted $D_{r\theta}$), we get a cylindrical coordinate system in which
we maintain the notation and relations of polar coordinates, i.e.,

$$x = r\cos\theta, \quad y = r\sin\theta, \quad z = z, \quad r^2 = x^2 + y^2, \quad dA_{r\theta} = r\, dr\, d\theta.$$

Example 4.5. The triple integral of Example 4.4 in cylindrical coordinates
is

$$\iint_{D_{r\theta}} \int_{r^2/2}^{\sqrt{3-r^2}} f(r\cos\theta, r\sin\theta, z)\, dz\, dA_{r\theta}$$

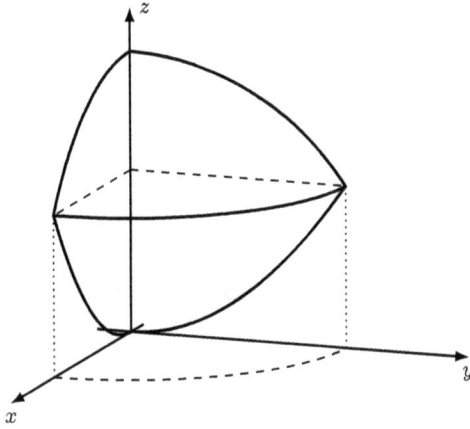

Fig. 4.5   First-octant portion of the volume region of Example 4.4.

$$= \int_0^{2\pi} \int_0^{\sqrt{2}} \int_{r^2/2}^{\sqrt{3-r^2}} f(r\cos\theta, r\sin\theta, z)\, dz\, r\, dr\, d\theta. \qquad \square$$

In cylindrical coordinates it makes sense to calculate triple integrals for the types of domains E suggested in Fig. 4.6.

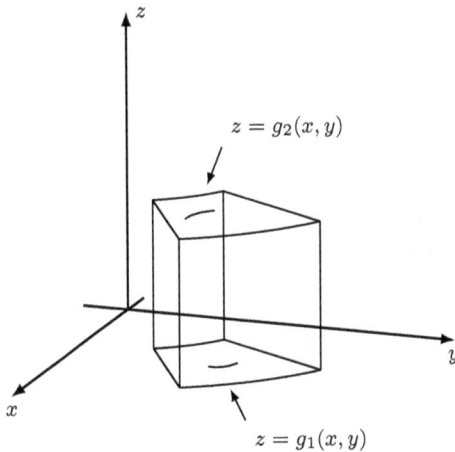

Fig. 4.6   Portion of a hollow cylinder bounded above and below by surfaces $z = g_2(x, y)$ and $z = g_1(x, y)$, respectively.

Example 4.6. Consider the solid bounded by the surfaces $x^2 + y^2 = 1$, $x^2 + y^2 = 4$, $\theta = 0$, $\theta = \pi/2$, $z = 1$, and $z = 3$. In this case $f(x, y, z) = 1$ and the volume is

$$V = \int_0^{\pi/2} \int_1^2 \int_1^3 dz\, r\, dr\, d\theta = \frac{\pi}{2}. \qquad \square$$

Two Other Domain Types

A domain of Type 1 has something like a natural axis, namely the z-axis. By rotating a Type 1 domain such that its new axis is along the x-axis or y-axis, we produce other types of domains for which integration formulas can be written simply through changes in notation.

Type 2: $dA_{zx} = dz\, dx$,

$$\iiint_E f(x, y, z)\, dV = \iint_{D_{zx}} \int_{g_1(x,z)}^{g_2(x,z)} f(x, y, z)\, dy\, dA_{zx}.$$

Type 3: $dA_{yz} = dy\, dz$,

$$\iiint_E f(x, y, z)\, dV = \iint_{D_{yz}} \int_{g_1(y,z)}^{g_2(y,z)} f(x, y, z)\, dx\, dA_{yz}.$$

Here D_{zx} and D_{yz} are the orthogonal projections of E onto the zx-plane and yz-plane, respectively, and straight lines parallel to the corresponding axes intersect the boundary of E at exactly two points for interior points of the projections of E onto the corresponding coordinate planes. Of course the functions g_1 and g_2, representing portions of the boundaries, should be at least piecewise continuous.

Example 4.7. Taking E as a domain in $\mathbb{R}^3$ in the first octant and bounded by the plane through the points $A = (2, 0, 0)$, $B = (0, 2, 0)$, $C = (0, 0, 2)$, we calculate

$$I = \iiint_E (x + y + z)\, dV.$$

First it is necessary to understand the shape of the domain. This may enable us to subdivide it into portions of the three types. See Fig. 4.7, left.

In fact E is of all three types, as any straight line parallel to any axis intersects its boundary exactly twice. Let us select Type 1 where D_{xy} is

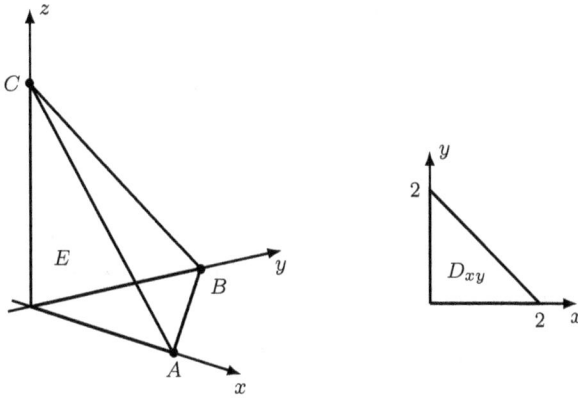

Fig. 4.7 Example 4.7.

the triangle ABO and $g_1(x, y) = 0$. From the equation of the plane ABC, which is $x + y + z = 2$, we get $z = g_2(x, y) = 2 - x - y$ and hence

$$I = \iint_{D_{xy}} \int_0^{2-x-y} (x + y + z) \, dz \, dA$$

$$= -\frac{1}{2} \iint_{D_{xy}} (x + y - 2)(x + y + 2) \, dA.$$

Calculation of the double integral is aided by Fig. 4.7, right. A 2D Type 1 approach yields

$$I = -\frac{1}{2} \int_0^2 \int_0^{2-x} (x + y - 2)(x + y + 2) \, dy \, dx$$

$$= \frac{1}{6} \int_0^2 (x^3 - 12x + 16) \, dx = 2.$$

Alternatively, from the analytic description of E,

$$E = \{(x, y, z) \colon 0 \le x \le 2,\ 0 \le y \le 2 - x,\ 0 \le z \le 2 - x - y\},$$

we could write

$$I = \int_0^2 \int_0^{2-x} \int_0^{2-x-y} (x + y + z) \, dz \, dy \, dx$$

and calculate step by step. $\qquad\qquad\qquad\qquad\qquad\qquad\qquad\qquad$ □

A Type 2 or Type 3 domain can be converted to a Type 1 domain through rotation of the coordinate system. This can be done formally by cyclic substitution. For example, if the domain is of Type 2 then the y-axis should be taken over to the z-axis through the change of variables

$$y \mapsto z, \qquad z \mapsto x, \qquad x \mapsto y.$$

Cyclic substitution is necessary if we wish to preserve the orientation of the coordinate system; otherwise, some expressions like the components of a cross product can change sign.

Example 4.8. Consider

$$\iiint_E y \, dV,$$

where E is bounded by the planes $x = 0$, $x = 1$, $z = 0$, $z = 1$, $y = x+z$, and $y = 10 - x - z$. The cyclic substitution mentioned above yields precisely the formulation of Example 4.2. $\qquad\qquad\qquad\qquad\square$

4.4 Change of Variables in Triple Integrals

The finite partitions of a segment $[a, b]$ for a definite integral are arbitrary. But the partitions of a body in $\mathbb{R}^3$ must be such that we can calculate the integral. We cannot employ cells for which we are unable to calculate the volumes, at least approximately. On the other hand, it is clear that partitions involving only rectangular parallelepipeds will not suffice for bodies of complex shape. In a calculus textbook we must restrict ourselves to partitions that can be dealt with using relatively simple methods. By staying with problems related to curvilinear coordinates in $\mathbb{R}^3$, we come to the basic change-of-variables problem for triple integrals.

We have seen one possible change of variables in triple integrals. This was for integrals in cylindrical coordinates for domains of Type 1. The corresponding iterated integrals are

$$\iiint_E f(x, y, z) \, dV = \int_{\theta_1}^{\theta_2} \int_{r_1(\theta)}^{r_2(\theta)} \int_{g_1(r,\theta)}^{g_2(r,\theta)} f(r \cos\theta, r \sin\theta, z) \, r \, dz \, dr \, d\theta.$$

As in polar coordinates, the area element is $dA = r \, dr \, d\theta$ and the volume element is $dV = r \, dr \, d\theta \, dz$. The scale factor r is required because the volume of a "rectangular parallelepiped" in the (r, θ, z) system depends on its position in $\mathbb{R}^3$ — namely, on its distance from the z-axis.

Coordinate translation and rotations do not affect the elementary volume dV. So we need not place a scale factor at the new $dV' = dx'\, dy'\, dz'$.

A general coordinate transformation in $\mathbb{R}^3$ is given by three equations

$$x = x(u, v, w), \qquad y = y(u, v, w), \qquad z = z(u, v, w), \qquad (4.7)$$

where (u, v, w) are called curvilinear coordinates (they may be rectilinear). The transformation $(x, y, z) \mapsto (u, v, w)$ should be one-to-one in the domain of integration, except maybe for some singular points or lines of measure zero like the z-axis of cylindrical coordinates or the origin of spherical coordinates. The transformation (4.7), also expressible as

$$\mathbf{r} = \mathbf{r}(u, v, w),$$

should be continuously differentiable with a continuously differentiable inverse. To find dV we need the tools of calculus.

Rather than trying to substitute (4.7) directly into the triple integral, we shall go back to approximating the mass of E using rectangular parallelepiped cells of partitions of E in the uvw coordinates. To establish the shape of E under the transformation, it is useful to note that the boundary of E in the xyz system is mapped to the boundary of E in the uvw system.

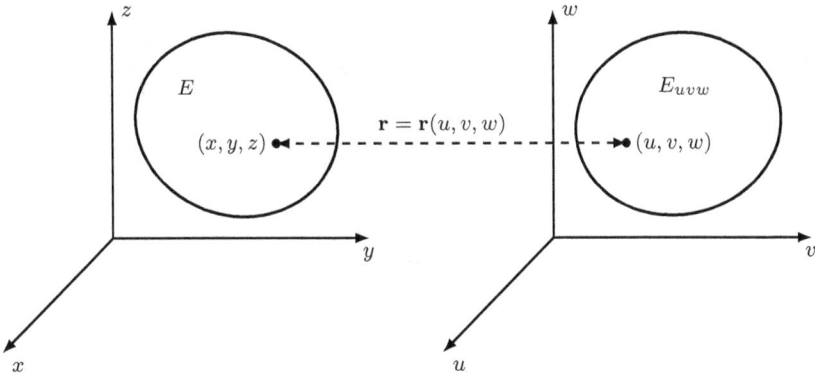

Fig. 4.8 $E \subset \mathbb{R}^3$ in Cartesian xyz and curvilinear uvw coordinates.

In Fig. 4.8 we draw the uvw coordinate system as if it were Cartesian. The straight lines parallel to the u-, v-, and w-axes are curves in the xyz system, and are called curvilinear coordinate lines. The images of the planes parallel to the uvw coordinate planes are curvilinear coordinate surfaces in the xyz system. These surfaces in cylindrical coordinates are illustrated in Figure 4.9.

(a)   (b)   (c)

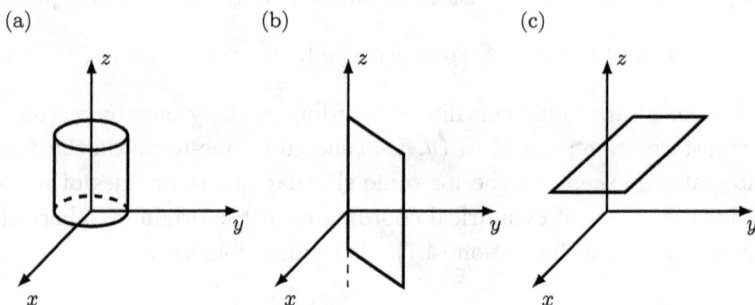

Fig. 4.9   A cylinder $r =$ constant, a half-plane $\theta =$ constant, and a plane $z =$ constant.

The partitions we shall use in the uvw coordinates, depicted as orthogonal, are rectangular (Fig. 4.10). But the volume of a cell with sides

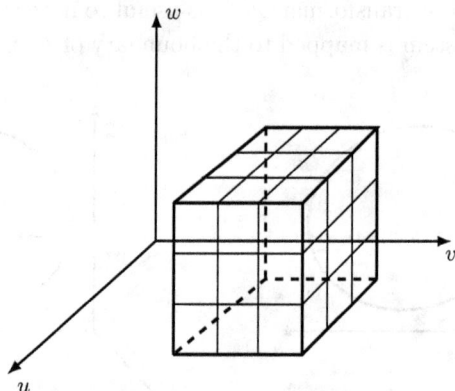

Fig. 4.10   Rectangular partition in uvw coordinates.

$\Delta u, \Delta v, \Delta w$ is not the product $\Delta u\, \Delta v\, \Delta w$ — even if the system is actually orthogonal. To find the mass of such a cell using the formula $\Delta m = \rho\, \Delta V$, we must use the volume ΔV of the cell in xyz coordinates whose image under transformation (4.7) is a rectangular parallelepiped with sides $\Delta u, \Delta v, \Delta w$ in uvw coordinates (Fig. 4.11).

So we find the mass of E with density $f(x(u, v, w), y(u, v, w), z(u, v, w))$. To avoid discussing what happens at the boundary of E in Riemann sums,

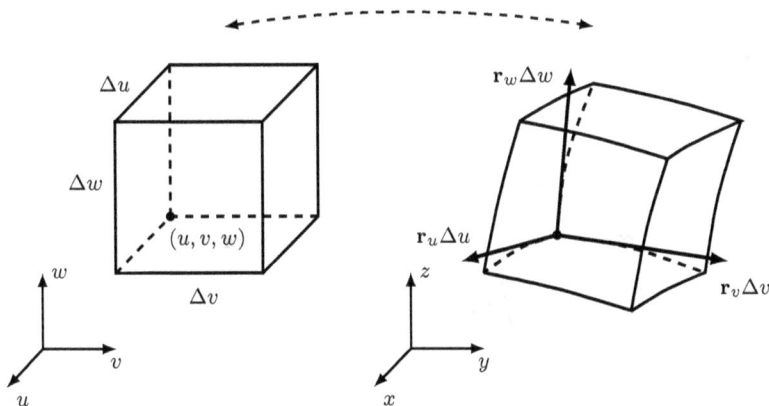

Fig. 4.11 *Left*, cell with sides $\Delta u, \Delta v, \Delta w$ in the uvw coordinates. *Right*, its preimage, a curvilinear "parallelepiped" in the xyz coordinates. Tangent vectors $\mathbf{r}_u(u,v,w)\Delta u$, $\mathbf{r}_v(u,v,w)\Delta v$, and $\mathbf{r}_w(u,v,w)\Delta w$ define the sides of a parallelepiped approximating the preimage cell.

let us extend f by zero outside of E (here we use the fact that the sum of terms with "parallelograms" covering the boundary tends to zero in the limit). We start with a rectangular partition in the uvw system (Fig. 4.10). An elementary parallelepiped we again tag with indices ijk. Its image in the xyz system is a curvilinear "parallelepiped" whose sides become "more and more straight" as $\Delta u_i, \Delta v_j, \Delta w_k$, the lengths of the sides of the ijk-parallelepiped in the uvw system, tend to zero. These sides are approximated by the tangents whose tangent vectors are

$$\mathbf{r}_u(u_i, v_j, w_k)\,\Delta u_i, \qquad \mathbf{r}_v(u_i, v_j, w_k)\,\Delta v_j, \qquad \mathbf{r}_w(u_i, v_j, w_k)\,\Delta w_k.$$

Refer again to Fig. 4.11.

The volume of the "tangent" parallelepiped is

$$\Delta V_{ijk} = \left| \mathbf{r}_u(u_i, v_j, w_k) \times \mathbf{r}_v(u_i, v_j, w_k) \cdot \mathbf{r}_w(u_i, v_j, w_k) \right| \Delta u_i \, \Delta v_j \, \Delta w_k.$$

The volume of the ijk curvilinear parallelepiped is $\Delta V_{ijk} + \varepsilon_{ijk}$. It can be proved that for a smooth transformation (4.7) of uvw to xyz coordinates, taking the limit of the corresponding Riemann sums, the terms having factor ε_{ijk} approach zero. So we can approximate the mass of E by

$$M \approx \sum_{i,j,k} f(x(u_i, v_j, w_k), y(u_i, v_j, w_k), z(u_i, v_j, w_k))$$

$$\times \left| \mathbf{r}_u(u_i, v_j, w_k) \times \mathbf{r}_v(u_i, v_j, w_k) \cdot \mathbf{r}_w(u_i, v_j, w_k) \right| \Delta u_i \, \Delta v_j \, \Delta w_k.$$

The expression on the right is a Riemann sum for the integral which gives the mass of M as

$$M = \iiint_{E_{uvw}} f(x(u,v,w), y(u,v,w), z(u,v,w))$$
$$\times |\mathbf{r}_u(u,v,w) \times \mathbf{r}_v(u,v,w) \cdot \mathbf{r}_w(u,v,w)| \, du \, dv \, dw.$$

Clearly we can repeat this when f is not a mass density function. So

$$\iiint_E f(x,y,z) \, dV = \iiint_{E_{uvw}} f(x(u,v,w), y(u,v,w), z(u,v,w))$$
$$\times |J(u,v,w)| \, du \, dv \, dw,$$

where

$$J(u,v,w) = \mathbf{r}_u \times \mathbf{r}_v \cdot \mathbf{r}_w = \begin{vmatrix} x_u & y_u & z_u \\ x_v & y_v & z_v \\ x_w & y_w & z_w \end{vmatrix}.$$

is called the *Jacobian* of the transformation $\mathbf{r} = \mathbf{r}(u,v,w)$. The value $|J|$ gives the ratio between the values of corresponding infinitesimal volumes in E_{uvw} and E.

Example 4.9. The Jacobian of the transformation between Cartesian and cylindrical coordinates is

$$J = \begin{vmatrix} x_r & y_r & z_r \\ x_\theta & y_\theta & z_\theta \\ x_z & y_z & z_z \end{vmatrix} = \begin{vmatrix} \cos\theta & \sin\theta & 0 \\ -r\sin\theta & r\cos\theta & 0 \\ 0 & 0 & 1 \end{vmatrix} = r(\cos^2\theta + \sin^2\theta) = r,$$

hence

$$\iiint_E f(x,y,z) \, dV = \iiint_{E_{r\theta z}} f(r\cos\theta, r\sin\theta, z) \, dV_{r\theta z}$$

where $dV_{r\theta z} = r \, dr \, d\theta \, dz$. □

Example 4.10. The Jacobian of the transformation between Cartesian and spherical coordinates is

$$J = \begin{vmatrix} x_\rho & y_\rho & z_\rho \\ x_\phi & y_\phi & z_\phi \\ x_\theta & y_\theta & z_\theta \end{vmatrix} = \rho^2 \sin\phi,$$

hence

$$\iiint_E f(x, y, z)\, dV = \iiint_{E_{\rho\phi\theta}} f(\rho \sin\phi\cos\theta, \rho\sin\phi\sin\theta, \rho\cos\phi)\, dV_{\rho\phi\theta}$$

where

$$dV_{\rho\phi\theta} = \rho^2 \sin\phi\, d\rho\, d\phi\, d\theta.$$

When spherical coordinates are used and a triple integral is done as an iterated integral, the inner-most integration is usually over ρ and the outer-most over θ. The triple integral in spherical coordinates is used most frequently when E is restricted by inequalities of the form $\rho_1 \le \rho \le \rho_2$, $\phi_1 \le \phi \le \phi_2$, $\theta_1 \le \theta \le \theta_2$, i.e., when E is a rectangle in the $\rho\phi\theta$ system. For instance, the region described by $a \le \rho \le b$, $0 \le \phi \le \pi$, $0 \le \theta \le 2\pi$ is called a *spherical shell*. If the mass density of such a shell varies as the inverse square of the radial coordinate, i.e., as $k\rho^{-2}$, then the total mass of the shell is

$$\int_0^{2\pi} \int_0^{\pi} \int_a^b k\rho^{-2}\rho^2 \sin\phi\, d\rho\, d\phi\, d\theta = 4\pi k(b-a). \qquad \square$$

4.5 Applications of Triple Integrals

All the continuum theories (the theories of elasticity, plasticity, viscoelasticity, electromagnetism, general relativity, mathematical biology, etc.) contain good portions of material based on triple integrals. We mention only a few simple applications.

We start with some facts already established. The volume of E is

$$V(E) = \iiint_E dV.$$

If $\rho = \rho(x, y, z)$ is the mass density function on E, then the mass M of E is

$$M = \iiint_E \rho(x, y, z)\, dV.$$

Now we come to the center of mass. Recall that the mass center of a system of particles m_k with coordinates (x_k, y_k, z_k) is

$$x_c = \frac{\sum_{k=1}^{n} m_k x_k}{\sum_{k=1}^{n} m_k}, \qquad y_c = \frac{\sum_{k=1}^{n} m_k y_k}{\sum_{k=1}^{n} m_k}, \qquad z_c = \frac{\sum_{k=1}^{n} m_k z_k}{\sum_{k=1}^{n} m_k}.$$

Let us write out an approximation for x_c based on a rectangular partition of E:

$$x_c \approx \frac{\displaystyle\sum_{i,j,k} \rho(x_i, y_j, z_k) x_i \, \Delta V_{ijk}}{\displaystyle\sum_{i,j,k} \rho(x_i, y_j, z_k) \, \Delta V_{ijk}}.$$

In the limit as

$$\max_{i,j,k}(\Delta x_i^2 + \Delta y_j^2 + \Delta z_k^2)^{1/2} \to 0$$

we get

$$x_c = \frac{\displaystyle\iiint_E x\rho(x, y, z) \, dV}{\displaystyle\iiint_E \rho(x, y, z) \, dV}.$$

Analogously

$$y_c = \frac{\displaystyle\iiint_E y\rho(x, y, z) \, dV}{\displaystyle\iiint_E \rho(x, y, z) \, dV}, \qquad z_c = \frac{\displaystyle\iiint_E z\rho(x, y, z) \, dV}{\displaystyle\iiint_E \rho(x, y, z) \, dV}.$$

The expressions in the numerators of the last three formulas are called the *moments* with respect to the corresponding coordinate planes. They are denoted

$$M_{yz} = \iiint_E x\rho(x, y, z) \, dV,$$

$$M_{zx} = \iiint_E y\rho(x, y, z) \, dV,$$

$$M_{xy} = \iiint_E z\rho(x, y, z) \, dV.$$

The subscripts refer to the corresponding coordinate planes.

In a similar manner we have the second-order moments that constitute the inertia tensor of E. The inertial moments about the Cartesian coordinate lines,

$$I_x = \iiint_E (y^2 + z^2)\rho(x, y, z) \, dV,$$

$$I_y = \iiint_E (x^2 + z^2) \rho(x, y, z) \, dV,$$

$$I_z = \iiint_E (x^2 + y^2) \rho(x, y, z) \, dV,$$

are used to study the rotation of a solid E under force moments. More complete discussion of the tensor can be found in textbooks on classical mechanics.

4.6 Problems

Evaluation of Triple Integrals

4.1 Use the definition of the triple integral to evaluate

(a)

$$I = \iiint_E x \, dz \, dy \, dx \quad \text{where} \quad E = [0, 1]^3,$$

(b)

$$I = \iiint_E xyz \, dz \, dy \, dx \quad \text{where} \quad E = [0, 1] \times [0, 2] \times [0, 3].$$

4.2 Verify:

(a)

$$\int_0^1 \int_0^1 \int_0^1 xyz(x + y + z) \, dz \, dy \, dx = \frac{1}{4}$$

(b)

$$\int_0^1 \int_0^1 \int_0^1 e^{x+y+z} \, dz \, dy \, dx = (e - 1)^3$$

(c)

$$\int_0^1 \int_0^\pi \int_0^{\sqrt{\cos\theta}} r^3 z \, dr \, d\theta \, dz = \frac{\pi}{16}$$

(d)

$$\int_0^1 \int_0^\pi \int_1^{\cos\theta+z} r^2 \sin\theta \, dr \, d\theta \, dz = -\frac{1}{6}$$

(e)

$$\int_0^1 \int_0^{2\pi} \int_1^{z\cos\theta} r^3 \sin\theta \, dr \, d\theta \, dz = 0$$

4.3 Find

(a)
$$I_a = \int_0^1 \int_1^2 \int_2^3 x^2 y z^3 \, dz \, dy \, dx,$$

(b)
$$I_b = \int_0^1 \int_1^2 \int_2^3 e^x y z^3 \, dz \, dy \, dx,$$

(c)
$$I_c = \int_0^1 \int_1^2 \int_0^\pi e^x y \sin z \, dz \, dy \, dx,$$

(d)
$$I_d = \int_0^1 \int_0^\pi \int_0^\pi e^x \cos y \sin z \, dz \, dy \, dx,$$

(e)
$$I_e = \int_0^a \int_0^x \int_0^y xyz \, dz \, dy \, dx.$$

4.4 Evaluate the integral of

(a) the function $f(x, y, z) = xyz$ over the volume region bounded by the planes $x = 0$, $y = 0$, $z = 0$, and $x + y + z = 1$;

(b) the function $f(x, y, z) = z$ over the volume region bounded by the sphere $\rho = a$ for $z \geq 0$.

4.5 After changing variables to cylindrical coordinates, evaluate

$$I = \int_{-1}^1 \int_{-\sqrt{1-x^2}}^{\sqrt{1-x^2}} \int_0^{\sqrt{x^2+y^2}} z^2 \, dz \, dy \, dx.$$

Calculation of Volume

4.6 Find the volumes of the regions bounded by the given surfaces:

(a) $y = x^2$, $y = 1$, $z = 0$, and $z = 4 - x - y$
(b) $z = 1 - x^2 - y^2$ and $z = 0$
(c) $z = x^2 + y^2$, $x^2 + y^2 = 1$, and $z = 0$

4.7 Find the volume of the region described by the inequalities $0 \leq z \leq xy$ and $y \leq x \leq 1$.

4.8 Find the volumes of the regions bounded by the given surfaces:

(a) $x + y \leq 1$, $z \leq x^2 + y^2$, and $x, y, z \geq 0$
(b) the sphere $\rho = a$ and the half-cone $\phi = \alpha$

(c) the sphere $\rho = a$ and the two half-planes $\theta = 0$ and $\theta = \theta_0$

(d) the planes $x + y + z = 1$, $x = 0$, $y = 0$, $z = 0$

(e) the hyperboloid $x^2/a^2 + y^2/b^2 - z^2/c^2 = 1$ and the planes $z = \pm h$

(f) the half-cone $\phi = \pi/4$ and the plane $z = z_0 > 0$

(g) the paraboloid $z = x^2 + y^2$, the parabolic cylinder $y = x^2$, and the planes $z = 0$ and $y = 1$

4.9 Find the volume of the ellipsoid (1.33).

4.10 Find the volume of the solid formed by rotating about the z-axis the planar region described by

(a) $y = 0$, $0 \leq x \leq a$, and $0 \leq z \leq \sqrt{x}$;

(b) $y = 0$, $0 \leq x \leq a$, and $\sqrt{x} \leq z \leq \sqrt{a}$.

4.11 Develop a formula for the volume of a solid of revolution obtained by revolving the curve $y = f(x)$, $a \leq x \leq b$, about the x-axis. Apply it to the parabola $y^2 = 4ax$ for $0 \leq x \leq h$.

4.12 Calculate the volume of the tetrahedron formed by the coordinate planes and the plane tangent to the surface $xyz = a^3$ at a point (x_0, y_0, z_0) in the first octant. Show that this volume does not depend on the choice of the point (x_0, y_0, z_0) on the surface.

4.13 Find the volume of the portion of the cylinder $x^2 + y^2 = 1$ that is cut out by the cylinder $x^2 + z^2 = 1$.

Triple Integration and the Nabla Operator

4.14 Calculate

$$I = \iiint_G \nabla \cdot \mathbf{A} \, dV$$

where

(a) $\mathbf{A}(x, y, z) = x^2 \mathbf{i} + y^2 \mathbf{j}$ and $G = [0, 1] \times [0, 2] \times [0, 3]$;

(b) $\mathbf{A}(x, y, z) = x \mathbf{i} + z^2 \mathbf{k}$ and G is the first-octant region enclosed by the plane $x + y + z = 1$ and the three coordinate planes.

4.15 Calculate

$$I = \iiint_G |\nabla f|^2 \, dV$$

where

(a) $f(x, y, z) = x \cos y$ and $G = [0, 1] \times [-1, 1] \times [-2, 2]$,

(b) $f(x, y, z) = x + 3y + z$ and G is the region $x^2 + y^2 + z^2 \leq 1$,

(c) $f(x, y, z) = -x + 4z$ and G is the region $x^2 + y^2 \leq 1$, $0 \leq z \leq 1$.

4.16 Calculate

$$I = \iiint_G \nabla^2 f \, dV$$

if

(a) $f(x, y, z) = x^2 y + y^2 z + z^2 x$ and $G = [0, 1] \times [-1, 1] \times [-2, 2]$,

(b) $f(x, y, z) = x^2 + y^2 + z^2$ and G is the region $x^2 + y^2 + z^2 \le 1$,

(c) $f(x, y, z) = 4z^3$ and G is the region $x^2 + y^2 \le 1$, $0 \le z \le 1$.

Other Applications

4.17 The mass density inside a ball of radius a is $k\rho^2$ where ρ is the radial distance from the center of the ball. Find the total mass of the ball.

4.18 The mass density inside the rectangular parallelepiped $[0, a] \times [0, b] \times [0, c]$ is $\rho(x, y, z) = (xy)^2 z$. Find the center of mass.

4.19 The mass density inside the cylindrical volume $r \le a$, $0 \le z \le h$ is given by $\rho(r, \theta, z) = z^2$. Find the total mass of the cylinder and the z-coordinate of the center of mass.

4.20 A solid of uniform unit density takes the form of the tetrahedron bounded by the planes $x + y + z = 1$, $x = 0$, $y = 0$, and $z = 0$. Find the center of mass.

4.21 The average or mean value of a function $f(x, y, z)$ over a volume region R is given by

$$\overline{f} = \frac{1}{V} \iiint_R f(x, y, z) \, dx \, dy \, dz$$

where V is the volume of R. Find the average value of $f(x, y, z) = xyz(x^2 + y^2 + z^2)$ over the cubical region $R = [0, 2]^3$.

A Useful Fact

4.22 Let $f(x, y, z)$ be a continuous function satisfying

$$\iiint_E f(x, y, z) \, dV = 0$$

for every region E. Show that $f(x, y, z) \equiv 0$.

Chapter 5

Line Integrals

The line integrals can be directly represented through ordinary definite integrals. There are different types of line integrals along a curve L. One type,

$$\int_L F(x, y, z) \, ds,$$

where s is the length parameter, provides various characteristics of a curve such as its length or the mass of a cable along L. Another type,

$$\int_L \mathbf{F} \cdot d\mathbf{r},$$

yields the work of a force field along L. It participates in Green's and Stokes' theorems, and in the derivation of boundary conditions for the problems of mathematical physics. For a plane curve $y = f(x)$ on $[a, b]$, an example of the first type of line integral is

$$\int_L F(x, y) \, ds = \int_a^b f(x, y(x)) \left\{ 1 + [f'(x)]^2 \right\}^{1/2} dx$$

where $ds = \left\{ 1 + [f'(x)]^2 \right\}^{1/2} dx$ is the differential of the length parameter given in the x variable. With $\mathbf{F} = P\mathbf{i} + Q\mathbf{j}$ the second type looks like

$$\int_L \mathbf{F} \cdot d\mathbf{r} = \int_L P(x, y) \, dx + Q(x, y) \, dy.$$

As with all mathematical objects, line integrals are extended to abstract fields of mathematics.

5.1 Prelude: The Length of a Space Curve

It is not a simple problem to find the exact value of curve length. The oldest nontrivial problem of this type is finding the circumference of a circle. In ancient Greece they bounded this length by calculating the perimeters of inscribed and circumscribed regular polygons (Fig. 5.1). The idea of seeking

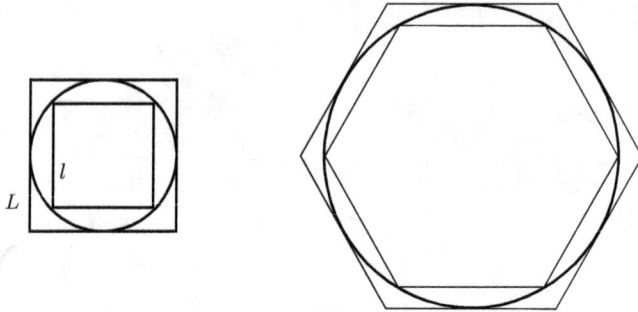

Fig. 5.1 Bounding the circumference of a circle. *Left*, inscribed and circumscribed squares. The circumference of the circle lies between the numbers $4l$ and $4L$. *Right*, the bounds can be tightened by increasing the number of sides of the inscribed and circumscribed polygons.

bounds and then applying a limit passage is a central idea in calculus. It serves to define derivatives, definite integrals, etc. Unfortunately for an arbitrary curve we cannot find approximating polygons whose perimeters give upper and lower bounds for the length. We shall not go to the definition of "rectifiable curves" for which length can be defined, opting instead for a more evident approach suitable for smooth curves. It is based on the idea of approximating the curve length with a few tangent segments.

Let us start with a parametric description of a curve L:

$$\mathbf{r} = x(t)\,\mathbf{i} + y(t)\,\mathbf{j} + z(t)\,\mathbf{k}.$$

Assume $t \in [a, b]$ and that the functions x, y, z are piecewise continuously differentiable with bounded first derivatives. First we partition $[a, b]$ into n subintervals by points $a = t_1 < t_2 < \cdots < t_{n+1} = b$ with small $\Delta t_k = t_{k+1} - t_k$. This partitions L into n small sections ΔL_k as shown in Fig. 5.2. Now we approximate each section ΔL_k by the segment connecting the endpoints of ΔL_k. Writing $\Delta t_k = t_{k+1} - t_k$, we introduce a small vector

$$\Delta \mathbf{r}_k = \mathbf{r}(t_{k+1}) - \mathbf{r}(t_k)$$

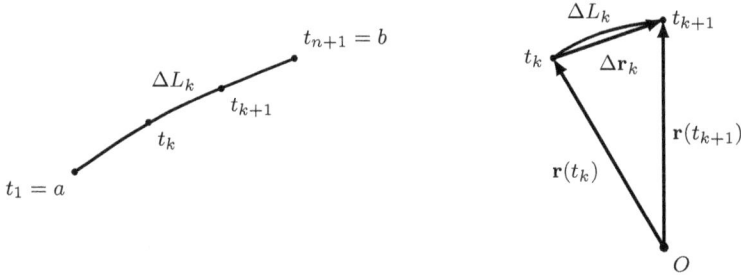

Fig. 5.2 Calculating the length of a space curve. *Left*, a partition for L; *right*, a more detailed view of ΔL_k.

$$= \mathbf{r}(t_k + \Delta t_k) - \mathbf{r}(t_k)$$

$$= \Delta x_k(\Delta t_k)\,\mathbf{i} + \Delta y_k(\Delta t_k)\,\mathbf{j} + \Delta z_k(\Delta t_k)\,\mathbf{k}.$$

A broken line approximation to the length $s(L)$ of L is

$$s_n(L) = \sum_{k=1}^{n} |\Delta \mathbf{r}_k|.$$

But now we introduce another approximation with use of the first differentials, the components of $\Delta \mathbf{r}_k$, which are

$$\Delta x_k = dx_k + \varepsilon_{k1}\,\Delta t_k = x'(t_k)\,\Delta t_k + \varepsilon_{k1}\,\Delta t_k, \quad \text{etc.,}$$

where $\varepsilon_{ki} \to 0$ as $\Delta t_k \to 0$. So

$$|\Delta \mathbf{r}_k| = \{[x'(t_k)]^2 + [y'(t_k)]^2 + [z'(t_k)]^2\}^{1/2}\,\Delta t_k + \varepsilon_k\,\Delta t_k$$

with $\varepsilon_k \to 0$ as $\Delta t_k \to 0$. We have

$$s_n(L) = \sum_{k=1}^{n}\{[x'(t_k)]^2 + [y'(t_k)]^2 + [z'(t_k)]^2\}^{1/2}\,\Delta t_k + \sum_{k=1}^{n}\varepsilon_k\,\Delta t_k. \quad (5.1)$$

As $x'(t), y'(t), z'(t)$ are piecewise continuous on $[a, b]$, it can be proved that $\max_k |\varepsilon_k| \to 0$ when $\max_k \Delta t_k \to 0$ and

$$\left|\sum_{k=1}^{n}\varepsilon_k\,\Delta t_k\right| \le \max_k |\varepsilon_k| \sum_{k=1}^{n}\Delta t_k = \max_k |\varepsilon_k| \cdot (b - a) \to 0$$

when $\max_k \Delta t_k \to 0$. The first sum in (5.1) is a Riemann sum for the integral

$$\int_a^b \{[x'(t)]^2 + [y'(t)]^2 + [z'(t)]^2\}^{1/2}\,dt.$$

Since the integrand is piecewise continuous and bounded on $[a, b]$, the integral exists and

$$s(L) = \lim_{n \to \infty} s_n(L)$$

$$= \int_a^b \{[x'(t)]^2 + [y'(t)]^2 + [z'(t)]^2\}^{1/2} \, dt. \tag{5.2}$$

Example 5.1. Let us check (5.2) for a circle of radius r. In polar coordinates, the circle is represented as

$$x = r\cos t, \qquad y = r\sin t, \qquad t \in [0, 2\pi).$$

Taking $z(t) = 0$ in (5.2) we get

$$s(L) = \int_0^{2\pi} (r^2 \sin^2 t + r^2 \cos^2 t)^{1/2} \, dt = 2\pi r. \qquad \square$$

Example 5.2. The position vector

$$\mathbf{r}(t) = r(t - \sin t)\,\mathbf{i} + r(1 - \cos t)\,\mathbf{j} \qquad t \in [0, 2\pi]$$

traces out one arc of a cycloid in the xy-plane. Equation (5.2) gives

$$s(L) = \int_0^{2\pi} [(r - r\cos t)^2 + (r\sin t)^2]^{1/2} \, dt = 8r. \qquad \square$$

When, instead of a general parameter t, we use the length s of the curve from a fixed point of L to the point corresponding to the value of the parameter s, we call s the *length parameter*. This choice has some advantages, in particular that $|\mathbf{r}'(s)| = 1$.

5.2 Introducing the Line Integral

We start by calculating the mass of a cable that lies along a contour $\mathbf{r} = \mathbf{r}(t)$. Let the line density $\gamma = \gamma(t)$ be piecewise continuous on $[a, b]$. Partitioning $[a, b]$ by points $a = t_1, t_2, \ldots, t_{n+1} = b$, we approximate the cable mass:

$$M \approx \sum_{k=1}^n \gamma(t_k) \, \Delta s_k, \tag{5.3}$$

where Δs_k is the length of that portion of the curve corresponding to the segment $[t_k, t_{k+1}]$. As the principal part of Δs_k is given by the expression

$$\{[x'(t_k)]^2 + [y'(t_k)]^2 + [z'(t_k)]^2\}^{1/2} \, \Delta t_k,$$

for small Δt_k we have

$$M \approx \sum_{k=1}^{n} \gamma(t_k) \left\{ [x'(t_k)]^2 + [y'(t_k)]^2 + [z'(t_k)]^2 \right\}^{1/2} \Delta t_k.$$

The right-hand side is a Riemann sum for the integral

$$\int_a^b \gamma(t) \left\{ [x'(t)]^2 + [y'(t)]^2 + [z'(t)]^2 \right\}^{1/2} dt.$$

Since t_k defines the point $(x(t_k), y(t_k), z(t_k))$ on the cable, we can also write $\gamma(t_k)$ as $\gamma(x(t_k), y(t_k), z(t_k))$ and get

$$M \approx \sum_{k=1}^{n} \gamma(x(t_k), y(t_k), z(t_k)) \left\{ [x'(t_k)]^2 + [y'(t_k)]^2 + [z'(t_k)]^2 \right\}^{1/2} \Delta t_k.$$

If γ, x, y, z are bounded piecewise continuous functions of t on $[a, b]$, the Riemann sum on the right converges to an integral giving the total mass of the cable:

$$M = \int_a^b \gamma(x(t), y(t), z(t)) \left\{ [x'(t)]^2 + [y'(t)]^2 + [z'(t)]^2 \right\}^{1/2} dt. \qquad (5.4)$$

Equation (5.3) prompts us to denote

$$M = \int_L \gamma(x, y, z) \, ds. \qquad (5.5)$$

Note that if $\gamma(t) = 1$ the result is the length of the curve $s(L) = \int_L ds$.

Example 5.3. The mass density of a helical spring is $\gamma(x, y, z) = kz$ where $k > 0$ is a constant. The spring is described parametrically by $x = a \cos t$, $y = a \sin t$, $z = t$ for $0 \le t \le 2\pi n$ where n is an integer. By (5.4) we have

$$M = \int_0^{2\pi n} kt \, (a^2 \sin^2 t + a^2 \cos^2 t + 1)^{1/2} \, dt$$

$$= k(a^2 + 1)^{1/2} \int_0^{2\pi n} t \, dt = 2\pi^2 n^2 k (a^2 + 1)^{1/2}. \qquad \square$$

We can replace the mass density γ by an abstract piecewise continuous function f given over L and repeat the above steps to get the definition of the abstract line integral:

$$\int_L f(x, y, z) \, ds = \int_a^b f(x(t), y(t), z(t)) \left\{ [x'(t)]^2 + [y'(t)]^2 + [z'(t)]^2 \right\}^{1/2} dt.$$

$$(5.6)$$

When interpreting the formal notation on the left-hand side, it is important to remember that $a < b$. The reader should remember that, as for all types of integrals in this book, $\int_L f(x, y, z)\, ds$ is only a formal notation which should be deciphered with an appropriate formula, here by means of (5.6).

Example 5.4. Let us calculate

$$I = \int_L (x + y + z)\, ds$$

over L given by the formulas $x = 1$, $y = 2$, $z = t$ for $t \in [1, 2]$. See Fig. 5.3.

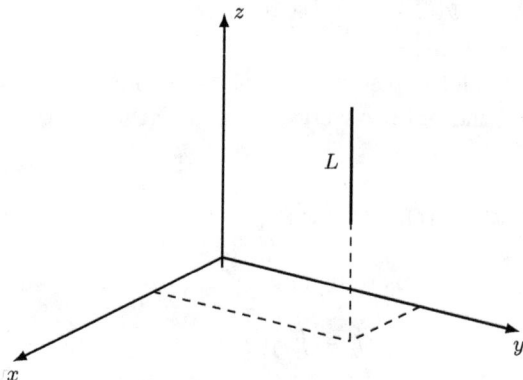

Fig. 5.3   Curve for Example 5.4.

Here

$$\{[x'(t)]^2 + [y'(t)]^2 + [z'(t)]^2\}^{1/2} = (0 + 0 + 1)^{1/2} = 1$$

so $ds = dt$. Thus

$$I = \int_1^2 (1 + 2 + t)\, dt = 5.5. \qquad \square$$

The role of parameter t is frequently played by x, y or z. For example, if a plane curve L is given by $y = g(x)$ for $x \in [a, b]$, then we can take $t = x$. Hence the parametric representation of L becomes $x = x$, $y = g(x)$ and we have

$$\{[x'(t)]^2 + [y'(t)]^2 + [z'(t)]^2\}^{1/2} = \{1 + [g'(x)]^2\}^{1/2}$$

so that

$$\int_L f(x,y)\, ds = \int_a^b f(x, g(x))\, \{1 + [g'(x)]^2\}^{1/2}\, dx. \tag{5.7}$$

The main properties of the line integral are shared with the ordinary definite integral. For any constant c we have

$$\int_L [cf_1(x,y,z) + f_2(x,y,z)]\, ds = c \int_L f_1(x,y,z)\, ds + \int_L f_2(x,y,z)\, ds.$$

If L is a union of two curves L_1 and L_2 that can intersect in a few points, then

$$\int_L f(x,y,z)\, ds = \int_{L_1} f(x,y,z)\, ds + \int_{L_2} f(x,y,z)\, ds.$$

There are also some useful inequality properties. If $f(x,y,z) \geq 0$ on L, then

$$\int_L f(x,y,z)\, ds \geq 0.$$

If $f(x,y,z) \leq g(x,y,z)$ on L, then

$$\int_L f(x,y,z)\, ds \leq \int_L g(x,y,z)\, ds.$$

If $m \leq f(x,y,z) \leq M$ on L, then

$$m\, s(L) \leq \int_L f(x,y,z)\, ds \leq M\, s(L).$$

Example 5.5. Suppose L can be decomposed into two curves L_1 and L_2 both lying in the xy-plane. Curve L_1 is the segment from $(1, -1)$ to $(1, 0)$, and L_2 is first quadrant portion of the unit circle. For L_1 a natural parameterization is $x = x(y) = 1$, $y = y$. Here $\{[x'(y)]^2 + 1\}^{1/2} = 1$ and we have

$$\int_{L_1} f(x,y)\, ds = \int_{-1}^0 f(1, y)\, dy.$$

For L_2 we use the polar coordinate representation $x = \cos\theta$, $y = \sin\theta$ with $\theta \in [0, \pi/2]$. As $[x'(\theta)]^2 + [y'(\theta)]^2 = 1$, we have

$$\int_{L_2} f(x,y)\, ds = \int_0^{\pi/2} f(\cos\theta, \sin\theta)\, d\theta.$$

Summation gives the line integral of $f(x,y)$ over L. $\qquad\qquad\square$

As additional applications of line integration, we could mention calculation of the cable moments

$$M_{yz} = \int_L x\,\gamma(x,y,z)\,ds,$$

$$M_{xz} = \int_L y\,\gamma(x,y,z)\,ds,$$

$$M_{xy} = \int_L z\,\gamma(x,y,z)\,ds,$$

and the center of mass coordinates

$$x_c = M_{yz}/M, \qquad y_c = M_{xz}/M, \qquad z_c = M_{xy}/M.$$

5.3   Another Type of Line Integral

Some physical problems involve finding the work done by a force acting along a curve, or the work of some field of forces along a curve. The form of the resulting integral differs from that of the line integral considered in Section 5.2 (although it can be reduced to this form).

Recall that the work done by a force $\mathbf{F} \in \mathbb{R}^3$ in moving a particle along a directed segment represented by vector $\mathbf{d}$ is $\mathbf{F} \cdot \mathbf{d}$. To approximate the work done by a force field $\mathbf{F}$ along a curve L in $\mathbb{R}^3$, we return to the idea of a broken line approximation:

$$W_n \approx \sum_{k=1}^{n} \mathbf{F}(x(t_k), y(t_k), z(t_k)) \cdot \Delta \mathbf{r}_k,$$

where the summand serves to approximate the work done by $\mathbf{F}$ over ΔL_k. Writing $\mathbf{F} = P\mathbf{i} + Q\mathbf{j} + R\mathbf{k}$ we also have

$$W_n \approx \sum_{k=1}^{n} [P(x(t_k), y(t_k), z(t_k))\,\Delta x_k$$

$$+ Q(x(t_k), y(t_k), z(t_k))\,\Delta y_k + R(x(t_k), y(t_k), z(t_k))\,\Delta z_k].$$

Using the differential approximations for Δx_k, Δy_k, Δz_k, we get

$$\sum_{k=1}^{n} \big[P(x(t_k), y(t_k), z(t_k))\,x'(t_k) + Q(x(t_k), y(t_k), z(t_k))\,y'(t_k)$$

$$+ R(x(t_k), y(t_k), z(t_k))\,z'(t_k) \big]\,\Delta t_k + \sum_{k=1}^{n} \tilde{\varepsilon}_k\,\Delta t_k$$

with $\tilde{\varepsilon}_k \to 0$ as $\Delta t_k \to 0$. We recognize the first term as the summation of three Riemann sums for ordinary integrals. Further, if P, Q, R are bounded piecewise continuous functions of (x, y, z) and hence of t, and if x', y', z' are piecewise continuous, this sum converges to the integral

$$\int_a^b \left[P(x(t), y(t), z(t)) \, x'(t) + Q(x(t), y(t), z(t)) \, y'(t) \right.$$
$$\left. + R(x(t), y(t), z(t)) \, z'(t) \right] dt. \tag{5.8}$$

Since $x'(t) \, dt = dx$, $y'(t) \, dt = dy$, and $z'(t) \, dt = dz$, this integral is expressed as an integral along L without explicitly showing the parameter t. It is also conventional to suppress the brackets. The work done by $\mathbf{F}$ along L is

$$W = \int_L P(x, y, z) \, dx + Q(x, y, z) \, dy + R(x, y, z) \, dz. \tag{5.9}$$

Of course P, Q, R can also be abstract functions, giving us a second type of line integral. It is important to maintain the connection between the notation on the right-hand side of (5.9) and the version (5.8) that shows the parameter explicitly. If L is a union of several distinct portions L_k, then

$$\int_L P \, dx + Q \, dy + R \, dz = \sum_k \int_{L_k} P_k \, dx + Q_k \, dy + R_k \, dz.$$

Let us emphasize the difference between the two types of line integrals. When we calculate $\int_L f \, ds$, then ds — the differential of the length parameter — should be "positive" and this is guaranteed by the fact that $a < b$ for the integration limits. Clearly we can use $b < a$ provided we compensate with a negative sign inside the integral with respect to the parameter t. Hence it makes no difference which endpoint of L is taken as the initial one. This is not true for the integral $\int_L P \, dx + Q \, dy + R \, dz$, hence the orientation of the contour must be made clear. Reversal in orientation changes the sign of the integral. This fact is clear if we think of the value of the integral as the work of a force field $\mathbf{F}$ along the path L. So when calculating a line integral of second type, we should show the orientation of L: that is, its initial and terminal points.

When treating line integrals over planar contours in the xy-plane, we formally omit the z dependence.

Example 5.6. We compute

$$I = \int_L y \, dx + z \, dy + x \, dz$$

along the twisted cubic $x = t$, $y = t^2$, $z = t^3$ from $t = 0$ to $t = 1$ (see Fig. 5.4):

$$I = \int_0^1 (t^2 + 2t^4 + 3t^3)\, dt = \frac{89}{60}. \qquad \square$$

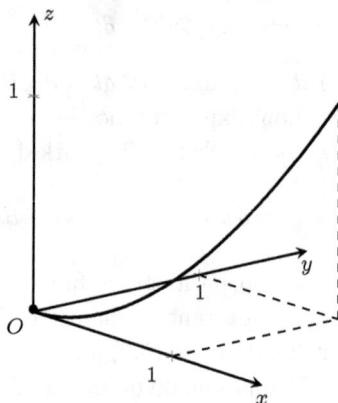

Fig. 5.4   Twisted cubic curve of Example 5.6

The second type of line integral also arises when we carry out iterative double integrals. Consider the integral

$$\int_a^b \int_{h_1(x)}^{h_2(x)} \frac{\partial f(x,y)}{\partial y}\, dy\, dx.$$

over the domain $D \subset \mathbb{R}^2$ of Fig. 5.5. The inner integral taken at a fixed $x \in [a,b]$ can be computed explicitly:

$$\int_a^b f(x,y)\Big|_{y=h_1(x)}^{h_2(x)}\, dx = \int_a^b f(x, h_2(x))\, dx - \int_a^b f(x, h_1(x))\, dx.$$

On the right we see the difference of two line integrals

$$\int_{L_2} f(x,y)\, dx = \int_a^b f(x, h_2(x))\, dx, \quad \int_{L_1} f(x,y)\, dx = \int_a^b f(x, h_1(x))\, dx$$

of the second type. We shall encounter such a line integral in Chapter 7.

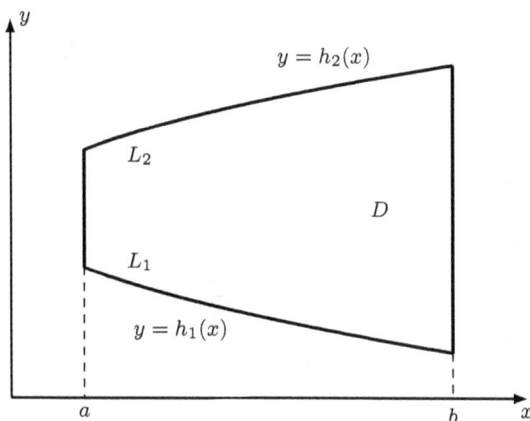

Fig. 5.5 Double integral giving rise to two line integrals.

Example 5.7. We calculate

$$I = \int_L x\,dx + y\,dy + z^2\,dz$$

over L defined by the equations $x = 1$, $y = t$, $z = t^3$ with initial point $(1,1,1)$ and terminal point $(1,0,0)$:

$$I = \int_1^0 \left[x(t)\,\frac{dx}{dt} + y(t)\,\frac{dy}{dt} + z^2(t)\,\frac{dz}{dt} \right] dt$$

$$= -\int_0^1 [1 \cdot 0 + t \cdot 1 + (t^3)^2 \cdot 3t^2]\,dt = -\frac{5}{6}. \qquad \square$$

5.4 Fundamental Theorem for Line Integrals

The centerpiece of integral calculus, the Newton–Leibnitz formula

$$\int_a^b f'(x)\,dx = f(b) - f(a) \tag{5.10}$$

can be extended to line integrals in a special case.

Theorem 5.1. *Let* $\mathbf{r} = \mathbf{r}(t) = x(t)\,\mathbf{i} + y(t)\,\mathbf{j} + z(t)\,\mathbf{k}(t)$ *be a piecewise continuously differentiable vector function with bounded first derivatives in* $\mathbb{R}^3$ *(or* $\mathbb{R}^2$*) representing a curve* L *with parameter* $t \in [a,b]$*. Let* $f = f(\mathbf{r}) \equiv$

$f(x, y, z)$ be continuously differentiable in a neighborhood of L. Then

$$\int_L \nabla f \cdot d\mathbf{r} = f(\mathbf{r}(b)) - f(\mathbf{r}(a)). \tag{5.11}$$

Remark 5.1. In the one-dimensional version $\nabla f \cdot d\mathbf{r}$ reduces to $f'(x) \, dx$, so (5.11) becomes (5.10). $\qquad\qquad\square$

Proof. We represent (5.11) parametrically:

$$\int_L \nabla f \cdot d\mathbf{r} = \int_a^b \left(\frac{\partial f}{\partial x} \frac{dx}{dt} + \frac{\partial f}{\partial y} \frac{dy}{dt} + \frac{\partial f}{\partial z} \frac{dz}{dt} \right) dt$$

$$= \int_a^b \frac{df(\mathbf{r}(t))}{dt} \, dt$$

$$= f(\mathbf{r}(b)) - f(\mathbf{r}(a)).$$

We have used the chain rule for the derivative of f with respect to t, along with the Newton–Leibnitz formula. $\qquad\qquad\square$

Note that the right side of (5.11) involves only the endpoints of L. This shows that if we have f defined in an open domain $D \subset \mathbb{R}^3$ where f is continuously differentiable, then for any two piecewise continuously differentiable curves L_1 and L_2 in D with the same endpoints

$$\int_{L_1} \nabla f \cdot d\mathbf{r} = \int_{L_2} \nabla f \cdot d\mathbf{r}. \tag{5.12}$$

Now let L be a smooth, finite and *closed* contour so that $\mathbf{r}(a) = \mathbf{r}(b)$. By cutting L at a point $t = c$ where $a < c < b$, we divide L into two parts L_1 and L_2 corresponding to $t \in [a, c]$ and $t \in [c, b]$ respectively. Reversing the orientation of L_2 and denoting the result by $-L_2$, we have by (5.12)

$$\int_L \nabla f \cdot d\mathbf{r} = \int_{L_1} \nabla f \cdot d\mathbf{r} + \int_{-L_2} \nabla f \cdot d\mathbf{r}$$

$$= \int_{L_1} \nabla f \cdot d\mathbf{r} - \int_{L_2} \nabla f \cdot d\mathbf{r}$$

$$= 0. \tag{5.13}$$

In physics (and frequently in mathematics) the line integral over a closed contour L is denoted

$$\int_L \mathbf{F} \cdot d\mathbf{r} = \oint_L \mathbf{F} \cdot d\mathbf{r}.$$

For more general vector fields $\mathbf{F}$, however, such integrals need not vanish.

Conservative Fields

Now we touch on the law of conservation of mechanical energy and its relation to the fundamental theorem. When a point mass m moves in the homogeneous gravitational field with $g = 9.8$ m/s^2, elementary physics dictates that total mechanical energy is conserved if there is no air resistance. That is, at any time t we have

$$\tfrac{1}{2}mv^2 + mgh = \text{constant}$$

or, in other words, $K + P = \text{constant}$ where the kinetic energy $K = \frac{1}{2}mv^2$ depends on the particle velocity $v(t)$, and the potential energy $P = mgh$ depends on height $h(t)$ above a reference level. In three dimensions the gravitational force acting on the body is

$$\mathbf{F} = -mg\,\mathbf{k}, \tag{5.14}$$

which is the gradient of the function $f = -mgz$.

We say that f is a *potential function* of an arbitrary vector field $\mathbf{F}$ if $\nabla f = \mathbf{F}$.

Equation (5.14) gives a simple approximation to the Earth's gravitational field. More accurate results are available from Newton's universal law of gravitation

$$\mathbf{F} = -G\frac{mM}{r^3}\mathbf{r}$$

where $\mathbf{r}$ is the position vector relative to the center of the Earth, M is the mass of the Earth, $r = |\mathbf{r}|$, and G is the gravitational constant. The reader should verify that $f = \gamma mM/r$ is a potential for $\mathbf{F}$.

The situation with the Coulomb forces of electrostatics is analogous. Indeed force fields having potentials play important roles in physics. Let us extend this situation to abstract vector fields.

Definition 5.1. A field $\mathbf{F}$ is *conservative* in $E \subset \mathbb{R}^3$ if there is a potential function f in the variables x, y, z such that for $\mathbf{r} \in E$ we have

$$\mathbf{F}(\mathbf{r}) = \nabla f(\mathbf{r}). \tag{5.15}$$

Conservation of Mechanical Energy

Let us show that for the motion of a point mass m, relation (5.15) implies conservation of mechanical energy in the form

$$K(t) + P(t) = \text{constant}$$

with potential energy $P = -f$. We start with Newton's second law $m\mathbf{r}''(t) = \mathbf{F}(\mathbf{r}(t)) = \nabla f(\mathbf{r}(t))$ at time instant t:

$$m\mathbf{r}''(t) = \nabla f(\mathbf{r}(t)). \tag{5.16}$$

Dot-multiplying both sides by $\mathbf{r}'(t)$ to get

$$m\,\mathbf{r}''(t) \cdot \mathbf{r}'(t) = \nabla f(\mathbf{r}(t)) \cdot \mathbf{r}'(t), \tag{5.17}$$

we proceed to transform the left side:

$$m\,\mathbf{r}''(t) \cdot \mathbf{r}'(t) = m[x''(t)x'(t) + y''(t)y'(t) + z''(t)z'(t)]$$

$$= \frac{m}{2}\frac{d}{dt}[x'(t)^2 + y'(t)^2 + z'(t)^2].$$

Then, with $v^2(t) = x'(t)^2 + y'(t)^2 + z'(t)^2$, equation (5.17) takes the form

$$\frac{d}{dt}\frac{mv(t)^2}{2} = \nabla f(\mathbf{r}(t)) \cdot \mathbf{r}'(t).$$

Now we integrate this over the trajectory L during the time interval $[t_1, t_2]$:

$$\int_{t_1}^{t_2} \frac{d}{dt}\frac{mv(t)^2}{2}\,dt = \int_{t_1}^{t_2} \nabla f(\mathbf{r}(t)) \cdot \mathbf{r}'(t)\,dt.$$

We can directly integrate on the left and use the fundamental theorem on the right to get

$$\left.\frac{mv(t)^2}{2}\right|_{t_1}^{t_2} = f(\mathbf{r}(t_2)) - f(\mathbf{r}(t_1)).$$

The result,

$$(K + P)|_{t=t_2} = (K + P)|_{t=t_1},$$

where $K = \frac{1}{2}mv^2$ and $P = -f$, is easily extended to a finite system of particles.

Hence for a particle moving in a conservative force field, energy conservation holds and the work integral $\int_{t_1}^{t_2} \mathbf{F} \cdot \mathbf{r}'(t)\,dt$ is path independent between fixed endpoints. It is not difficult to prove the converse.

Theorem 5.2. *Let $\int_L \mathbf{F} \cdot d\mathbf{r}$ be independent of the path connecting any pair of fixed points in an open connected domain $D \subset \mathbb{R}^3$ over which $\mathbf{F} = P\mathbf{i} + Q\mathbf{j} + R\mathbf{k}$ is continuous. Then $\mathbf{F}$ is a conservative field in D.*

Proof. As D is connected, we can join any $(x, y, z) \in D$ to an arbitrary but fixed $(a, b, c) \in D$ through a curve L. Point (a, b, c) is the initial point of L. Since $\int_L \mathbf{F} \cdot d\mathbf{r}$ is independent of the path from (a, b, c) to (x, y, z), the integral becomes a function of (x, y, z):

$$\int_L \mathbf{F} \cdot d\mathbf{r} = f(x, y, z).$$

Let us show that f is a potential function for $\mathbf{F}$, i.e., $\mathbf{F}(x, y, z) = \nabla f(x, y, z)$. We shall demonstrate that $P = \partial f/\partial x$; the other two relations, $Q = \partial f/\partial y$ and $R = \partial f/\partial z$, can be shown similarly. By definition,

$$\frac{\partial f(x, y, z)}{\partial x} = \lim_{\Delta x \to 0} \frac{f(x + \Delta x, y, z) - f(x, y, z)}{\Delta x}.$$

To determine $f(x + \Delta x, y, z)$ we take the path consisting of L and a straight segment $C_{\Delta x}$ from (x, y, z) to $(x + \Delta x, y, z)$. See Fig. 5.6. Then

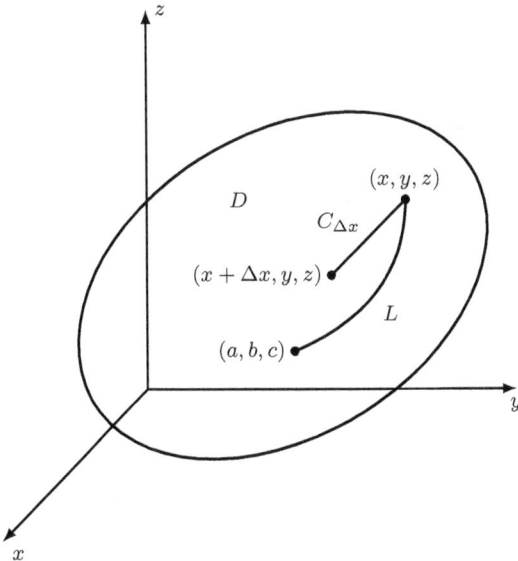

Fig. 5.6   Curve consisting of curve L from point (a, b, c) to (x, y, z) in D and a straight segment $C_{\Delta x}$ from (x, y, z) to $(x + \Delta x, y, z)$.

$$f(x+\Delta x,y,z)-f(x,y,z)=\int_L \mathbf{F}\cdot d\mathbf{r}+\int_{C_{\Delta x}}\mathbf{F}\cdot d\mathbf{r}-\int_L \mathbf{F}\cdot d\mathbf{r}=\int_{C_{\Delta x}}\mathbf{F}\cdot d\mathbf{r}.$$

On $C_{\Delta x}$ we have $dy=0=dz$. Moreover P is continuous at (x,y,z), so the last integral reduces to the ordinary integral

$$\int_x^{x+\Delta x} P(t,y,z)\,dt = (P(x,y,z)+\varepsilon)\,\Delta x$$

with $\varepsilon \to 0$ as $\Delta x \to 0$. Dividing through by Δx and taking the limit, we get the needed equality $P(x,y,z)=\partial f(x,y,z)/\partial x$ at any point of D.   □

Thus, on an open connected domain $D \subset \mathbb{R}^3$, a continuous vector field $\mathbf{F}$ is conservative if and only if the work integral $\int_L \mathbf{F}\cdot d\mathbf{r}$ is independent of the form of L connecting the fixed endpoints for any L in D. In physics, they prefer to speak of equality to zero of the work integral over any closed L in D.

Theorem 5.3. *If for any closed curve $L \subset D$ we have*

$$\oint_L \mathbf{F}\cdot d\mathbf{r}=0,$$

then $\mathbf{F}$ is conservative in D.

The proof follows from the fact that for any two curves L_1 and L_2 connecting two fixed points, the pair of curves L_1 and $-L_2$ constitutes a closed contour. So we have equality of the work integrals over L_1 and L_2.

A Simple Necessary Condition for F to be Conservative

How can we check whether $\mathbf{F}$ is conservative? There is a simple necessary condition for this if $\mathbf{F}$ is continuously differentiable in D. Let $\nabla f = \mathbf{F}$ so that f has all the second derivatives continuous in an open set D. By Clairaut's theorem the mixed second derivatives of f in D do not depend on the order of differentiation. In terms of the components of $\mathbf{F}$, this means that in D we have the three equalities

$$\left.\begin{array}{l} P_y(x,y,z)=Q_x(x,y,z), \\ P_z(x,y,z)=R_x(x,y,z), \\ Q_z(x,y,z)=R_y(x,y,z). \end{array}\right\} \qquad (5.18)$$

For the two-dimensional case this reduces to the single equality

$$P_y(x,y)=Q_x(x,y). \qquad (5.19)$$

Of course we should like to know whether (5.19) is sufficient for $\mathbf{F}$ to be conservative. In general it is not, but it is if D is open and simply-connected in $\mathbb{R}^2$ (meaning that D has no "holes" or that any closed contour can be reduced continuously to a point inside D) and $\mathbf{F} = P\mathbf{i} + Q\mathbf{j}$ is continuously differentiable. This follows from Green's theorem (proved later) which reduces to the equality

$$\oint_{\partial E} P\,dx + Q\,dy = \iint_E (Q_x - P_y)\,dA$$

for any $E \subset D$ and positively-oriented boundary ∂E of E. By (5.19) the right-hand integral vanishes for any E and

$$\oint_{\partial E} P\,dx + Q\,dy = 0$$

for any closed contour ∂E. Consequently $\mathbf{F}$ is conservative.

Example 5.8. The two-dimensional field

$$\mathbf{F}(x,y) = \cos(xy)(y\,\mathbf{i} + x\,\mathbf{j})$$

is conservative, as is easily verified using (5.19). $\qquad\qquad\qquad\square$

A similar result holds in the 3D case when $\mathbf{F}$ is continuously differentiable and satisfies (5.18) or, equivalently, $\nabla \times \mathbf{F} = \mathbf{0}$.

It is possible to find the potential function under the above conditions if the components of $\mathbf{F}$ satisfy (5.19). Such a procedure is applied in particle mechanics, electrostatics, the theory of exact first-order ordinary differential equations, etc.

The Potential for $\mathbf{F} = P\mathbf{i} + Q\mathbf{j}$ *Under Condition* (5.19)

Let us find the potential f in the general 2D case where it must satisfy

$$f_x(x,y) = P(x,y), \qquad f_y(x,y) = Q(x,y).$$

This is a system of two partial differential equations. Although the reader likely has no knowledge of how to solve such systems, this one is rather simple. Let us first integrate one of the equations, for example the first one. The integration is done for an arbitrary but fixed value of y:

$$f(x,y) = \int P(x,y)\,dx + c.$$

The indefinite "constant" c depends on y. We write $c = g(y)$ and hence

$$f(x,y) = \int P(x,y)\,dx + g(y).$$

To determine g, let us substitute f into the second equation of (5.19):

$$\frac{\partial}{\partial y}\left(\int P(x,y)\,dx + g(y)\right) = Q(x,y).$$

Since P is continuously differentiable we can rewrite this as

$$g'(y) = Q(x,y) - \int \frac{\partial P(x,y)}{\partial y}\,dx. \tag{5.20}$$

It is easily verified that under the condition (5.19) the right side does not depend on x; this is done by differentiating it with respect to x:

$$\frac{\partial}{\partial x}\left(Q(x,y) - \int \frac{\partial P(x,y)}{\partial y}\,dx\right) = \frac{\partial Q(x,y)}{\partial x} - \frac{\partial P(x,y)}{\partial y} = 0.$$

So we can integrate (5.20) with respect to y, getting

$$g(y) = \int \left(Q(x,y) - \int \frac{\partial P(x,y)}{\partial y}\,dx\right) dy.$$

We have omitted a constant of integration, as we require just one function f. In 3D we can integrate (5.18) in a similar fashion.

Example 5.9. Let us find the potential for $\mathbf{F}$ of Example 5.8. Here $P(x,y) = y\cos(xy)$ and $Q(x,y) = x\cos(xy)$. We start with

$$f_x(x,y) = y\cos(xy), \qquad f_y(x,y) = x\cos(xy),$$

and integrate the first equation:

$$f(x,y) = \int y\cos(xy)\,dx + g(y) = \sin(xy) + g(y).$$

Substituting this into the second equation for f_y we get $x\cos(xy) + g'(y) = x\cos(xy)$, hence $g'(y) = 0$. So $g(y) = 0$ and $f(x,y) = \sin(xy)$. $\square$

5.5   Problems

Evaluation of Line Integrals

5.1 Calculate the following line integrals:

(a)
$$I = \int_L x\, dy + y\, dx$$

where L is the segment of the line $y = x$ from $(0,0)$ to $(1,1)$

(b)
$$I = \int_L x\, dy - y\, dx$$

where L is the portion of the parabola $y = x^2$ from $(0,0)$ to $(1,1)$

(c)
$$I = \int_L (xy - 1)\, dx + x^2 y\, dy$$

where L is the first-quadrant portion of the ellipse $x = \cos t$, $y = 3\sin t$

(d)
$$I = \int_L (y+1)\, dx + (x+y+1)\, dy$$

where L is the parabola $y = 3x^2/2$ between the points $(0,0)$ and $(2,6)$

(e)
$$I = \int_L (x^2 + y^2)\, dx + xy\, dy$$

where L is the portion of the circle $x^2 + y^2 = 1$ between the points $(1,0)$ and $(0,1)$

5.2 Line integrate the field $\mathbf{A}(\mathbf{r}) = z\,\mathbf{i} + x\,\mathbf{j} + y\,\mathbf{k}$ along the helix given by

$$\mathbf{r}(t) = \cos t\,\mathbf{i} + \sin t\,\mathbf{j} + t\,\mathbf{k} \qquad (0 \le t < 2\pi).$$

5.3 Line integrate the constant 1 along the curve given by $x = e^{-t}\cos t$, $y = e^{-t}\sin t$, $z = e^{-t}$ for $0 \le t \le 2\pi$.

5.4 Compute
$$I = \int_L P(x,y)\, dx + Q(x,y)\, dy$$

where L is the path given by $\mathbf{r}(t) = \cos t\,\mathbf{i} + \sin t\,\mathbf{j}$ for $0 \le t \le \pi/2$, taking

(a) $P(x,y) = x$ and $Q(x,y) = y$,

(b) $P(x,y) = 3x^2 y^2$ and $Q(x,y) = 2x^3 y$,

Fundamentals of Multivariable Calculus

(c) $P(x,y) = 2xy$ and $Q(x,y) = x^2 + y^2$,

(d) $P(x,y) = ye^x$ and $Q(x,y) = xe^y$.

5.5 In each case letting the parameter t run from 0 to 1, compute

$$I = \int_L P(x,y,z)\,dx + Q(x,y,z)\,dy + R(x,y,z)\,dz$$

for the given integrand functions and parameterization of L:

(a) $P(x,y,z) = xy$, $Q(x,y,z) = -yz$, $R(x,y,z) = e^z$; $\mathbf{r}(t) = 2t\,\mathbf{i} - t\,\mathbf{j} + t^2\,\mathbf{k}$

(b) $P(x,y,z) = x$, $Q(x,y,z) = x + z$, $R(x,y,z) = z$; $\mathbf{r}(t) = t\,\mathbf{i} + t^2\,\mathbf{j} + t^3\,\mathbf{k}$

(c) $P(x,y,z) = xe^y$, $Q(x,y,z) = ye^z$, $R(x,y,z) = ze^x$; $\mathbf{r}(t) = t^3\,\mathbf{i} + t^2\,\mathbf{j} + t\,\mathbf{k}$

5.6 For each of the following vector fields, compute

$$I = \int_L \mathbf{F} \cdot d\mathbf{r}$$

where L is the line segment that runs from point A to point B:

(a) $\mathbf{F} = xyz\,\mathbf{i} + \sin(yz)\,\mathbf{j} + xz\,\mathbf{k}$; $A = (1,1,1)$, $B = (2,1,3)$

(b) $\mathbf{F} = ye^z\,\mathbf{i} + ze^x\,\mathbf{j} + xe^y\,\mathbf{k}$; $A = (1,2,3)$, $B = (2,3,4)$

(c) $\mathbf{F} = (y+z)\,\mathbf{i} + (z+x)\,\mathbf{j} + (x+y)\,\mathbf{k}$; $A = (0,0,0)$, $B = (2,2,2)$

Properties of Line Integrals

5.7 Verify the following properties of line integrals over C:

(1)
$$\int_C (\mathbf{A} + \mathbf{B}) \cdot d\mathbf{r} = \int_C \mathbf{A} \cdot d\mathbf{r} + \int_C \mathbf{B} \cdot d\mathbf{r}.$$

(2) For any constant λ,
$$\int_C \lambda \mathbf{A} \cdot d\mathbf{r} = \lambda \int_C \mathbf{A} \cdot d\mathbf{r}.$$

(3) If $C = C_1 \cup C_2$ with C_1, C_2 having the same orientation as C, then
$$\int_C \mathbf{A} \cdot d\mathbf{r} = \int_{C_1} \mathbf{A} \cdot d\mathbf{r} + \int_{C_2} \mathbf{A} \cdot d\mathbf{r}.$$

(4) If C has length L and $|\mathbf{A}| \le M$ on C, then
$$\left| \int_C \mathbf{A} \cdot d\mathbf{r} \right| \le ML.$$

General Expressions for Line Integrals

5.8 A curve L on the plane is $y = g(x)$ for $x \in [a, b]$. Take $t = x$. Verify that the line integrals take the form:

$$\int_L f(x, y) \, ds = \int_a^b f(x, g(x)) \left\{ 1 + [g'(x)]^2 \right\}^{1/2} dx,$$

$$\int_L \mathbf{F}(x, y) \cdot \mathbf{r} = \int_a^b [P(x, g(x)) + Q(x, g(x))] \, dx, \quad \mathbf{F} = P \mathbf{i} + Q \mathbf{j}.$$

Present similar formulas if L is given by equation $x = h(y)$ on $[c, d]$.

5.9 Extend the expressions of Problem 5.8 for a space curve L given by equations $x = x$, $y = y(x)$, $z = z(x)$ for $x \in [a, b]$.

5.10 A curve L on the unit cylinder $x^2 + y^2 = 1$ is given by the equation $z = h(\theta)$ for $\theta \in [a, b]$. Derive an expression for the line integral $\int_L f(x, y, z) \, ds$.

5.11 (a) A curve L on the unit sphere $x^2 + y^2 + z^2 = 1$ is given by the equation $\phi = h(\theta)$ for $\theta \in [a, b]$. Derive an expression for the line integral $\int_L f(x, y, z) \, ds$.
(b) Apply the result of part (a) to the function $f \equiv 1$ and the curve described by $h(\theta) = \theta$ for $\theta \in [\pi/4, \pi/2]$.

Calculation of Length

5.12 Find the length of each given plane curve for $0 \le x \le a$, assuming $a > 0$. If the integral doesn't seem to exist in closed form, then set $a = 1$ and try to approximate it numerically.

(a) $y = x^{3/2}$	(g) $y = 1 + \cosh x$
(b) $y = 2x$	(h) $y = e^x$
(c) $y = x^2$	(i) $y = e^{-x}$
(d) $y = x^{1/2}$	(j) $y = \sinh x$
(e) $y = x^3$	(k) $y = \sin x$
(f) $y = \cosh x$	(l) $y = \cos x$

5.13 Find the length of the space curve

(a) $\mathbf{r}(t) = t \mathbf{i} + t^2 \mathbf{j} + (2t^3/3) \mathbf{k}$ for $t \in [0, 3]$,
(b) $\mathbf{r}(t) = 3t \mathbf{i} + t \mathbf{j} + 2t \mathbf{k}$ for $t \in [0, 1]$.

5.14 Show that the two surfaces $2ay = x^2$ and $6a^2 z = x^3$ intersect in a curve whose length, measured from the coordinate origin to a point (x, y, z) on the curve, is $x + z$.

5.15 Argue intuitively that the arc length along a plane curve $r = f(\theta)$ between $\theta = \alpha$ and $\theta = \beta$ is given by

$$L = \int_\alpha^\beta \{[f'(\theta)]^2 + [f(\theta)]^2\}^{1/2}\, d\theta.$$

Then, assuming $0 \leq \theta \leq 2\pi$, compute the length of each curve below:

(a) $r = 1 + \cos\theta$ (d) $r = \cos\theta$

(b) $r = \theta$ (e) $r = \theta^2$

(c) $r = e^\theta$ (f) $r = 1 + \sin\theta$

Some Applications

5.16 A flexible cable hangs in the shape of the catenary $y = \cosh x$ for $0 \leq x \leq a$. Find its length.

5.17 A wire of nonuniform mass density $\gamma(x, y) = y$ takes the shape of the curve $y = x^3$, $z = 0$, for $0 \leq x \leq 1$. Find its mass.

5.18 A wire of uniform mass density γ_0 takes the shape of the semicircle $y = (1 - x^2)^{1/2}$, $|x| \leq 1$, $z = 0$. Find its center of mass.

5.19 A particle moves along the cycloid given by $x = t - \sin t$, $y = 1 - \cos t$ for $0 \leq t \leq 2\pi$. What is the total distance traveled?

5.20 Calculate the work done by the force $\mathbf{F} = \mathbf{F}(x, y, z)$ along the specified path:

(a) $\mathbf{F} = x^2\,\mathbf{i} + yx\,\mathbf{j} + z\,\mathbf{k}$, along the straight segment from the origin to the point $(1, -1, 2)$

(b) $\mathbf{F} = x\,\mathbf{i} + yz\,\mathbf{j}$, along the hyperbola $xy = 1$ from the point $(1, 1, 0)$ to the point $(2, 1/2, 0)$

(c) $\mathbf{F} = y\,\mathbf{i} + x\,\mathbf{j}$, along the parabola $y = x^2$ from the origin to the point $(2, 4, 0)$

(d) $\mathbf{F} = xy\,\mathbf{i}$, along the curve $y = \cos x$ from the point $(0, 1, 0)$ to the point $(\pi/2, 0, 0)$

Chapter 6

Surface Integrals

6.1 Introduction

Surface integrals are of two types. The first,

$$\iint_S f(x, y, z)\, dS,$$

is used to find the area and mass characteristics of a massive surface S. The second,

$$\iint_S \mathbf{F} \cdot d\mathbf{S},$$

yields the flux of a vector field $\mathbf{F}$ through an oriented surface S; it appears in Stokes' and Gauss' theorems, and in transformations of the variational formulations of boundary value problems in mathematical physics. Surface integrals reduce to ordinary double integrals that can be calculated via Fubini's theorem. They are important practically and theoretically.

Let us return to the notion of surface. A simple example is a portion S of a plane. Introducing Cartesian coordinates over S, we place the points of S into one-to-one correspondence with coordinate pairs. By deforming S we obtain a more general surface. The Cartesian coordinate lines (which are, of course, straight lines parallel to the coordinate axes on the plane) map to certain curves. These curves provide a curvilinear coordinate framework on the surface if the mapping of points under the deformation is one-to-one and continuous (or, better yet, smooth). However on a surface there could be u- and v-coordinate lines of other origin, and the surface in 3D can be described with three parametric equations

$$x = x(u, v), \qquad y = y(u, v), \qquad z = z(u, v),$$

209

or expressed in the vector form

$$\mathbf{r} = \mathbf{r}(u, v) = x(u, v)\,\mathbf{i} + y(u, v)\,\mathbf{j} + z(u, v)\,\mathbf{k}. \qquad (6.1)$$

A simpler description of a surface is

$$z = f(x, y) \qquad (6.2)$$

where the role of (u, v) is played by the pair (x, y). With this form of surface equation many textbook problems can be solved. It seems to be all that is needed to describe various surfaces. However, we should consider the idea of a surface having *sides*. A portion S of a plane has two sides. This means that if we move the surface normal over S, and do so in a continuous fashion, we cannot move this normal to the other side of the plane. Perhaps it seems that the same would hold for any surface that is a deformed piece of a plane. It certainly would for a sphere or a cylindrical surface. But certain deformations of a planar surface can lead to unexpected results, and we may find ourselves able to move the normal vector continuously over the surface in such a way that we return to the same point but with the normal vector in the opposite direction. The first such surface to be discovered was the *Möbius strip* of Problem 1.54. Although this and other surfaces like the "Klein bottle" are of great interest in topology, we shall exclude them from further discussion and consider only two-sided surfaces given by (6.1). It is worth noting that even if one cannot represent a surface using (6.1), it may still be possible to write down equations like (6.1) for some neighborhood of each point. For smooth surfaces one can even select some discrete number of such neighborhoods that cover the bounded surface. Clearly this is still a surface, but in such a case the applicable term is *manifold*. We shall not go into that complicated theory, however; we restrict ourselves to cases for which (6.1) or (6.2) will suffice.

Our goal is to construct a surface integral that can be applied directly to physical problems. We begin with the problem of surface area.

6.2 Area of a Surface

Recalling what it took to obtain curve length, we certainly do not expect that surface area will be simpler to obtain. First we restrict ourselves to a smooth geometric surface S given by (6.1). Our approach is to partition S into cells by finite numbers of coordinate lines, getting small curvilinear "quadrangles." We approximate the cell area by that of its orthogonal projection onto the tangent plane at a quadrangle point. The sum of these areas approximates $A(S)$. Finally, the limit (if it exists) defines $A(S)$ for a smooth surface. It can be proved that the limit of the approximations, as for the double integral, does not depend on the type of smooth partitioning or the selection of points inside the partition elements. As usual we rely on intuition and avoid deeper theoretical questions.

Recall that on S the inner coordinate lines are defined by fixing one of the coordinates u, v at a time:

$$\mathbf{r} = \mathbf{r}(u, v_0), \qquad \mathbf{r} = \mathbf{r}(u_0, v),$$

with respective parameters u, v. The lines

$$\mathbf{r} = \mathbf{r}(u, v_0 + \Delta v), \quad \mathbf{r} = \mathbf{r}(u, v_0), \quad \mathbf{r} = \mathbf{r}(u_0, v), \quad \mathbf{r} = \mathbf{r}(u_0 + \Delta u, v),$$

with fixed u_0 or v_0 and small fixed $\Delta u, \Delta v$ constitute the boundary of a small "curvilinear parallelogram" $ABCD$. Here each of the four equations is of the type $\mathbf{r} = \mathbf{r}(t)$ where t is u or v. See Fig. 6.1.

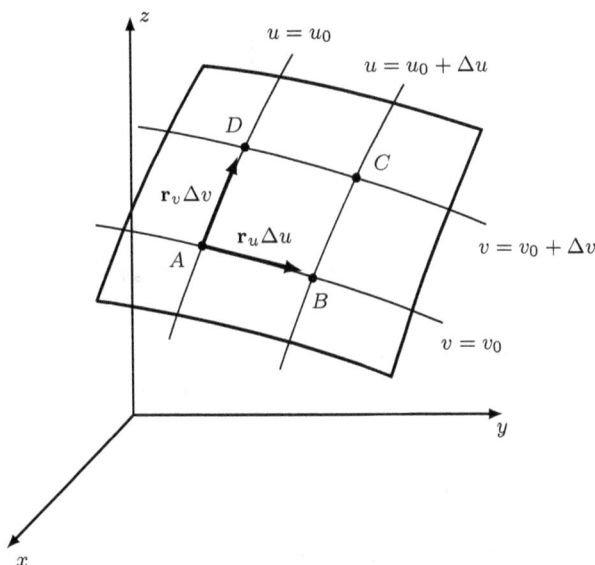

Fig. 6.1 A "curvilinear parallelogram" $ABCD$, a cell, on surface S.

The tangent vectors to the coordinate lines, which "approximate" two sides of $ABCD$, are $\mathbf{r}_u(u_0, v_0) \Delta u$ and $\mathbf{r}_v(u_0, v_0) \Delta v$. If $\mathbf{r} = \mathbf{r}(u, v)$ is continuously differentiable on the domain D_{uv} in the uv-plane where S is determined, and Δu and Δv are small, then $ABCD$ is almost flat and its area ΔS is approximately that of the "tangent" parallelogram defined by the above tangent vectors. It can be proved that this area

$$\Delta \tilde{S} = |\mathbf{r}_u(u_0, v_0)\, \Delta u \times \mathbf{r}_v(u_0, v_0)\, \Delta v|$$

is such that

$$\left| \frac{\Delta S - \Delta \tilde{S}}{\Delta \tilde{S}} \right| \to 0 \quad \text{as} \quad [(\Delta u)^2 + (\Delta v)^2]^{1/2} \to 0.$$

So $\Delta S = \Delta \tilde{S} + \varepsilon\, \Delta \tilde{S}$, where $\varepsilon \to 0$ as $[(\Delta u)^2 + (\Delta v)^2]^{1/2} \to 0$.

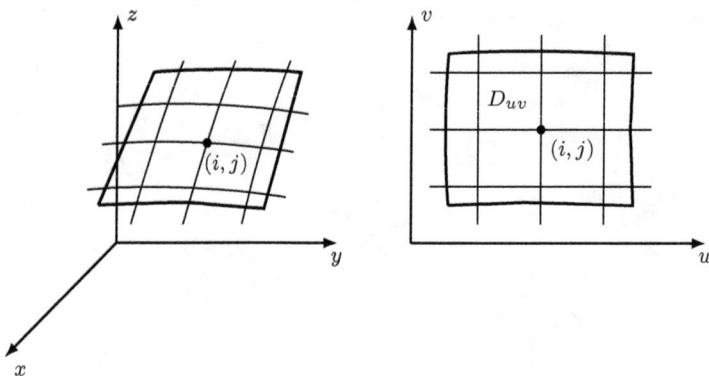

Fig. 6.2 Rectangular partition of D_{uv} and its corresponding preimage under a smooth transformation.

Thus, partitioning S by coordinate lines (Fig. 6.2 left), which corresponds to the rectangular partition of D_{uv} in the uv-plane (Fig. 6.2 right), we get the area of S:

$$A(S) = \sum_{i,j} \Delta S_{ij} + \sum_{\text{bdry}} \Delta S_{ij}$$

$$= \sum_{i,j} \Delta \tilde{S}_{ij} + \sum_{i,j} \varepsilon_{ij}\, \Delta \tilde{S}_{ij} + \sum_{\text{bdry}} \Delta S_{ij}$$

$$= \sum_{i,j} |\mathbf{r}_u(u_i, v_j) \times \mathbf{r}_v(u_i, v_j)\, \Delta u_i\, \Delta v_j| + \sum_{i,j} \varepsilon_{ij}\, \Delta \tilde{S}_{ij} + \sum_{\text{bdry}} \Delta S_{ij},$$

where (u_i, v_j) are nodal points of the partition of D_{uv} and $\sum_{\text{bdry}} \Delta S_{ij}$ is the area of the cells covering the boundary. If

$$\max_{i,j} [(\Delta u_i)^2 + (\Delta v_j)^2]^{1/2} \to 0,$$

the second and third terms in $A(S)$ tend to zero and the first term, which takes the form of a Riemann sum, yields a double integral:

$$A(S) = \iint_{D_{uv}} |\mathbf{r}_u(u,v) \times \mathbf{r}_v(u,v)| \, du \, dv. \qquad (6.3)$$

We assume the boundary of D_{uv} is piecewise smooth to avoid trouble in the limit passage with partition elements containing boundary points.

Example 6.1. Let us use (6.3) to calculate the area of the conical surface $z = (x^2 + y^2)^{1/2}$ lying in the first octant and above the disk $x^2 + y^2 \leq 1$. See Fig. 6.3.

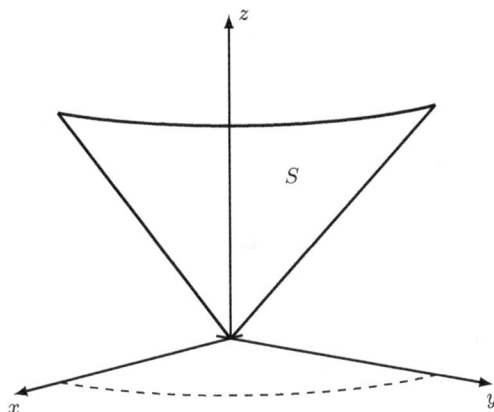

Fig. 6.3   Surface of Example 6.1.

With $x = u$, $y = v$, we can represent S as

$$\mathbf{r}(u,v) = u\,\mathbf{i} + v\,\mathbf{j} + (u^2 + v^2)^{1/2}\,\mathbf{k} \qquad (u, v \geq 0, \; u^2 + v^2 \leq 1).$$

We have

$$\mathbf{r}_u = \mathbf{i} + \frac{u}{(u^2 + v^2)^{1/2}}\,\mathbf{k}, \qquad \mathbf{r}_v = \mathbf{j} + \frac{v}{(u^2 + v^2)^{1/2}}\,\mathbf{k},$$

hence $|\mathbf{r}_u \times \mathbf{r}_v| = \sqrt{2}$ and

$$A(S) = \iint_{D_{uv}} \sqrt{2}\, dv\, du = \int_0^{\pi/4} \int_0^1 \sqrt{2}\, r\, dr\, d\theta = \frac{\sqrt{2}\pi}{4}. \qquad \Box$$

The integral (6.3) is denoted by

$$\iint_S dS \quad \text{with ``elementary area''} \ dS = |\mathbf{r}_u \times \mathbf{r}_v| \, du \, dv.$$

But

$$\mathbf{r}_u \times \mathbf{r}_v = \begin{vmatrix} \mathbf{i} & \mathbf{j} & \mathbf{k} \\ x_u & y_u & z_u \\ x_v & y_v & z_v \end{vmatrix} = \mathbf{i}\,\frac{\partial(y,z)}{\partial(u,v)} + \mathbf{j}\,\frac{\partial(z,x)}{\partial(u,v)} + \mathbf{k}\,\frac{\partial(x,y)}{\partial(u,v)}$$

so that

$$|\mathbf{r}_u \times \mathbf{r}_v| = \left\{ \left[\frac{\partial(y,z)}{\partial(u,v)}\right]^2 + \left[\frac{\partial(z,x)}{\partial(u,v)}\right]^2 + \left[\frac{\partial(x,y)}{\partial(u,v)}\right]^2 \right\}^{1/2}$$

and

$$A(S) = \iint_{D_{uv}} \left\{ \left[\frac{\partial(y,z)}{\partial(u,v)}\right]^2 + \left[\frac{\partial(z,x)}{\partial(u,v)}\right]^2 + \left[\frac{\partial(x,y)}{\partial(u,v)}\right]^2 \right\}^{1/2} dA,$$

where $dA = du\,dv$ is the "area" of an elementary rectangle in the "orthonormal" uv-plane. Hence we can calculate the area using techniques for double integrals.

In particular let us consider the representation

$$z = f(x,y) \quad \text{or} \quad \mathbf{r} = x\,\mathbf{i} + y\,\mathbf{j} + f(x,y)\,\mathbf{k}.$$

The uv-plane is now the xy-plane and we have

$$\mathbf{r}_x \times \mathbf{r}_y = \begin{vmatrix} \mathbf{i} & \mathbf{j} & \mathbf{k} \\ 1 & 0 & f_x \\ 0 & 1 & f_y \end{vmatrix} = -\mathbf{i}\,f_x - \mathbf{j}\,f_y + \mathbf{k}$$

so that $|\mathbf{r}_x \times \mathbf{r}_y| = (1 + f_x^2 + f_y^2)^{1/2}$ and

$$A(S) = \iint_{D_{xy}} (1 + f_x^2 + f_y^2)^{1/2}\, dA, \qquad dA = dx\,dy, \qquad (6.4)$$

where D_{xy} is the orthogonal projection of S onto the xy-plane. So here

$$dS = (1 + f_x^2 + f_y^2)^{1/2}\, dx\,dy.$$

But clearly we are free to employ polar coordinates when convenient.

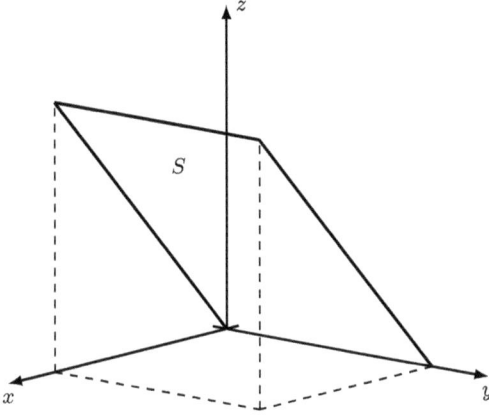

Fig. 6.4 Surface of Example 6.2.

Example 6.2. We find the area of that part S of the plane $z = ax$ $(a > 0)$ whose projection on the xy-plane is $[0, 1]^2$. See Fig. 6.4.

The angle α of inclination of the plane to the xy-plane satisfies $\tan \alpha = z_x = a$. The square $[0, 1]^2$ is the projection of S onto the xy-plane so the needed area is $1/\cos \alpha = (1 + a^2)^{1/2}$. We get the same result by (6.4):

$$\iint_D (1 + f_x^2 + f_y^2)^{1/2} \, dA = \int_0^1 \int_0^1 (1 + a^2)^{1/2} \, dx \, dy = (1 + a^2)^{1/2}. \qquad \square$$

Remark 6.1. The angle α in Example 6.2 is also the angle between the unit normal $\mathbf{n}$ and the unit vector $\mathbf{k}$ along the z-axis. That is, $\alpha = \widehat{\mathbf{n}, \mathbf{k}}$. The z-component of $\mathbf{n}$ is frequently denoted $n_z = \cos \alpha$. The relation between $A(S)$ and the area A of the projection of plane S onto the xy-plane is

$$A = n_z A(S).$$

The same relation holds for any plane $z = ax + by + c$ with $\mathbf{n} \cdot \mathbf{k} > 0$. Moreover, if we take the area dS of a very small ("infinitesimal") part of an arbitrary surface $z = f(x, y)$, where $\mathbf{n} \cdot \mathbf{k} > 0$, and compare it with the area $dA = dx \, dy$ of its projection onto the xy-plane, we get a similar relation

$$dA = n_z \, dS$$

since the infinitesimal part is approximated by part of the tangent plane at a point of the surface. The relation can be also written as

$$dS = (1 + f_x^2 + f_y^2)^{1/2} \, dA. \qquad \square$$

Example 6.3. We seek the area of the paraboloid $z = x^2 + y^2$ inside the cylinder $x^2 + y^2 = 1$. See Fig. 6.5. Taking x, y as inner coordinates, we

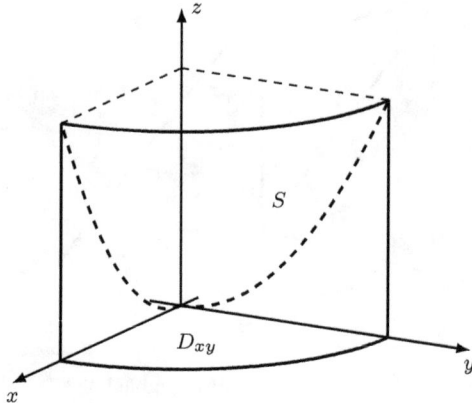

Fig. 6.5   A quarter of the surface for Example 6.3.

calculate

$$A = \iint_S (1 + z_x^2 + z_y^2)^{1/2} \, dA = \iint_{D_{xy}} (1 + 4x^2 + 4y^2)^{1/2} \, dx \, dy.$$

Since the projection D_{xy} of S onto the xy-plane is a quarter of the unit disk, we switch to polar coordinates:

$$A = \iint_{D_{r\theta}} (1 + 4r^2)^{1/2} \, dA = 4 \int_0^{\pi/2} \int_0^1 (1 + 4r^2)^{1/2} \, r \, dr \, d\theta.$$

Changing variables to $u = 1 + 4r^2$, we get $du = 8r \, dr$ and

$$A = \frac{1}{2} \int_0^{\pi/2} \int_0^1 u^{1/2} \, du \, d\theta = \frac{\pi}{6}. \qquad \Box$$

Example 6.4. Figure 6.6 shows the surface S cut off from the sphere $x^2 + y^2 + z^2 = 1$ in the first octant by the cylinder $x^2 + (y - 1/2)^2 = (1/2)^2$. Rewriting the cylinder equation as $x^2 + y^2 = y$, we see that the sphere is intersected in the curve $z = (1 - y)^{1/2}$. Partial differentiation gives

$$z_x = \frac{-x}{(1 - x^2 - y^2)^{1/2}}, \qquad z_y = \frac{-y}{(1 - x^2 - y^2)^{1/2}},$$

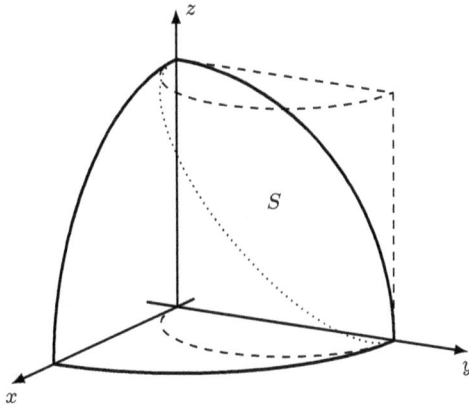

Fig. 6.6   Surface S for Example 6.4.

and the area of S is

$$A = \int_0^1 \int_0^{\sqrt{y(1-y)}} \left(1 + \frac{x^2}{1 - x^2 - y^2} + \frac{x^2}{1 - x^2 - y^2} \right) dx\, dy$$

$$= \int_0^1 \int_0^{\sqrt{y(1-y)}} \frac{dx\, dy}{(1 - x^2 - y^2)^{1/2}}$$

$$= \int_0^1 \sin^{-1} \left(\frac{y}{y+1} \right)^{1/2} dy$$

$$= \frac{\pi}{2} - 1.$$

As an exercise, the reader may wish to calculate the area of that portion of the cylinder cut off by the sphere. ☐

Cylindrical Surfaces

An object of common occurrence is the circular cylinder

$$x^2 + y^2 = a^2.$$

Natural coordinates for such a surface, given in vector form as

$$\mathbf{r} = \mathbf{r}(\theta, z) = a \cos\theta\, \mathbf{i} + a \sin\theta\, \mathbf{j} + z\, \mathbf{k},$$

are clearly (θ, z). As $\mathbf{r}_\theta = -a \sin \theta \, \mathbf{i} + a \cos \theta \, \mathbf{j}$ and $\mathbf{r}_z = \mathbf{k}$, we get

$$\mathbf{r}_\theta \times \mathbf{r}_z = \begin{vmatrix} \mathbf{i} & \mathbf{j} & \mathbf{k} \\ -a \sin \theta & a \cos \theta & 0 \\ 0 & 0 & 1 \end{vmatrix} = a \cos \theta \, \mathbf{i} + a \sin \theta \, \mathbf{j}$$

and hence $|\mathbf{r}_\theta \times \mathbf{r}_z| = a$, the integrand quantity for the area of the cylindrical surface:

$$A(S) = \iint_S a \, dA, \qquad dA = d\theta \, dz, \qquad dS = a \, d\theta \, dz.$$

Of course the extent of S along the z-axis must be restricted to obtain a finite result.

Example 6.5. We calculate the surface area of the wedge considered in Example 4.3 on p. 171. Eliminating x from the two equations $x^2 + y^2 = 1$ and $z = x$, we find that the upper edge of the surface is given by $z^2 = 1 - y^2 = 1 - \sin^2 \theta = \cos^2 \theta$. The area sought is

$$\int_0^{\pi/2} \int_0^{\cos \theta} 1 \, dz \, d\theta = \int_0^{\pi/2} \cos \theta \, d\theta = 1. \qquad \square$$

Spherical Surfaces

Now suppose S is a part of the sphere of radius a:

$$x^2 + y^2 + z^2 = a^2.$$

We take

$$\mathbf{r} = \mathbf{r}(\phi, \theta) = a \sin \phi \cos \theta \, \mathbf{i} + a \sin \phi \sin \theta \, \mathbf{j} + a \cos \phi \, \mathbf{k}$$

and compute

$$\mathbf{r}_\phi \times \mathbf{r}_\theta = \begin{vmatrix} \mathbf{i} & \mathbf{j} & \mathbf{k} \\ a \cos \phi \cos \theta & a \cos \phi \sin \theta & -a \sin \phi \\ -a \sin \phi \sin \theta & a \sin \phi \cos \theta & 0 \end{vmatrix}$$

$$= a^2 (\sin^2 \phi \cos \theta \, \mathbf{i} - \sin^2 \phi \sin \theta \, \mathbf{j} + \sin \phi \cos \phi \, \mathbf{k}).$$

So

$$|\mathbf{r}_\phi \times \mathbf{r}_\theta| = a^2 \sin \phi$$

(note that $\sin \phi \geq 0$ when $0 \leq \phi \leq \pi$, which gives $(\sin^2 \phi)^{1/2} = \sin \phi$). Thus for a portion S of a sphere of radius a the area is

$$A(S) = \iint_S a^2 \sin \phi \, dA, \qquad dA = d\phi \, d\theta, \qquad dS = a^2 \sin \phi \, d\phi \, d\theta.$$

Now we are prepared to introduce the surface integral.

6.3 Surface Integral

Once again wishing to introduce a physically meaningful integral, we assume S is given by

$$\mathbf{r} = \mathbf{r}(u, v)$$

and that ρ is a mass density over S. Let us try to define the total M mass of S. In Fig. 6.7 we show D_{uv}, the domain of (u, v) for which $\mathbf{r}(u, v)$ locates points of S, a rectangular partition of S in the uv-plane, and the induced partition of S. Since the area of a small "parallelogram" located at (i, j) is

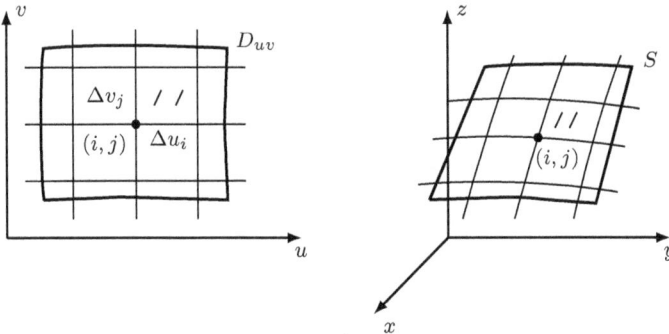

Fig. 6.7 Toward a definition of surface integral. Small "parallelogram" located at (i, j) is hatched in the right-hand picture.

$$\Delta S_{ij} \approx |\mathbf{r}_u(u_i, v_j) \times \mathbf{r}_v(u_i, v_j)| \, \Delta u_i \, \Delta v_j,$$

its mass is

$$\Delta m_{ij} \approx \rho(x(u_i, v_j), y(u_i, v_j), z(u_i, v_j)) \, \Delta S_{ij}$$

where (u_i, v_j) is a node point of the hatched rectangle tagged with (i, j) in the left half of the figure. Summing over i, j in such a way that each such rectangle contains points of D_{uv}, we get

$$M \approx \sum_{i,j} \rho(x(u_i, v_j), y(u_i, v_j), z(u_i, v_j)) \, |\mathbf{r}_u(u_i, v_j) \times \mathbf{r}_v(u_i, v_j)| \, \Delta u_i \, \Delta v_j.$$

Taking the limit over a sequence of partitions as

$$\max_{i,j} [(\Delta u_i)^2 + (\Delta v_j)^2]^{1/2} \to 0,$$

and assuming this limit exists independent of how the partitions are selected, we obtain

$$M = \iint_{D_{uv}} \rho(x(u, v), y(u, v), z(u, v)) \, |\mathbf{r}_u(u, v) \times \mathbf{r}_v(u, v)| \, dA, \quad dA = du \, dv.$$

Existence is guaranteed if the integrand is bounded and piecewise continuous on the closure of D_{uv}. Since our construction made use of no special properties of mass density, we may change ρ to an abstract function f. Although the resulting integral is usually denoted

$$\iint_S f(x, y, z) \, dS,$$

we must keep in mind that this is merely shorthand for

$$\iint_{D_{uv}} f(x(u, v), y(u, v), z(u, v)) \, |\mathbf{r}_u(u, v) \times \mathbf{r}_v(u, v)| \, dA, \quad dA = du \, dv.$$

For a surface represented as $z = z(x, y)$, we work with the projection D_{xy} of S onto the xy-plane:

$$\iint_S f(x, y, z) \, dS = \iint_{D_{xy}} f(x, y, z(x, y)) \, [1 + z_x^2(x, y) + z_y^2(x, y)]^{1/2} \, dx \, dy.$$

Other useful formulas are as follows. For a finite portion S_{cyl} of a cylinder $x^2 + y^2 = a^2$, we obtain

$$\iint_{S_{\text{cyl}}} f(x, y, z) \, dS = a \iint_{D_{\theta z}} f(a \cos \theta, a \sin \theta, z) \, dA, \quad dA = d\theta \, dz$$

$$(6.5)$$

where $D_{\theta z}$ is the image of S_c in the θz-plane. For a portion S_{sph} of a sphere $x^2 + y^2 + z^2 = a^2$, we get $\iint_{S_{\text{sph}}} f(x, y, z) \, dS$ to be

$$a^2 \iint_{D_{\phi \theta}} f(a \sin \phi \cos \theta, a \sin \phi \sin \theta, a \cos \phi) \sin \phi \, dA, \quad dA = d\phi \, d\theta. \quad (6.6)$$

Note that in the above formulas for surface integrals, dA must be positive.

Example 6.6. Let us integrate the function $f(x, y, z) = z$ over the upper half of the unit sphere $z = (1 - x^2 - y^2)^{1/2}$. By (6.6) we obtain

$$\iint_S f(x, y, z) \, dS = 1^2 \int_0^{2\pi} \int_0^{\pi/2} 1 \cdot \cos\phi \sin\phi \, d\phi \, d\theta = \pi. \qquad \square$$

In summary, we see that the surface integral

$$\iint_S f(x, y, z) \, dS$$

yields the mass of S if $f = \rho$ is the surface density, or the area of S if $f = 1$ over S. When $f = \rho$, we also have the coordinates (x_c, y_c, z_c) of the center of mass for S:

$$x_c = \frac{1}{M} \iint_S x\rho(x, y, z) \, dS,$$

$$y_c = \frac{1}{M} \iint_S y\rho(x, y, z) \, dS,$$

$$z_c = \frac{1}{M} \iint_S z\rho(x, y, z) \, dS.$$

Now we proceed to surface integrals of a second type.

6.4 Flux of a Vector Field Across a Surface

Let us start with the rate of flow of a liquid having density $\rho = $ constant and flowing with speed v in a tube of cross-sectional area A:

$$\rho v A.$$

If we place an imaginary tube having section S of area A into water where the velocity is $\mathbf{v}$ and the tube axis is along a unit vector $\mathbf{n}$ (Fig. 6.8), the flow rate across S is $\rho \mathbf{v} \cdot \mathbf{n} A$.

Since a surface S has two sides, there are two possible unit normals $\mathbf{n}_1, \mathbf{n}_2$. We should decide (by engineering reasons) which to select in a certain problem. However there is a convention that for a closed surface like a sphere, the positive orientation has the normal vector chosen outward from the enclosed region. In Fig. 6.9 we show two such normals for the boundary $S_1 + S_2$ of a ball with a spherical cavity.

For a non-closed surface there is no certain rule for surface orientation. However for a surface given by the equation $z = f(x, y)$ the conventional positive orientation has $\mathbf{n} = \mathbf{n}_+$ where $\mathbf{n}_+ \cdot \mathbf{k} > 0$ (see Fig. 6.10). In this

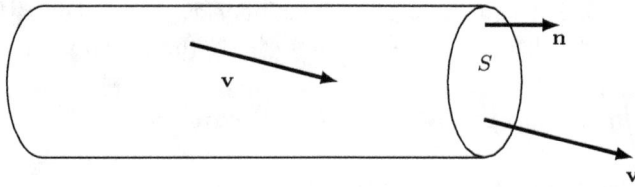

Fig. 6.8   Imaginary tube immersed in a fluid flow field.

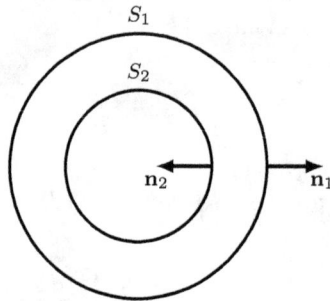

Fig. 6.9   Ball with a spherical cavity; note the two-part boundary surface.

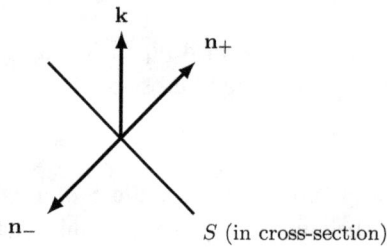

Fig. 6.10   Conventional orientations for a surface of the form $z = f(x, y)$.

case, representing the surface equation in the "level form"

$$z - f(x, y) = 0$$

and using the gradient, we get

$$\mathbf{n}_+ = \frac{-f_x\,\mathbf{i} - f_y\,\mathbf{j} + \mathbf{k}}{(1 + f_x^2 + f_y^2)^{1/2}} = -\mathbf{n}_-.$$

When S is given in parametric form $\mathbf{r} = \mathbf{r}(u, v)$, a unit normal is

$$\mathbf{n} = \frac{\mathbf{r}_u \times \mathbf{r}_v}{|\mathbf{r}_u \times \mathbf{r}_v|} \tag{6.7}$$

and $(\mathbf{r}_u, \mathbf{r}_v, \mathbf{n})$ will constitute a coordinate basis. However, $\mathbf{n}$ will not be outward normal from a closed surface if the order of u, v is selected inappropriately. So depending on the engineering meaning of a problem we may have to use $-\mathbf{n}$.

Now let us approximate the flow rate for a liquid, having density $\rho = \rho(x, y, z)$ and velocity $\mathbf{v} = \mathbf{v}(x, y, z)$, across a surface S with unit normal $\mathbf{n}$ that changes from point to point. For this, we introduce a partition on S as we did in Fig. 6.3, tagging a partition cell with subscripts i, j. Across the cell we have the flow rate

$$\Delta F_{ij} \approx \rho \mathbf{v} \cdot \mathbf{n} \, \Delta S_{ij}\big|_{\mathbf{r}=\mathbf{r}(u_i, v_j)}$$

and summing these over i, j over S we get

$$\sum_{i,j} \rho \mathbf{v} \cdot \mathbf{n} \, \Delta S_{ij}\bigg|_{\mathbf{r}=\mathbf{r}(u_i, v_j)}$$

which is the Riemann sum for the integral

$$\iint_S \rho \mathbf{v} \cdot \mathbf{n} \, dS.$$

This yields the mass flux through S.

To pass to the abstract case, we merely replace $\rho \mathbf{v}$ by an abstract vector function $\mathbf{F} = \mathbf{F}(x, y, z)$ and write $d\mathbf{S} = \mathbf{n} \, dS$. By definition the flux of $\mathbf{F}$ across S is the surface integral

$$\iint_S \mathbf{F}(x, y, z) \cdot d\mathbf{S}.$$

For practical purposes, we have

$$d\mathbf{S} = \mathbf{n} \, dS = \frac{\mathbf{r}_u \times \mathbf{r}_v}{|\mathbf{r}_u \times \mathbf{r}_v|} |\mathbf{r}_u \times \mathbf{r}_v| \, du \, dv = \mathbf{r}_u \times \mathbf{r}_v \, du \, dv.$$

Then

$$\iint_S \mathbf{F} \cdot d\mathbf{S} = \iint_{D_{uv}} \mathbf{F} \cdot (\mathbf{r}_u \times \mathbf{r}_v) \, dA$$

$$= \iint_{D_{uv}} \begin{vmatrix} P & Q & R \\ x_u & y_u & z_u \\ x_v & y_v & z_v \end{vmatrix} dA, \qquad dA = du \, dv. \tag{6.8}$$

We should verify the appropriate orientation of $\mathbf{r}_u \times \mathbf{r}_v$, inserting a negative sign if needed. Expanding the determinant in the integrand and remembering the Jacobian notation, we get

$$\iint_S \mathbf{F} \cdot d\mathbf{S} = \iint_{D_{uv}} \left[P\frac{\partial(y,z)}{\partial(u,v)} + Q\frac{\partial(z,x)}{\partial(u,v)} + R\frac{\partial(x,y)}{\partial(u,v)} \right] dA, \quad dA = du\, dv$$

(6.9)

since

$$d\mathbf{S} = \mathbf{i}\,\frac{\partial(y,z)}{\partial(u,v)}\, du\, dv + \mathbf{j}\,\frac{\partial(z,x)}{\partial(u,v)}\, du\, dv + \mathbf{k}\,\frac{\partial(x,y)}{\partial(u,v)}\, du\, dv.$$

(6.10)

For S given by $z = f(x,y)$ and positively oriented,

$$d\mathbf{S} = \mathbf{n}\, dS = \frac{-f_x\mathbf{i} - f_y\mathbf{j} + \mathbf{k}}{(f_x^2 + f_y^2 + 1)^{1/2}}\,(f_x^2 + f_y^2 + 1)^{1/2}\, dA$$

$$= (-f_x\mathbf{i} - f_y\mathbf{j} + \mathbf{k})\, dA, \quad dA = dx\, dy.$$

Hence

$$\iint_S \mathbf{F} \cdot d\mathbf{S} = \iint_{D_{xy}} (-Pf_x - Qf_y + R)\, dA, \quad dA = dx\, dy,$$

(6.11)

where D_{xy} is the orthogonal projection of S onto the xy-plane. In the solution of problems (and in the proofs of some theorems) this formula is preferred for its simplicity.

Example 6.7. Let us find the flux of the vector field $\mathbf{F}(x,y,z) = \mathbf{k}$ through the upper half of the unit sphere $z = (1 - x^2 - y^2)^{1/2}$. Here $P = 0$, $Q = 0$, and $R = 1$. With use of polar coordinates, formula (6.11) yields

$$\iint_S \mathbf{F} \cdot d\mathbf{S} = \int_{-1}^{1}\int_{-\sqrt{1-x^2}}^{\sqrt{1-x^2}} dy\, dx = \int_0^{2\pi}\int_0^1 r\, dr\, d\theta = \pi. \quad \square$$

Example 6.8. For a cylinder $x^2 + y^2 = a^2$, the outward unit normal is $\mathbf{n} = (x\mathbf{i} + y\mathbf{j})/a = \cos\theta\,\mathbf{i} + \sin\theta\,\mathbf{j}$. Using (6.5) we obtain the flux of the field $\mathbf{F} = P\mathbf{i} + Q\mathbf{j} + R\mathbf{k}$ across a finite portion S_c of the cylinder:

$$\iint_{S_{cyl}} \mathbf{F} \cdot d\mathbf{S} = a \iint_{D_{\theta z}} (P\cos\theta + Q\sin\theta)\, d\theta\, dz,$$

(6.12)

where $D_{\theta z}$ is the image of S_c in the θz-plane. $\quad \square$

Example 6.9. For a sphere $x^2 + y^2 + z^2 = a^2$, the outward unit normal is $\mathbf{n} = (x\,\mathbf{i} + y\,\mathbf{j} + z\,\mathbf{k})/a = \cos\phi\cos\theta\,\mathbf{i} + \cos\phi\sin\theta\,\mathbf{j} + \cos\phi\,\mathbf{k}$. Using (6.6), we find the flux of the field $\mathbf{F} = P\,\mathbf{i} + Q\,\mathbf{j} + R\,\mathbf{k}$ across a finite portion S_{sph} of the sphere:

$$\iint_{S_{\text{sph}}} \mathbf{F} \cdot d\mathbf{S} = a^2 \iint_{D_{\phi\theta}} (P\cos\phi\cos\theta + Q\cos\phi\sin\theta + R\cos\phi)\sin\phi\, d\phi\, d\theta.$$

$\square$

Example 6.10. Let M be the mass of a particle at the origin. By Newton's law, the gravitational force acting on a unit mass at point $\mathbf{r} = (x, y, z)$ is

$$\mathbf{F} = \frac{\gamma M}{|\mathbf{r}|^3}\,\mathbf{r}.$$

We seek the flux of $\mathbf{F}$ through the sphere of radius a centered at the origin. On the sphere, $|\mathbf{r}| = a$ and $\mathbf{F} = (\gamma M/a^3)\mathbf{r}$. We get

$$\iint_S \mathbf{F} \cdot d\mathbf{S} = \iint_S \frac{\gamma M}{a^3}\,\mathbf{r} \cdot d\mathbf{S} = \frac{\gamma M}{a^2} \iint_S dS = \frac{\gamma M}{a^2}\, 4\pi a^2 = 4\pi\gamma M,$$

independent of a.

A similar result holds for the Coulomb force; the flux does not depend on the radius of the sphere. In Chapter 7 we shall see that the result still holds even if we deform the sphere. $\square$

6.5 Problems

Evaluation of Surface Integrals

6.1 Integrate the function $f(x, y, z) = xy^2 z^3$ over each of the following rectangular surfaces:

(a) $0 \le x \le 1,\ 0 \le y \le 2,\ z = 3$

(b) $0 \le x \le 2,\ y = -1,\ 0 \le z \le 1$

(c) $x = 3,\ 0 \le y \le 1,\ -1 \le z \le 1$

6.2 Integrate each of the following functions over the circular disk described by $x^2 + y^2 \le 1,\ z = 0$:

(a) $f(x, y, z) = k$, a constant

(b) $f(x, y, z) = xy(z + 1)$

(c) $f(x, y, z) = \sin^2 x \cos y \sin z$

6.3 Calculate

(a)
$$I = \iint_S (x^2 + y^2)\, dS$$

where S is the sphere $r = R$, and

(b)
$$I = \iint_S [(1 - y^2)^{1/2} + x] z\, dS$$

where S is the portion of the cylinder $x^2 + y^2 = 1$ lying between the planes $z = 0$ and $z = 1$.

6.4 Calculate

(a)
$$I = \iint_S x\, dS$$

where S is the portion of the plane $x + y + z = 1$ enclosed by the cylinder $x^2 + y^2 = 1$;

(b)
$$I = \iint_S xyz\, dS$$

where S is given by $z = (1 - x^2 - y^2)^{1/2}$ for $x, y \geq 0$;

(c)
$$I = \iint_S (1 + 4x^2 + 4y^2)^{1/2}\, dS$$

where S is the portion of the paraboloid $z = 1 - x^2 - y^2$ that lies above the plane $z = 0$.

6.5 For a surface given by $z = g(x, y)$, another formulation of the surface integral is
$$\iint_S f(x, y, z)\, dS = \iint_{D_{xy}} f(x, y, g(x, y))\, |\mathbf{r}_x \times \mathbf{r}_y|\, dA$$

where $\mathbf{r} = x\,\mathbf{i} + y\,\mathbf{j} + g(x, y)\,\mathbf{k}$.

(a) Use this to find the area of the portion of the plane $z = ax + by$ above the region bounded by $y = 0$, $x = x_0$, and $x = ky$, taking all the constants to be positive.

(b) For the surface S of part (a), evaluate the integral
$$I = \iint_S (x + y + z)\, dA.$$

Surface Area Calculation

6.6 Find the area of the cone described by $r(u, v) = u \cos v \, \mathbf{i} + u \sin v \, \mathbf{j} + u \, \mathbf{k}$ for $0 \le u \le 1$ and $0 \le v < 2\pi$.

6.7 Find the area of that portion of

(a) the plane $x + 2y + 3x = 12$ lying within the cylinder $x^2 + y^2 = 9$,

(b) the hyperbolic paraboloid $z = xy$ lying within the cylinder $x^2 + y^2 = a^2$,

(c) the cylinder $r = a$ for $0 < \theta < \pi/3$ and $0 \le z \le 1$,

(d) the annulus $1 \le r \le 2$, $z = 1$ for $\pi/2 \le \theta \le \pi$, and

(e) the cone $z^2 = x^2 + y^2$ in the first octant and bounded by the plane $x + y = 1$.

6.8 Let $a, b, c, d > 0$. Calculate the area of that portion of the plane $ax + by + cz = d$ bounded by the positive coordinate axes.

6.9 Find the area of

(a) the paraboloid $z = x^2 + y^2$ for $x^2 + y^2 \le 1$;

(b) the plane $x + y + z = 1$ in the first octant;

(c) the surface $z = xy$ for $x^2 + y^2 \le 1$, $x \ge 0$, $y \ge 0$.

6.10 Use intuitive reasoning to write down a formula for the area of the surface of revolution generated by

(a) a curve $y = f(x)$, defined for $x \in [a, b]$, revolved about the x-axis;

(b) a curve $z = g(y)$, defined for $y \in [c, d]$, revolved about the y-axis;

(c) a curve $x = h(z)$, defined for $z \in [e, f]$, revolved about the z-axis.

6.11 Find the area of the surface obtained by revolving the parabola $y^2 = 4ax$, $0 \le x \le h$, about the x-axis.

6.12 Find the surface area of the prolate spheroid $r^2 + z^2/4 = a^2$.

6.13 The areas S_x and S_y of the surfaces of revolution about the x and y axes when a curve is given parametrically as $x = x(t)$ and $y = y(t)$ are

$$ S_x = 2\pi \int_{t_1}^{t_2} y(t) \left[\left(\frac{dx}{dt} \right)^2 + \left(\frac{dy}{dt} \right)^2 \right]^{1/2} dt, $$

$$ S_y = 2\pi \int_{t_1}^{t_2} x(t) \left[\left(\frac{dx}{dt} \right)^2 + \left(\frac{dy}{dt} \right)^2 \right]^{1/2} dt. $$

Find S_y for the elliptical curve given by $x = a \cos t$ and $y = b \sin t$ where $-\pi/2 \le t \le \pi/2$ and $a \ge b$.

Flux Calculation

6.14 Find the flux of the constant field $\mathbf{A} = \mathbf{k}$ through

(a) the square $z = 0,\ 0 \le x \le 1,\ 0 \le y \le 1$,

(b) the square $x = 0,\ 0 \le y \le 1,\ 0 \le z \le 1$,

(c) the hemisphere $r = a,\ 0 \le \phi \le \pi/2,\ 0 \le \theta \le 2\pi$,

(d) the disk $z = 0,\ r \le a$, and

(e) the first-octant portion of the plane $x + y + z = 1$.

6.15 Compute the flux of $\mathbf{A}$ through S if

(a) $\mathbf{A} = x\mathbf{j} + y\mathbf{k}$ and S is the planar surface $x + y + z = 2$ for $x \ge 0,\ y \ge 0,$ $z \ge 0$;

(b) $\mathbf{A} = x\mathbf{i} + y\mathbf{j} + z\mathbf{k}$ and S is the hemisphere $x^2 + y^2 + z^2 = 1$ for $z \le 0$;

(c) $\mathbf{A} = x\mathbf{i} + y\mathbf{j}$ and S is the cylindrical surface $x^2 + y^2 = 1$ for $0 \le z \le 1$.

6.16 Compute

$$I = \iint_S \nabla f \cdot d\mathbf{S},$$

where

(a) $f(x,y,z) = x(yz)^2$ and S is the surface $x = 0,\ 0 \le y \le 1,\ -1 \le z \le 0$;

(b) $f(x,y,z) = x^2 + z^2$ and S is the surface $x^2 + y^2 = 1$ for $x \ge 0$ and $1 \le z \le 2$;

(c) $f(x,y,z) = \cos z$ and S is the surface $x^2 + y^2 \le 1,\ z = \pi/2$;

(d) $f(x,y,z) = 3x$ and S is the surface $\theta = \pi/4,\ 1 \le r \le 2,\ 0 \le z \le 1$.

6.17 Compute

$$I = \iint_S (\nabla \times \mathbf{A}) \cdot d\mathbf{S},$$

where

(a) $\mathbf{A} = 3z\,\mathbf{i} + \mathbf{j} + 2x\,\mathbf{k}$ and S is the surface described by $x = 1,\ 0 \le y \le 1$, and $0 \le z \le 1$;

(b) $\mathbf{A} = (z+y)\,\mathbf{i} + (x+z)\,\mathbf{j} + (y+x)\,\mathbf{k}$ and S is the surface described by $y = -1,$ $0 \le x \le 1$, and $0 \le z \le 1$;

(c) $\mathbf{A} = 2zy\,\mathbf{i} + (x+y)^2\,\mathbf{j} + e^{(y-x)}\,\mathbf{k}$ and S is the described by surface $z = 2,$ $0 \le x \le 1$, and $0 \le y \le 1$.

Chapter 7

Principal Theorems of Multivariable Calculus

We have considered relatively simple problems concerning some character-istics of bodies and fields. But various integrals are used in mathematical physics to establish important variational principles as well as formulas for the finite element method. The formulas here discussed are called theorems because they are deduced under some restrictions that must be formulated each time. They are Green's theorem, Stokes' theorem, and the divergence theorem. All are wide extensions of the Newton–Leibnitz formula

$$\int_a^b f'(x)\,dx = f(b) - f(a).$$

7.1 Green's Theorem

Green's theorem underpins the other theorems of this chapter. Its main content is an integration formula relating a double integral over a domain D to a line integral over the boundary L of D. The line integral is of the second type, for which the orientation of the contour is obligatory. So let us discuss how to orient L. We start with a simple contour L, say a circle. The orientation of a circle C in the plane is positive when we traverse C counterclockwise as in Fig. 7.1.

This definition can be adapted to any simple non-self-intersecting closed contour that can be deformed continuously into a circle. Examples include squares and triangles. To extend the definition to a contour enclosing an-other contour, we first draw two circles as in Fig. 7.2. Let us introduce a cut along a diameter, dividing D into a union of subdomains D_1 and D_2. Each of these is Type 1, with a simple boundary contour that we can regard as positively oriented. Arrows indicate the positive orientations for the boundaries L_1 and L_2. Over the segments AB and DE the routes

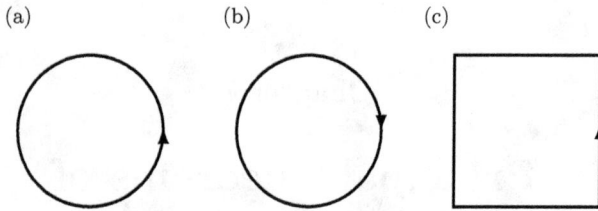

Fig. 7.1   Oriented closed contours.   (a) Positive orientation for circle.   (b) Negative orientation for circle.   (c) Positive orientation for square.

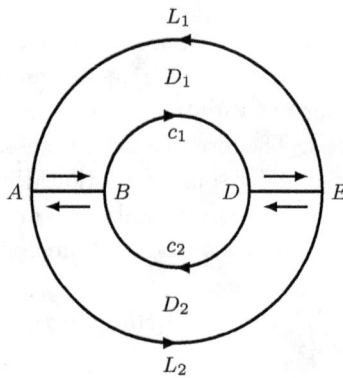

Fig. 7.2   Contour containing another contour.

"go" in opposite directions, resulting in cancellation of the corresponding Type 2 line integral contributions. The parts not eliminated are the outer and inner circles. Note that the outer circle is traversed counterclockwise but the inner circle is traversed clockwise. Mathematicians elaborated a simple rule for such "multiply-connected" domains with "holes" present: the boundary contour is positively oriented if a "person" going around the contour with head "above" finds the domain on the left.

Theorem 7.1 (Green). *Let the boundary L of a domain $D \subset \mathbb{R}^2$ be a piecewise smooth contour positively oriented with respect to D. Let the functions P, Q be continuous with continuous partial derivatives P_y and Q_x on a bounded open domain containing $D \cup L$. Then*

$$\iint_D (Q_x - P_y)\, dA = \int_L P\, dx + Q\, dy. \tag{7.1}$$

The term "piecewise smooth contour" means a contour described by one or a few pairs of functions $x = x(t)$, $y = y(t)$ that are piecewise continuously differentiable.

Proof. Clearly (7.1) involves two independent equalities

$$\iint_D Q_x \, dA = \int_L Q \, dy, \qquad \iint_D P_y \, dA = -\int_L P \, dx.$$

It suffices to prove the second one. We start with a domain $D = D_1$ of Type 1 with boundary L_1. See Fig. 7.3 for the notation. By Fubini's theorem

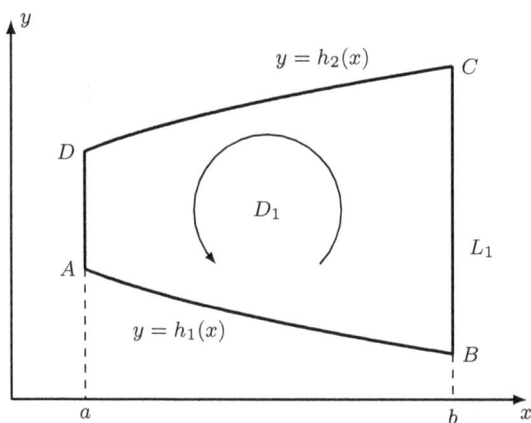

Fig. 7.3 First step in the proof of Green's theorem: L_1 consists of four simple lines, AB, BC, CD, and DA. The needed integral is the sum of four integrals over these lines.

$$\iint_{D_1} P_y(x,y) \, dA = \int_a^b \int_{h_1(x)}^{h_2(x)} P_y(x,y) \, dy \, dx$$

$$= \int_a^b [P(x, h_2(x)) - P(x, h_1(x))] \, dx.$$

Since the line integrals over AB, BC, CD, DA constituting L_1 are

$$\int_{AB} P(x,y) \, dx = \int_a^b P(x, h_1(x)) \, dx, \qquad \int_{BC} P(x,y) \, dx = 0,$$

$$\int_{CD} P(x,y) \, dx = -\int_a^b P(x, h_2(x)) \, dx, \qquad \int_{DA} P(x,y) \, dx = 0,$$

we have the needed equality over D_1.

Under the conditions of the theorem it can be proved that D can be represented as a finite set of subdomains $D_1, \ldots, D_k$ all of Type 1 and that may intersect only via their boundaries $L_1, \ldots, L_k$. Thus

$$\iint_D P_y(x, y)\, dA = -\sum_{i=1}^{k} \int_{L_i} P(x, y)\, dx.$$

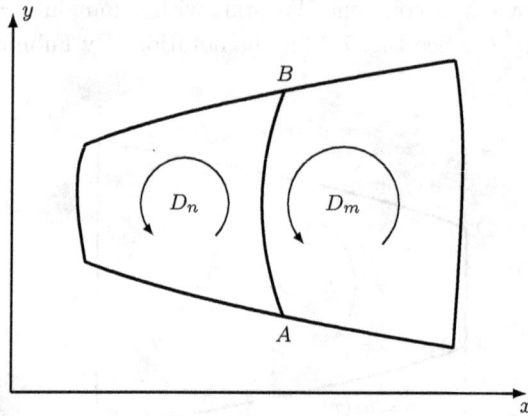

Fig. 7.4   Two adjacent subdomains of D. On two subdomains D_m and D_n having a shared boundary portion AB, the orientations of AB are opposite. Hence the corresponding parts of the line integrals mutually cancel.

Along any boundary L_k that is common to two subdomains, cancellation occurs (Fig. 7.4) and we are led to the required statement for P. The statement for Q is analogous.   □

Reversal in the orientation of L produces a sign change on the right-hand side of (7.1).

Our proof holds when D can be represented as a finite union of domains of Type 1, and simultaneously as a finite union of domains of Type 2, which is valid in the conditions of Green's theorem.

Example 7.1. Let us compute both sides of Green's theorem for $P = x^2 y$, $Q = xe^y$, and $D = [0, 1] \times [0, 1]$:

$$\int_L P\, dx + Q\, dy = -\int_0^1 x^2\, dx + \int_0^1 e^y\, dy = e - \frac{4}{3},$$

$$\iint_D (Q_x - P_y)\, dA = \int_0^1 \int_0^1 (e^y - x^2)\, dy\, dx = e - \frac{4}{3}. \qquad \Box$$

Example 7.2. This time we shall make use of polar coordinates. Let $P = xy$ and $Q = x^2y^2$, and consider the quarter annulus shown in Fig. 7.5. We have $Q_x = 2xy^2$, $P_y = x$, and therefore

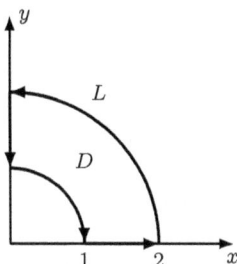

Fig. 7.5   Domain for Example 7.2.

$$\iint_D (Q_x - P_y)\, dA = \int_0^{\pi/2} \int_1^2 [2r\cos\theta(r\sin\theta)^2 - r\cos\theta] r\, dr\, d\theta$$

$$= \int_0^{\pi/2} \int_1^2 r\cos\theta[2r^2\sin^2\theta - 1] r\, dr\, d\theta$$

$$= \frac{9}{5}.$$

The other side of Green's theorem is

$$\int_L P\, dx + Q\, dy = \int_0^{\pi/2} [(2\cos\theta \cdot 2\sin\theta)(-2\sin\theta\, d\theta)$$

$$+ (2\cos\theta \cdot 2\sin\theta)^2(2\cos\theta\, d\theta)]$$

$$- \int_0^{\pi/2} [(\cos\theta \cdot \sin\theta)(-\sin\theta\, d\theta)$$

$$+ (\cos\theta \cdot \sin\theta)^2(\cos\theta\, d\theta)]$$

$$= \int_0^{\pi/2} \sin^2\theta \cos\theta(-7 + 31\cos^2\theta)\, d\theta$$

$$= \frac{9}{5}. \qquad \Box$$

7.2 Other Forms of Green's Equality

Stokes' Form

Let $\mathbf{F} = P\mathbf{i} + Q\mathbf{j}$ and $d\mathbf{r} = dx\,\mathbf{i} + dy\,\mathbf{j}$. Then $P\,dx + Q\,dy = \mathbf{F} \cdot d\mathbf{r}$ and Green's formula becomes

$$\iint_D (Q_x - P_y)\, dA = \int_L \mathbf{F} \cdot d\mathbf{r}.$$

We can consider $\mathbf{F}$ as a function given in $\mathbb{R}^3$, i.e., as $\mathbf{F}(x,y) = P(x,y)\,\mathbf{i} + Q(x,y)\,\mathbf{j} + 0\,\mathbf{k}$ and ∇ as a three-dimensional operator. Then $(\nabla \times \mathbf{F}) \cdot \mathbf{k} = Q_x - P_y$ and

$$\iint_D (\nabla \times \mathbf{F}) \cdot \mathbf{k}\, dA = \int_L \mathbf{F} \cdot d\mathbf{r}. \qquad (7.2)$$

We shall encounter this in a more general form when studying Stokes' theorem for surface integrals.

Divergence Form

Replacing Q by P and $-P$ by Q in (7.1), we get another form of Green's equality:

$$\iint_D (P_x + Q_y)\, dA = \int_L -Q\, dx + P\, dy. \qquad (7.3)$$

It is similar to Gauss' formula (known in Russia as the Gauss–Ostrogradsky formula) and is related to the divergence theorem. We see that $\nabla \cdot \mathbf{F} = P_x + Q_y$. Now let us transform the integrand on the right-hand side.

Figure 7.6 shows a portion of L given by the formula $\mathbf{r}(t) = x(t)\,\mathbf{i} + y(t)\,\mathbf{j}$. The outward normal from D is $\mathbf{n}$. At point t a tangent vector to $\mathbf{r} = \mathbf{r}(t)$ is $\mathbf{r}'(t) = x'(t)\,\mathbf{i} + y'(t)\,\mathbf{j}$. Clearly $\mathbf{n}$ is perpendicular to $\mathbf{r}'(t)$. Taking $\mathbf{N} = y'(t)\,\mathbf{i} - x'(t)\,\mathbf{j}$ we get $\mathbf{N} \cdot \mathbf{r}'(t) = 0$. So $\mathbf{N}$ is also orthogonal to $\mathbf{r}'(t)$. As $|\mathbf{N}| = \mathbf{r}'(t)$ we have

$$\mathbf{n} = \frac{\mathbf{N}}{|\mathbf{N}|} = \frac{\mathbf{N}}{|\mathbf{r}'(t)|}.$$

(The reader should show that $\mathbf{n}$ is really outward normal if $\mathbf{r}'(t)$ is directed along the positive orientation of L.) Since $ds = |\mathbf{r}'(t)|\, dt$ is the elementary length of the part of L corresponding to dt,

$$\mathbf{n}\, ds = \left(\frac{y'(t)}{|\mathbf{r}'(t)|}\,\mathbf{i} - \frac{x'(t)}{|\mathbf{r}'(t)|}\,\mathbf{j} \right) |\mathbf{r}'(t)|\, dt = dy\,\mathbf{i} - dx\,\mathbf{j}$$

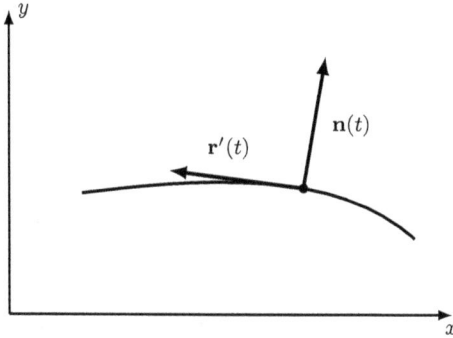

Fig. 7.6 Tangent and normal vectors on the boundary of D.

and thus $-Q\,dx + P\,dy = \mathbf{F}\cdot\mathbf{n}\,ds$. It follows that

$$\iint_D \nabla\cdot\mathbf{F}\,dA = \int_L \mathbf{F}\cdot\mathbf{n}\,ds. \qquad (7.4)$$

This is quite similar to Gauss' equality in $\mathbb{R}^3$, deduced later, with the volume replaced by a planar domain D and the boundary surface replaced by boundary contour L.

Equation (7.4) occurs frequently in mathematical physics. Written in coordinate-free form, it can be used in any curvilinear coordinate system.

Applications

Equation (7.4) can be considered as a two-dimensional generalization of the ordinary integration by parts formula. Indeed, introducing functions $P_k = P_k(x,y)$ and $Q_k = Q_k(x,y)$ for $k = 1,2$, and writing $P = P_1 P_2$ and $Q = Q_1 Q_2$, we get from (7.4)

$$\iint_D (P_{1x}P_2 + Q_{1y}Q_2)\,dA + \iint_D (P_{2x}P_1 + Q_{2y}Q_1)\,dA$$

$$= \int_L (P_1 P_2 + Q_1 Q_2)\,ds. \qquad (7.5)$$

Another simple application of (7.1) follows from the fact that $\iint_D dA$ is the area of D. If we take P, Q such that $Q_x - P_y = 1$, the right side of (7.1) yields this area. It is easy to find such combinations; they include (a) $Q = x$ and $P = 0$, (b) $Q = 0$ and $P = -y$, and (c) $Q = \frac{1}{2}x$ and $P = -\frac{1}{2}y$.

Hence the area of D can be found by the formulas

$$A(D) = \int_L x\,dy = -\int_L y\,dx = \frac{1}{2}\int_L x\,dy - y\,dx. \qquad (7.6)$$

Example 7.3. Let us calculate the area enclosed by the ellipse (1.8). Using its representation $x = a\cos t$ and $y = b\sin t$ for $t \in [0, 2\pi)$, we get

$$
\begin{aligned}
A(D) &= \frac{1}{2}\int_L x\,dy - y\,dx = \frac{1}{2}\int_L \left(x\frac{dy}{dt} - y\frac{dx}{dt} \right) dt \\
&= \frac{1}{2}\int_0^{2\pi} [(a\cos t)(b\cos t) - (b\sin t)(-a\sin t)]\,dt \\
&= \frac{1}{2}ab\int_0^{2\pi} dt = \pi ab. \qquad \qquad \square
\end{aligned}
$$

Some Formulas Pertaining to the Equilibrium Problem for a Planar Membrane

The equilibrium of a planar membrane under applied forces is described by Poisson's equation

$$-\frac{\partial^2 u}{\partial x^2} - \frac{\partial^2 u}{\partial y^2} = F, \qquad (x, y) \in D \subset \mathbb{R}^2, \qquad (7.7)$$

where u is the normal displacement at the membrane point (x, y) lying on D, and F is a normal force distributed over D.

First we consider the Dirichlet problem when the membrane edge ∂D is fixed:

$$u\big|_{\partial D} = 0. \qquad (7.8)$$

Let u be a solution of the *boundary value problem* (7.7)–(7.8). All the functions as well as ∂D are assumed to be smooth enough to justify all the needed manipulations.

Let v be an arbitrary function satisfying Dirichlet's condition $v|_{\partial D} = 0$. Let us multiply both sides of (7.7) by v and then integrate the equality over D:

$$-\iint_D (u_{xx} + u_{yy})v\,dA = \iint_D Fu\,dA.$$

Integrating on the left by parts and using (7.5), we get

$$\iint_D (u_x v_x + u_y v_y)\,dA = \iint_D Fv\,dA \qquad (7.9)$$

holding for any v satisfying $v|_{\partial D} = 0$. In mechanics this equation is used to formulate the *virtual displacement principle*: a function u satisfying Dirichlet's condition and the equation (7.9) for any smooth function v, also satisfying Dirichlet's condition, is a solution of the boundary value problem (7.7)–(7.8). The principle is used to define the concept of a *weak solution* to Dirichlet's problem for the membrane, which is one of the central points of modern mathematical physics.

It is worth noting that by setting $v = u$ in (7.9), from its terms we can construct the integral

$$P(u) = \frac{1}{2} \iint_D (u_x^2 + u_y^2)\, dA - \iint_D Fu\, dA,$$

which is the total potential energy of the membrane with fixed edge. In the calculus of variations it is shown that $P(u)$ attains its minimum over the set of functions satisfying Dirichlet's condition on the solution u of the problem, if it exists.

Similar results can be obtained for other types of boundary value problems for a membrane. For a membrane with the edge under load ϕ, which is given by the condition

$$\left.\frac{\partial u}{\partial n}\right|_{\partial D} = \phi \tag{7.10}$$

instead of (7.8), similar transformations lead to the equation

$$\iint_D (u_x v_x + u_y v_y)\, dA = \iint_D Fv\, dA + \int_{\partial D} \phi v\, ds \tag{7.11}$$

valid for arbitrary v, without given conditions on the boundary. The equation, holding for arbitrary v, defines a solution u of equation (7.7) and boundary condition (7.10). Also called the virtual displacement principle, it is the basis for posing weak setups of boundary value problems. Here the total potential energy transforms to

$$P_1(u) = \frac{1}{2} \iint_D (u_x^2 + u_y^2)\, dA - \iint_D Fu\, dA - \int_{\partial D} \phi u\, ds.$$

The principle of minimum total potential energy is valid for this problem and for various mixed problems of the membrane theory.

Weak formulation of problems in mechanics and physics plays an important role in engineering; it is the foundation of the finite element method, one of the principal methods used to find numerical solutions of various practical problems.

Now we start preparing to formulate Stokes' theorem.

7.3   A Curve on a Surface

Let a surface S be given by the equation $\mathbf{r} = \mathbf{r}(u, v)$ in a domain D_{uv} of the uv-plane (Fig. 7.7).

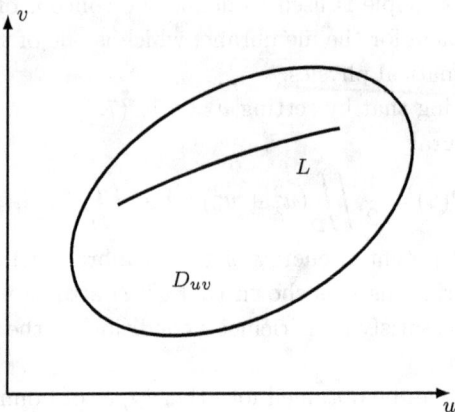

Fig. 7.7   Curve L on the uv-plane.

A curve L in D_{uv} is described by equations

$$u = u(t), \qquad v = v(t), \qquad t \in [a, b].$$

Its image $\mathbf{r} = \mathbf{r}(u(t), v(t))$ in $\mathbb{R}^3$, for $t \in [a, b]$, is a curve lying on S. By the chain rule

$$\mathbf{dr} = \mathbf{i}\, dx + \mathbf{j}\, dy + \mathbf{k}\, dz$$
$$= \mathbf{i}\,(x_u\, du + x_v\, dv) + \mathbf{j}\,(y_u\, du + y_v\, dv) + \mathbf{k}\,(z_u\, du + z_v\, dv)$$
$$= [\,(x_u\, u' + x_v\, v')\,\mathbf{i} + (y_u\, u' + y_v\, v')\,\mathbf{j} + (z_u\, u' + z_v\, v')\,\mathbf{k}\,]\, dt,$$

where in the last line we understand that $x_u = x_u(u(t), v(t))$, $u' = u'(t)$, and so on. For a vector field $\mathbf{F} = P\,\mathbf{i} + Q\,\mathbf{j} + R\,\mathbf{k}$ then, we have

$$\int_L \mathbf{F} \cdot d\mathbf{r} = \int_a^b [\,P(x_u u' + x_v v') + Q(y_u u' + y_v v') + R(z_u u' + z_v v')\,]\, dt,$$

where

$$P = P(x, y, z) = P(x(u(t), v(t)), y(u(t), v(t)), z(u(t), v(t)))$$

and so forth.

In the following section we use a special curve, the boundary ∂S of S, oriented as in Fig. 7.8.

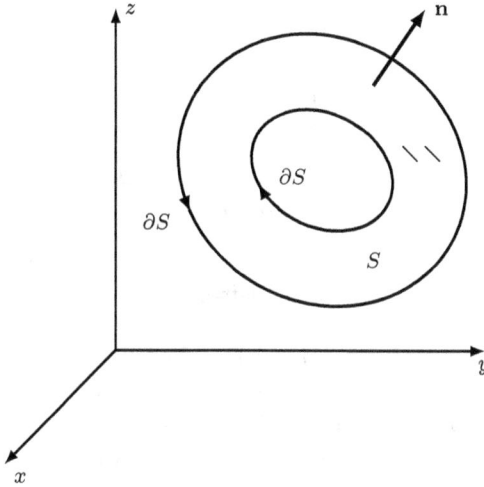

Fig. 7.8   Positive orientations of boundary contours.

7.4   Stokes' Theorem

The principal part of Stokes' theorem is the equality (7.12).

Theorem 7.2 (Stokes). *For a vector function* $\mathbf{F} = P\mathbf{i} + Q\mathbf{j} + R\mathbf{k}$ *continuously differentiable in a bounded open domain* $E \subset \mathbb{R}^3$ *containing* S *and its boundary, the equation*

$$\int_{\partial S} \mathbf{F} \cdot d\mathbf{r} = \iint_S \nabla \times \mathbf{F} \cdot d\mathbf{S} \tag{7.12}$$

holds under the following conditions:

(1) S is a two-sided and positively oriented (as shown in Fig. 7.8) smooth surface.

(2) ∂S, the boundary of S, is positively oriented and piecewise smooth.

Proof. In components, equation (7.12) (still not proven) is

$$\int_{\partial S} P\,dx + Q\,dy + R\,dz = \iint_{D_{uv}} \left[(R_y - Q_z)\frac{\partial(y,z)}{\partial(u,v)} \right] du\,dv$$

$$+ \iint_{D_{uv}} \left[(P_z - R_x)\frac{\partial(z,x)}{\partial(u,v)} \right] du\, dv$$

$$+ \iint_{D_{uv}} \left[(Q_x - P_y)\frac{\partial(x,y)}{\partial(u,v)} \right] du\, dv.$$

As P, Q, R are independent this constitutes three independent equations which can be proved similarly. So we focus on the equation for P:

$$\int_{\partial S} P\, dx = \iint_{D_{uv}} \left[P_z\frac{\partial(z,x)}{\partial(u,v)} - P_y\frac{\partial(x,y)}{\partial(u,v)} \right] du\, dv. \tag{7.13}$$

Because the contour ∂S is in one-to-one correspondence with the piecewise smooth boundary ∂D_{uv} of D_{uv}, the left-hand side is

$$\int_{\partial S} P\, dx = \int_{\partial D_{uv}} Px_u\, du + Px_v\, dv. \tag{7.14}$$

Moreover ∂D_{uv} is positively oriented with respect to the orientation of D_{uv} (the normal is out of the page in Fig. 7.9). An application of Green's theorem gives

$$\int_{\partial D_{uv}} Px_u\, du + Px_v\, dv = \iint_{D_{uv}} [(Px_v)_u - (Px_u)_v]\, du\, dv.$$

Now let us expand the integrand on the right side of the last equality:

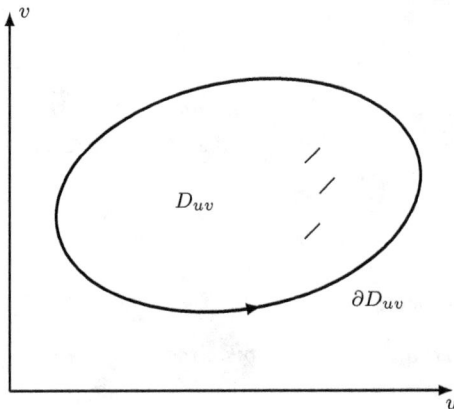

Fig. 7.9   Proof of Stokes' theorem.

$$(Px_v)_u - (Px_u)_v = Px_{vu} + P_u x_v - (Px_{uv} + P_v x_u)$$

$$= P x_{vu} + (P_x x_u + P_y y_u + P_z z_u) x_v$$
$$- [P x_{uv} + (P_x x_v + P_y y_v + P_z z_v) x_u]$$
$$= P_y (y_u x_v - y_v x_u) + P_z (z_u x_v - z_v x_u)$$
$$= -P_y \frac{\partial(x,y)}{\partial(u,v)} + P_z \frac{\partial(z,x)}{\partial(u,v)}.$$

Comparing this with the integrand in (7.14), we finish the proof. □

Corollary 7.1. *The value of*

$$\iint_S \nabla \times \mathbf{F} \cdot d\mathbf{S}$$

does not depend on the shape of smooth S having boundary ∂S if S lies inside an open domain G.

Let us note that for a plane surface S Stokes' equality is (7.4), which is one form of the Green's theorem equality.

Example 7.4. Let us evaluate both sides of (7.12) for $\mathbf{F} = y\,\mathbf{i} + z\,\mathbf{j} + x\,\mathbf{k}$ and ∂S described by $x = \cos t$, $y = \sin t$, $z = 1$ for $0 \le t < 2\pi$. We have

$$\int_{\partial S} \mathbf{F} \cdot d\mathbf{r} = \int_0^{2\pi} \left(y\frac{dx}{dt} + z\frac{dy}{dt} + x\frac{dz}{dt} \right) dt = -\int_0^{2\pi} \sin^2 t \, dt = -\pi.$$

Next

$$\nabla \times \mathbf{F} = \begin{vmatrix} \mathbf{i} & \mathbf{j} & \mathbf{k} \\ \dfrac{\partial}{\partial x} & \dfrac{\partial}{\partial y} & \dfrac{\partial}{\partial z} \\ y & z & x \end{vmatrix} = -\mathbf{i} - \mathbf{j} - \mathbf{k}, \qquad d\mathbf{S} = \mathbf{k}\, dS$$

and

$$\iint_S \nabla \times \mathbf{F} \cdot d\mathbf{S} = \iint_S (-1)\, dS = -\pi(1)^2 = -\pi$$

as expected. □

Stokes' theorem extends to the case when S can be divided into a few subdomains S_k, with possible intersections at their boundaries, and satisfying the conditions of the theorem on each S_k (Fig. 7.10). In this case we get

$$\int_{\partial S_k} \mathbf{F} \cdot d\mathbf{r} = \iint_{S_k} \nabla \times \mathbf{F} \cdot d\mathbf{S}$$

for each subsurface, and by summation over k

$$\sum_{k=1}^{n} \int_{\partial S_k} \mathbf{F} \cdot d\mathbf{r} = \sum_{k=1}^{n} \iint_{S_k} \nabla \times \mathbf{F} \cdot d\mathbf{S}.$$

On the right stands the integral over S. On the left stands the integral over ∂S, as the integrals over the internal cuts appear twice in opposite directions and mutually cancel. So (7.12) holds.

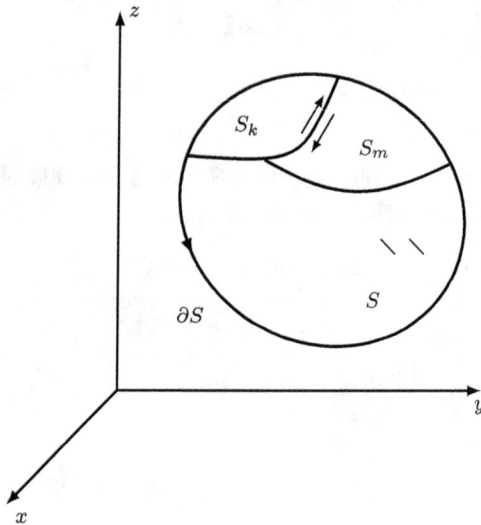

Fig. 7.10   Extension of Stokes' theorem.

An important consequence of Stokes' theorem is a condition under which a field $\mathbf{F}$ is conservative in an open set E. We know the necessary condition for this is $\nabla \times \mathbf{F} = \mathbf{0}$. Let this condition prevail in E, and let E be such that any closed contour L lying in E is the boundary of a piecewise smooth surface S contained in E. Then by Stokes' theorem

$$\int_{L} \mathbf{F} \cdot d\mathbf{r} = \iint_{S} \nabla \times \mathbf{F} \cdot d\mathbf{S} = 0.$$

As $L \subset E$ is arbitrary, $\mathbf{F}$ is a conservative field.

7.5 Divergence Theorem

This theorem is attributed to K. Gauss (in Russia to Gauss–Ostrogradsky), and one of its versions is due to Lagrange. We present it in a form more general than that covered by our proof.

Theorem 7.3. *Let* $\mathbf{F} = P\mathbf{i} + Q\mathbf{j} + R\mathbf{k}$ *and let the partial derivatives* P_x, Q_y, R_z *be continuous on an open domain including the closure of a bounded domain* $E \subset \mathbb{R}^3$ *with piecewise smooth positively oriented boundary surface* ∂E. *Then*

$$\iiint_E \nabla \cdot \mathbf{F}\, dV = \iint_{\partial E} \mathbf{F} \cdot d\mathbf{S}. \tag{7.15}$$

Proof. Recall that $\mathbf{F}$ and P_x, Q_y, R_z should be continuous on $E \cup \partial E$, and that "piecewise smooth" for ∂E means that, except for a few smooth lines, the surface ∂E is represented by continuously differentiable functions $\mathbf{r} = \mathbf{r}(u, v)$ and the normal $\mathbf{n}$ on ∂E is outward. Formula (7.15) consists of three independent formulas

$$\iiint_E P_x\, dV = \iint_{\partial E} P\mathbf{i} \cdot \mathbf{n}\, dS,$$

$$\iiint_E Q_y\, dV = \iint_{\partial E} Q\mathbf{j} \cdot \mathbf{n}\, dS,$$

$$\iiint_E R_z\, dV = \iint_{\partial E} R\mathbf{k} \cdot \mathbf{n}\, dS.$$

We shall prove the third one under the condition that E is a Type 1 domain. Let E satisfy $g_1(x, y) \le z \le g_2(x, y)$ and have projection D_{xy} onto the xy-plane. Then

$$\iiint_E R_z\, dV = \iint_{D_{xy}} \int_{g_1(x,y)}^{g_2(x,y)} R_z\, dz\, dx\, dy$$

$$= \iint_{D_{xy}} R(x, y, g_2(x, y))\, dx\, dy - \iint_{D_{xy}} R(x, y, g_1(x, y))\, dx\, dy. \tag{7.16}$$

In the notation of Fig. 7.11, the relation between an elementary area dS on the upper surface $z = g_2(x, y)$ and its projection onto D_{xy} is

$$dx\, dy = dS \cos \gamma = dS\, \mathbf{k} \cdot \mathbf{n}$$

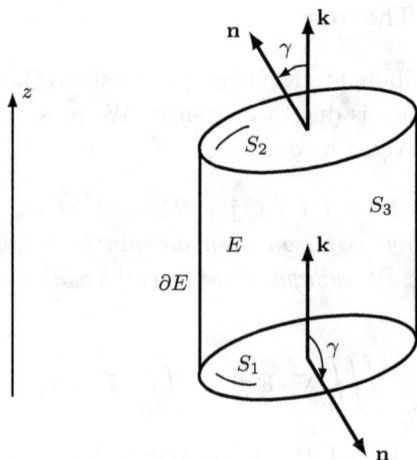

Fig. 7.11   Proof of the divergence theorem.

while on $z = g_1(x, y)$ where $\mathbf{k} \cdot \mathbf{n} < 0$ we have $dx\, dy = -dS\, \mathbf{k} \cdot \mathbf{n}$. Since $\partial E = S_1 + S_2 + S_3$ where $\mathbf{k} \cdot \mathbf{n} = 0$ on S_3, the right-hand side of (7.16) becomes

$$\iint_{\partial E} R\,\mathbf{k} \cdot \mathbf{n}\, dS$$

as needed.

We can extend the relation

$$\iiint_E R_z\, dV = \iint_{\partial E} R\,\mathbf{k} \cdot d\mathbf{S} \qquad (7.17)$$

to the case when E is a finite union of Type 1 domains with possible intersections only at their boundaries. Writing (7.17) for each domain E_k, we get

$$\iiint_E R_z\, dV = \sum_k \iiint_{E_k} R_z\, dV = \sum_k \iint_{\partial E_k} R\,\mathbf{k} \cdot d\mathbf{S}.$$

But each shared boundary portion E_k is encountered twice with normals directed oppositely; the resulting cancellation of terms gives

$$\sum_k \iint_{\partial E_k} R\,\mathbf{k} \cdot d\mathbf{S} = \iint_{\partial E} R\,\mathbf{k} \cdot d\mathbf{S}$$

as required.

The proofs for the terms

$$\iiint_E P_x \, dV, \qquad \iiint_E Q_y \, dV,$$

are analogous, but require that E can be represented as a finite union of domains of Types 2 and 3, respectively. So our proof is for the case where E can be represented in three forms: as unions of subdomains of all three types. $\qquad\square$

Writing $\mathbf{i} \cdot \mathbf{n} = n_x$, $\mathbf{j} \cdot \mathbf{n} = n_y$, and $\mathbf{k} \cdot \mathbf{n} = n_z$, we can also present (7.15) in the form

$$\iiint_E (P_x + Q_y + R_z) \, dV = \iint_{\partial E} (Pn_x + Qn_y + Rn_z) \, dS.$$

Further, note that if

$$P = P_1 P_2, \qquad Q = Q_1 Q_2, \qquad R = R_1 R_2$$

then we get an analogue to the formula for integration by parts:

$$\iiint_E (P_{1x} P_2 + Q_{1y} Q_2 + R_{1z} R_2) \, dV$$

$$= -\iiint_E (P_1 P_{2x} + Q_1 Q_{2y} + R_1 R_{2z}) \, dV$$

$$+ \iint_{\partial E} (P_1 P_2 n_x + Q_1 Q_2 n_y + R_1 R_2 n_z) \, dS. \qquad (7.18)$$

This is used in weak formulations of boundary value problems for the partial differential equations of mechanics.

Example 7.5. Let E be the tetrahedron bounded by the Cartesian coordinate planes and the plane $ax + by + cz = d$ for $a, b, c, d > 0$. The total outflux of $\mathbf{F} = P\mathbf{i} + Q\mathbf{j} + R\mathbf{k}$ over ∂E is

$$\iiint_E \nabla \cdot \mathbf{F} \, dV = \int_0^{d/a} \int_0^{(d-ax)/b} \int_0^{(d-ax-by)/c} (P_x + Q_y + R_z) \, dz \, dy \, dx.$$

We exploited the divergence theorem to avoid computation of four surface integrals. $\qquad\square$

Finally, suppose $\mathbf{F}$ is a conservative field with

$$\mathbf{F} = \nabla u$$

for some twice differentiable function u. Then $\nabla \cdot \mathbf{F} = \nabla^2 u$,

$$Pn_x + Qn_y + Rn_z = u_x \mathbf{i} \cdot \mathbf{n} + u_y \mathbf{j} \cdot \mathbf{n} + u_z \mathbf{k} \cdot \mathbf{n} = \nabla u \cdot \mathbf{n} = \frac{\partial u}{\partial n},$$

and

$$\iiint_E \nabla^2 u \, dV = \iint_{\partial E} \frac{\partial u}{\partial n} \, dS. \tag{7.19}$$

Other equalities for potential fields appear, for example, in the theory of weak solutions to Laplace's equation. From (7.18) we can deduce two equalities with $P_1 = u_x$, $Q_1 = u_y$, $R_1 = u_z$, and $P_2 = Q_2 = R_2 = v$. The first is

$$\iiint_E (u_{xx} + u_{yy} + u_{zz}) v \, dV = - \iiint_E (u_x v_x + u_y v_y + u_z v_z) \, dV$$

$$+ \iint_{\partial E} (u_x n_x + u_y n_y + u_z n_z) v \, dS$$

or

$$\iiint_E (\nabla^2 u) v \, dV = - \iiint_E \nabla u \cdot \nabla v \, dV + \iint_{\partial E} \frac{\partial u}{\partial n} v \, dS. \tag{7.20}$$

This is known as *Green's first identity*. Swapping u and v, we get

$$\iiint_E (\nabla^2 v) u \, dV = - \iiint_E \nabla v \cdot \nabla u \, dV + \iint_{\partial E} \frac{\partial v}{\partial n} u \, dS, \tag{7.21}$$

and by subtracting (7.21) from (7.20) we arrive at *Green's second identity*

$$\iiint_E (v \nabla^2 u - u \nabla^2 v) \, dV = \iint_{\partial E} \left(v \frac{\partial u}{\partial n} - u \frac{\partial v}{\partial n} \right) dS. \tag{7.22}$$

Finally, let us use the divergence theorem to further elucidate the meaning of $\nabla \cdot \mathbf{F}$. We apply the mean value theorem to Gauss' formula

$$\iiint_E \nabla \cdot \mathbf{F} \, dV = \iint_{\partial E} \mathbf{F} \cdot d\mathbf{S}.$$

If $\nabla \cdot \mathbf{F}(x, y, z)$ is continuous on the closed region E, there is a point $(\tilde{x}, \tilde{y}, \tilde{z}) \in E$ such that

$$\iiint_E \nabla \cdot \mathbf{F} \, dV = \nabla \cdot \mathbf{F}(\tilde{x}, \tilde{y}, \tilde{z}) \, V(E)$$

so

$$\nabla \cdot \mathbf{F}(\tilde{x}, \tilde{y}, \tilde{z}) = \frac{1}{V(E)} \iint_{\partial E} \mathbf{F} \cdot d\mathbf{S}. \tag{7.23}$$

Let us select a sequence of regions E_ε contracting down to the point (x, y, z) such that the diameter $\mathrm{diam}(E_\varepsilon) < \varepsilon$, the maximum distance between the points of E_ε. Then the limit passage in (7.23) brings us to

$$\nabla \cdot \mathbf{F}(x, y, z) = \lim_{\varepsilon \to 0} \frac{1}{V(E_\varepsilon)} \iint_{\partial E_\varepsilon} \mathbf{F} \cdot d\mathbf{S}.$$

In some physics books this formula is used to introduce the divergence operation. If $\nabla \cdot \mathbf{F}(x, y, z) \neq 0$ then for any small E_ε containing (x, y, z) we get

$$\iint_{\partial E_\varepsilon} \mathbf{F} \cdot d\mathbf{S} \neq 0,$$

so the flux of $\mathbf{F}$ through ∂E_ε is not zero and E_ε must contain a "source of $\mathbf{F}$."

Example 7.6. In Example 6.10 we established that the flux of the gravity force of a mass M through the sphere S_a centered at M is $4\pi\gamma M$ for any radius of the sphere. A similar result for the Coulomb electrostatic field

$$\mathbf{E} = \frac{\varepsilon Q}{|\mathbf{r}|^3} \mathbf{r}$$

is

$$\iint_{S_a} \mathbf{E} \cdot d\mathbf{S} = 4\pi\varepsilon Q.$$

Using the divergence theorem we will show that we can deform S_a into an arbitrary closed surface S provided we leave the charge Q inside. Let Q be at the coordinate origin surrounded by surface S, and let S_a be centered at the origin and inside S as well. Let V_a be the space between surfaces S and S_a. Then Gauss' theorem gives

$$\iiint_{V_a} \nabla \cdot \mathbf{E} \, dV = \iint_S \mathbf{E} \cdot d\mathbf{S} - \iint_{S_a} \mathbf{E} \cdot d\mathbf{S}.$$

It is easy to calculate that $\nabla \cdot \mathbf{E} = 0$ at all points $\mathbf{r} \neq \mathbf{0}$. So the integral on the left is zero and

$$\iint_S \mathbf{E} \cdot d\mathbf{S} = \iint_{S_a} \mathbf{E} \cdot d\mathbf{S} = 4\pi\varepsilon Q.$$

If Q is not at the origin and we transfer the origin to Q, the flux of $\mathbf{E}$ through the moved surface S does not change, and so for any position of Q inside S the flux through S is $4\pi\varepsilon Q$.

So if inside S we have n charges Q_k, then by additivity the total flux is

$$\sum_{k=1}^{n} \iint_S \mathbf{E}_k \cdot d\mathbf{S} = \sum_{k=1}^{n} 4\pi\varepsilon Q_k.$$

Hence the flux of n charges Q_k inside S is

$$\iint_S \mathbf{E} \cdot d\mathbf{S} = 4\pi\varepsilon \sum_{k=1}^{n} Q_k. \qquad \square$$

7.6   Problems

Green's Theorem

7.1   Use Green's theorem to calculate the area enclosed by

(a) the circle $r = a$,

(b) the rectangle $[a, b] \times [c, d]$,

(c) the triangle having vertices $(0,0)$, $(1,0)$, and $(1,1)$,

(d) the curves $y = x^2$, $y = 1$, and $x = 0$, and

(e) the curves $y = x^2$ and $y = x$ for $0 \le x \le 1$.

7.2   Compute both sides of Green's theorem for

(a) $P = xy$, $Q = 2x^2 y^2$, and D the rectangle $\{(x,y)\colon 0 \le x \le 1, 0 \le y \le 2\}$;

(b) $P = (x - y)^2$, $Q = (x + y)^2$, and D the first-quadrant region lying between the curves $y = x^2$ and $y = x$.

7.3   Evaluate the line integral

$$I = \int_L x^2 y\, dx + xy^2\, dy$$

where L is the boundary of the region $[-1, 1]^2$ in the xy-plane.

7.4   Show that the line integrals of the following vector fields vanish for every simple closed curve in the xy-plane:

(a) $\mathbf{A} = y\,\mathbf{i} + x\,\mathbf{j}$

(b) $\mathbf{A} = 3x^2 y^2\,\mathbf{i} + 2x^3 y\,\mathbf{j}$

(c) $\mathbf{A} = 2xy\,\mathbf{i} + (x^2 + y^2)\,\mathbf{j}$

(d) $\mathbf{A} = \cos x \sin y\,\mathbf{i} + \sin x \cos y\,\mathbf{j}$

Stokes' Theorem

7.5 Evaluate both sides of Stokes' theorem for

(a) the vector field $\mathbf{F} = y\,\mathbf{i} + z\,\mathbf{j} + x\,\mathbf{k}$ and the surface S given by $z = 1$, $0 \le x \le 1$, $0 \le y \le 1$;

(b) the vector field $\mathbf{F} = z\,\mathbf{i}$ and the surface S consisting of the first-octant portion of the plane $x + y + z = 1$;

(c) the vector field $\mathbf{F} = (y + z)\,\mathbf{i} + (z + x)\,\mathbf{j} + (x + y)\,\mathbf{k}$ and the closed contour C given in parametric form by $x = \cos t$, $y = \sin t$, $z = 1$ for $0 \le t < 2\pi$;

(d) the vector field $\mathbf{F} = (2z - y)\,\mathbf{i} + (x - z)\,\mathbf{j} + (y - 2x)\,\mathbf{k}$ and triangular contour with vertices at the points $(0,0,0)$, $(1,0,0)$, and $(1,1,0)$.

7.6 Calculate

$$I = \oint_C \mathbf{A} \cdot d\mathbf{r},$$

where $\mathbf{A} = y\,\mathbf{i} + z\,\mathbf{j} + x\,\mathbf{k}$ and C is the intersection of the sphere $r = R$ with the plane $x + y + z = 0$.

7.7 Find

$$I = \int_C (x - y)\,dx + (x + y)\,dy + (x - z)\,dz,$$

where C is the circle $x^2 + y^2 = 4$, $z = 1$.

Divergence Theorem

7.8 Evaluate both sides of the divergence theorem for the vector field

$$\mathbf{F}(x, y, z) = x^2\,\mathbf{i} + yz\,\mathbf{j} + y^2\,\mathbf{k}$$

and the unit cube $[0, 1]^3$.

7.9 Let E be the volume region bounded by the planes $x = 0$, $y = 0$, $z = 0$, and $x + y + z = 1$. Calculate the following integrals:

(a)

$$I_1 = \iint_{\partial E} x\,dy\,dz + y\,dz\,dx + z\,dx\,dy.$$

(b)

$$I_2 = \iint_{\partial E} \mathbf{F} \cdot d\mathbf{S} \text{ where } \mathbf{F} = xz\,\mathbf{i}.$$

7.10 Carry out the computation of Example 7.5 for the case $a = b = 2$, $c = 1$, $d = 8$, $P = x + y$, $Q = 2x - z$, $R = y$.

7.11 Find the flux of the field $\mathbf{A} = x\mathbf{i} + y\mathbf{j} + z\mathbf{k}$ through

(a) the portion of the ellipsoid (1.33) that lies in the first octant of three-dimensional space;

(b) the closed cylindrical can $r = a$, $0 \leq z \leq h$;

(c) the closed surface bounded by $z = 1 - (x^2 + y^2)^{1/2}$ and $z = 0$.

7.12 Find the net outflux of the field $\mathbf{F} = (x + y)\mathbf{i} + (y + z)\mathbf{j} + (z + x)\mathbf{k}$ from

(a) the ball $\rho \leq a$, and

(b) the cubical volume $|x| \leq 1$, $|y| \leq 1$, $|z| \leq 1$.

7.13 Use the divergence theorem to derive the formula

$$\iint_{\partial E} f \, d\mathbf{S} = \iiint_E \nabla f \, dV.$$

7.14 Show that if $\nabla^2 f = 0$ at all points (x, y, z) of a region E, then

$$\iint_{\partial E} \nabla f \cdot d\mathbf{S} = 0.$$

Deductions from Green's Identities

7.15 By setting $v = u$ in (7.20), show that

$$\iiint_E (u\nabla^2 u + |\nabla u|^2) \, dV = \iint_{\partial E} u \frac{\partial u}{\partial n} \, dS.$$

7.16 By setting $v = 1$ in (7.20), show that

$$\iiint_E \nabla^2 u \, dV = \iint_{\partial E} \frac{\partial u}{\partial n} \, dS.$$

7.17 Use (7.22) to show that if $\nabla^2 u = 0$ and $\nabla^2 v = 0$ then

$$\iint_{\partial E} v \frac{\partial u}{\partial n} \, dS = \iint_{\partial E} u \frac{\partial v}{\partial n} \, dS.$$

This is called *Green's reciprocal theorem*.

Appendix A

Vectors and Vector Fields in Curvilinear Coordinates

A.1 Introduction

Physics commences with fundamental notions about space and time. In this appendix we (fortunately or unfortunately) will not need the time concept. We accept Newton's point of view regarding an undefinable absolute space. But in this space of points without any content, empty space, we shall place solids and other "geometric" objects to be described using a reference frame and a coordinate system. And in physics, not here, these objects will have properties such as mass, temperature, charge, and so on; they will be subject to forces and torques, distributed or concentrated.

Mathematics works with models, not with real objects. So what is a model in mathematical physics? Everything happens in empty space: we could say "in vacuum" but this doesn't explain anything. In the model of space we study something like an instantaneous snapshot, as we do not introduce time, are not interested in inertia effects, etc. Since there are no special points or directions in pure space, we choose an origin point O and fix a reference frame. We should understand that the trajectory of a particle differs in reference frames that move relative to each other. As the frame we usually take a right-handed trihedron denoted $\mathbf{i}, \mathbf{j}, \mathbf{k}$. This trihedron is a bit strange. It is orthogonal (our space is Euclidean) and the vectors $\mathbf{i}, \mathbf{j}, \mathbf{k}$ have magnitude 1 but no physical units. Hence we can use them simultaneously as a basis for vectors of different natures, say for position vectors and forces. But then it becomes hard to explain logically what the unit vector $\mathbf{i}$ is for different spaces of vectors.

We shall erect a coordinate system with respect to the reference frame $\mathbf{i}, \mathbf{j}, \mathbf{k}$ and study various spatial material or geometrical relations for the objects embedded in the space.

To describe a point position we use two methods. One is through Cartesian coordinates (x, y, z). The other is through the position vector

$$\mathbf{r} = x\,\mathbf{i} + y\,\mathbf{j} + z\,\mathbf{k}$$

from the coordinate origin to the point; here x, y, z are called the components of $\mathbf{r}$. This vector can be considered as a vector from $\mathbb{R}^3$. So we have a one-to-one correspondence between the points of our space with a reference frame and the vectors of $\mathbb{R}^3$. In this way we describe objects from the space and say that we consider them in $\mathbb{R}^3$.

It should be noted that the vectors we use in physics frequently have additional properties that the vectors of linear algebra do not have. Furthermore, vectors do not exist in the space; they are external with respect to the space. Placing them onto a figure (as we do with forces, for instance) is something like placing a hologram over some real object.

First we study how to work with geometric vectors of $\mathbb{R}^3$. The set of vectors of $\mathbb{R}^3$ constitutes a linear (or vector) space. This means they satisfy seven or eight axioms of a vector space (depending on the textbook), and this should be verified if we wish to use all the results of linear algebra (mathematics is very formal). But we know that $\mathbb{R}^3$ obeys these axioms and will skip this boring work.

Vector of $\mathbb{R}^3$ in Non-Cartesian Frames

To use the machinery of Albert Einstein, we re-denote

$$\mathbf{i}_1 = \mathbf{i}, \qquad \mathbf{i}_2 = \mathbf{j}, \qquad \mathbf{i}_3 = \mathbf{k}, \qquad x_1 = x, \qquad x_2 = y, \qquad x_3 = z.$$

Then a general vector can be expanded in the frame as

$$\mathbf{x} = \sum_{k=1}^{3} x_k\,\mathbf{i}_k \equiv x_k\,\mathbf{i}_k \tag{A.1}$$

where on the right there is summation over the repeated index k. Suppose we take another basis $\mathbf{e}_k$ in $\mathbb{R}^3$. For $\mathbf{e}_k$ to be a basis, the mixed product

$$(\mathbf{e}_1 \times \mathbf{e}_2) \cdot \mathbf{e}_3 = V$$

must be nonzero. If $V > 0$ then V is volume of the parallelepiped with sides $\mathbf{e}_k$, and the basis is right-handed. If $V < 0$, the frame is left-handed.

For $V \neq 0$ any vector $\mathbf{y} \in \mathbb{R}^3$ can be expanded as

$$\mathbf{y} = y^k\,\mathbf{e}_k. \tag{A.2}$$

Here we use Einstein's rule again, but summation occurs only over an index that is part of a superscript-subscript pair. However, we retain the summation rule (A.1) when the basis is orthonormal. Such index positioning is used because we require the *dual* or *reciprocal basis* $\mathbf{e}^k$, in which vector components are tagged with subscripts. The dual basis makes it easy to express the components of vector expansions.

The reciprocal basis $\mathbf{e}^i$ is defined by the relations

$$\mathbf{e}_i \cdot \mathbf{e}^j = \delta_i^j = \begin{cases} 1, & i = j, \\ 0, & i \neq j, \end{cases} \tag{A.3}$$

where δ_i^j are called *Kronecker coefficients*. For these we have three equivalent notations:

$$\delta_i^j = \delta_{ij} = \delta^{ij}.$$

An orthonormal basis $\mathbf{i}_i$ satisfies

$$\mathbf{i}_i \cdot \mathbf{i}_j = \delta_{ij}.$$

This means that the basis $\mathbf{i}_i$ is reciprocal to itself, i.e.,

$$\mathbf{i}^k = \mathbf{i}_k.$$

By (A.3) each $\mathbf{e}^j$ is orthogonal to two other frame vectors $\mathbf{e}_i, \mathbf{e}_k$ with $i \neq j$ and $k \neq j$. It is easy to check directly that

$$\mathbf{e}^1 = \frac{\mathbf{e}_2 \times \mathbf{e}_3}{V}, \qquad \mathbf{e}^2 = \frac{\mathbf{e}_3 \times \mathbf{e}_1}{V}, \qquad \mathbf{e}^3 = \frac{\mathbf{e}_1 \times \mathbf{e}_2}{V}, \tag{A.4}$$

where

$$V = (\mathbf{e}_1 \times \mathbf{e}_2) \cdot \mathbf{e}_3. \tag{A.5}$$

Let $\mathbf{x} = x^k \mathbf{e}_k$. Dot multiplying by $\mathbf{e}^j$, we use (A.3) to get

$$\mathbf{x} \cdot \mathbf{e}^j = x^k \mathbf{e}_k \cdot \mathbf{e}^j = x^k \delta_k^j = x^j.$$

Similarly when $\mathbf{x} = x_k \mathbf{e}^k$ we obtain $\mathbf{x} \cdot \mathbf{e}_j = x_j$. So

$$x^j = \mathbf{x} \cdot \mathbf{e}^j, \qquad x_j = \mathbf{x} \cdot \mathbf{e}_j. \tag{A.6}$$

The components x^j of $\mathbf{x}$ are called *contravariant* and the components x_j *covariant*. For an orthonormal basis $x^i = x_i$.

There is no need to consider general transformations of frames, as we shall require only the transformation (A.4) (and its inverse, which we urge the reader to deduce).

We have used a fundamental property of objective vectors: they do not change under a change of basis. Such a vector is said to be *polar*. The reader may ask whether there exist vectors that do change. The answer is yes: they are not completely objective, but rather depend on an agreement regarding how to choose one of two possible directions; this is case for $\mathbf{a} \times \mathbf{b}$ whose direction depends on the frame orientation. Such vectors are said to be *axial*. If in basis transformations we do not change the basis orientation, axial vectors cannot be distinguished from polar vectors. However, under a change of basis orientation axial vectors they reverse direction. They are also called *pseudovectors*.

But it is well to keep in mind that, although we frequently use the cross product to find a vector orthogonal to two other vectors, often this operation $\mathbf{a} \times \mathbf{b}$ provides only a formal normal vector. If $\mathbf{a}, \mathbf{b}$ are vectors defining the sides of a parallelogram, the absolute value of their cross product is the enclosed area. If the vectors are forces, their cross product has no physical meaning. But it can be used to determine a direction normal to both forces.

For basis vectors $\mathbf{e}_k$ we define the *metric coefficients*

$$g_{km} = \mathbf{e}_k \cdot \mathbf{e}_m \tag{A.7}$$

that constitute a 3×3 symmetric matrix. The metric coefficients for $\mathbf{e}^k$ are

$$g^{km} = \mathbf{e}^k \cdot \mathbf{e}^m. \tag{A.8}$$

Recall that $\mathbf{e}^k \cdot \mathbf{e}_m = \delta^k_m$, which we could also call g^k_m. All these indexed "g" quantities constitute the components of the metric tensor, an entity that extends the idea of a vector. In mechanics there are plenty of tensor entities that are represented by 3×3 matrices in a basis, like the inertial tensor for a solid, the stress and strain tensors, etc. They transform according to certain rules different from those for vectors under change of basis.

But back to the metric coefficients. Dot multiplying the equality

$$x^k \mathbf{e}_k = x_m \mathbf{e}^m$$

(we could use k on both sides since k and m are "dummy" summation indices, but to avoid ambiguities we prefer to use different letters in such expressions) by $\mathbf{e}^i$ we get

$$x^k \mathbf{e}_k \cdot \mathbf{e}^i = x_m \mathbf{e}^m \cdot \mathbf{e}^i$$

so that

$$x^k \delta^i_k = x_m g^{mi}$$

and therefore

$$x^i = x_m g^{mi}.$$

Similarly dot multiplying $x^k \mathbf{e}_k = x_m \mathbf{e}^m$ by $\mathbf{e}_j$ we get

$$x^k g_{kj} = x_j.$$

These two operations are called raising and lowering of indices, respectively. We note without proof that

$$g_{kj} g^{jm} = \delta_k^m.$$

It follows that the matrices (g_{kj}) and (g^{jm}) are mutually inverse.

Now it is time to reconsider the cross product of two vectors $\mathbf{a} = a^j \mathbf{e}_j$ and $\mathbf{b} = b^k \mathbf{e}_k$. Here $\mathbf{c} = \mathbf{a} \times \mathbf{b}$ is represented in the right-handed trihedron $(\mathbf{e}_1, \mathbf{e}_2, \mathbf{e}_3)$. We recall that the cross product of two vectors is a pseudovector, but if the frame never changes orientation then pseudovectors behave as ordinary vectors.

Let $\mathbf{c} = c_m \mathbf{e}^m$. By the distributive property we have

$$c_m \mathbf{e}^m = a^j b^k \mathbf{e}_j \times \mathbf{e}_k.$$

Dot multiplying both sides by $\mathbf{e}_i$ we get

$$c_i = a^j b^k (\mathbf{e}_j \times \mathbf{e}_k) \cdot \mathbf{e}_i = a^j b^k \epsilon_{ijk},$$

where ϵ_{ijk} are called *Levi-Civita symbols*. These are the components of the Levi-Civita tensor presented in books on tensor calculus. They are nonzero only when i, j, k differ. For the three (i, j, k) triples

$$(1, 2, 3), \qquad (2, 3, 1), \qquad (3, 1, 2),$$

(the even permutations of $(1, 2, 3)$), we have $\epsilon_{ijk} = V$ where

$$V = (\mathbf{e}_j \times \mathbf{e}_k) \cdot \mathbf{e}_i.$$

For odd permutations of $(1, 2, 3)$, which are $(3, 2, 1)$, $(1, 3, 2)$, and $(2, 1, 3)$, the symbols ϵ_{ijk} are equal to $-V$. We also have $\epsilon^{ijk} = 1/V$ for even permutations of $(1, 2, 3)$, $\epsilon^{ijk} = -1/V$ for odd permutations of $(1, 2, 3)$, and $\epsilon^{ijk} = 0$ for other values of (i, j, k). In textbooks the reader can find various identities related to the Levi-Civita symbols and the Kronecker symbols. They prove useful in many types of algebraic transformation.

Now we are prepared to start the discussion of calculus in a space with curvilinear coordinates.

A.2   Curvilinear Coordinates in Space

Let us return to the position vector, changing (x, y, z) to (x^1, x^2, x^3):

$$\mathbf{r} = x^1 \, \mathbf{i}_1 + x^2 \, \mathbf{i}_2 + x^3 \, \mathbf{i}_3.$$

We put the point $\mathbf{r}$ into correspondence with a triple (u^1, u^2, u^3) called *curvilinear coordinates*. They are related to the Cartesian coordinates by functional relations

$$x^1 = x^1(u^1, u^2, u^3), \qquad x^2 = x^2(u^1, u^2, u^3), \qquad x^3 = x^3(u^1, u^2, u^3),$$

which are continuously differentiable, one-to-one correspondences except possibly at a few singular points or lines. If two of the three coordinates (u^1, u^2, u^3) are fixed, here to be marked with subscript zero, we get corresponding coordinate lines:

$$\mathbf{r} = \mathbf{r}(u^1, u_0^2, u_0^3), \quad \mathbf{r} = \mathbf{r}(u_0^1, u^2, u_0^3), \quad \text{or} \quad \mathbf{r} = \mathbf{r}(u_0^1, u_0^2, u^3).$$

By fixing only one of the three coordinates (u^1, u^2, u^3) we produce a coordinate surface with which we can create partitions of space suitable for defining the triple integral in (u^1, u^2, u^3) coordinates. We met these concepts in Chapter 1. But now we go into more detail regarding vectors tangent to the coordinate lines at each point.

First we should understand how to take derivatives in curvilinear coordinates. At each point $\mathbf{r}(u^1, u^2, u^3)$ we introduce a local vector frame. This frame, changing from point to point, is often used to expand vector quantities in the treatment of physical problems. Such quantities clearly do not belong to the space of points, and serious mathematical books define one or more tangent spaces to which vectors like forces can belong.

We start with the first differential of $\mathbf{r}$:

$$d\mathbf{r} = \frac{\partial \mathbf{r}}{\partial u^1} \, du^1 + \frac{\partial \mathbf{r}}{\partial u^2} \, du^2 + \frac{\partial \mathbf{r}}{\partial u^3} \, du^3. \tag{A.9}$$

Let us denote

$$\mathbf{r}_k = \frac{\partial \mathbf{r}(u^1, u^2, u^3)}{\partial u^k}$$

so that

$$d\mathbf{r} = \mathbf{r}_k \, du^k.$$

This gives us a frame trihedron $\mathbf{r}_1, \mathbf{r}_2, \mathbf{r}_3$ at each point if $(\mathbf{r}_1 \times \mathbf{r}_2) \cdot \mathbf{r}_3 \neq 0$. The vector $\mathbf{r}_k$ is tangent to the u^k-coordinate line. Note that k in the symbol $\partial / \partial u^k$ is equivalent to a subscript.

By (A.9), in Cartesian coordinates where $\mathbf{r} = x_k\,\mathbf{i}_k$, we have

$$d\mathbf{r} = \mathbf{i}_k\,dx_k$$

as expected.

Example A.1. In polar coordinates we have $u^1 = r$ and $u^2 = \theta$. From

$$\mathbf{r} = r\cos\theta\,\mathbf{i} + r\sin\theta\,\mathbf{j}, \qquad \mathbf{i}_1 = \mathbf{i}, \qquad \mathbf{i}_2 = \mathbf{j},$$

we obtain

$$\mathbf{r}_1 = \frac{\partial\mathbf{r}}{\partial u^1} = \frac{\partial(r\cos\theta\,\mathbf{i} + r\sin\theta\,\mathbf{j})}{\partial r} = \cos\theta\,\mathbf{i} + \sin\theta\,\mathbf{j},$$

$$\mathbf{r}_2 = \frac{\partial\mathbf{r}}{\partial u^2} = \frac{\partial(r\cos\theta\,\mathbf{i} + r\sin\theta\,\mathbf{j})}{\partial\theta} = -r\sin\theta\,\mathbf{i} + r\cos\theta\,\mathbf{j}.$$

Since $\mathbf{r}_1\cdot\mathbf{r}_2 = 0$, the frame vectors are orthogonal and the reciprocal vectors are codirected with the respective vectors $\mathbf{r}_1, \mathbf{r}_2$. From

$$\mathbf{r}_1\cdot\mathbf{r}^1 = 1, \qquad \mathbf{r}_2\cdot\mathbf{r}^2 = 1,$$

we get immediately

$$\mathbf{r}^1 = \mathbf{r}_1 = \cos\theta\,\mathbf{i} + \sin\theta\,\mathbf{j},$$

$$\mathbf{r}^2 = \frac{1}{r^2}\,\mathbf{r}_2 = -\frac{1}{r}\sin\theta\,\mathbf{i} + \frac{1}{r}\cos\theta\,\mathbf{j}.$$

The first differential is

$$d\mathbf{r} = (\cos\theta\,\mathbf{i} + \sin\theta\,\mathbf{j})\,dr + (-r\sin\theta\,\mathbf{i} + r\cos\theta\,\mathbf{j})\,d\theta. \qquad \square$$

Example A.2. By adding a term $z\mathbf{k}$ to the expression for $\mathbf{r}$ in polar coordinates, we get the frame vectors for cylindrical coordinates:

$$\mathbf{r}_1 = \frac{\partial\mathbf{r}}{\partial u^1} = \frac{\partial(r\cos\theta\,\mathbf{i} + r\sin\theta\,\mathbf{j} + z\,\mathbf{k})}{\partial r} = \cos\theta\,\mathbf{i} + \sin\theta\,\mathbf{j},$$

$$\mathbf{r}_2 = \frac{\partial\mathbf{r}}{\partial u^2} = \frac{\partial(r\cos\theta\,\mathbf{i} + r\sin\theta\,\mathbf{j} + z\,\mathbf{k})}{\partial\theta} = -r\sin\theta\,\mathbf{i} + r\cos\theta\,\mathbf{j},$$

$$\mathbf{r}_3 = \frac{\partial\mathbf{r}}{\partial u^3} = \frac{\partial(r\cos\theta\,\mathbf{i} + r\sin\theta\,\mathbf{j} + z\,\mathbf{k})}{\partial z} = \mathbf{k},$$

and

$$\mathbf{r}^1 = \mathbf{r}_1 = \cos\theta\,\mathbf{i} + \sin\theta\,\mathbf{j},$$

$$\mathbf{r}^2 = \frac{1}{r^2}\,\mathbf{r}_2 = -\frac{1}{r}\sin\theta\,\mathbf{i} + \frac{1}{r}\cos\theta\,\mathbf{j},$$

$$\mathbf{r}^3 = \mathbf{k}.$$

The first differential is

$$d\mathbf{r} = (\cos\theta\,\mathbf{i} + \sin\theta\,\mathbf{j})\,dr + (-r\sin\theta\,\mathbf{i} + r\cos\theta\,\mathbf{j})\,d\theta + \mathbf{k}\,dz. \qquad \square$$

Example A.3. Let us consider spherical coordinates. Here

$$(u^1, u^2, u^3) = (\rho, \phi, \theta)$$

and the position vector is

$$\mathbf{r} = \rho\sin\phi\cos\theta\,\mathbf{i} + \rho\sin\phi\sin\theta\,\mathbf{j} + \rho\cos\phi\,\mathbf{k}.$$

Calculations similar to those in the previous examples give

$$\mathbf{r}_1 = \sin\phi\cos\theta\,\mathbf{i} + \sin\phi\sin\theta\,\mathbf{j} + \cos\phi\,\mathbf{k}, \quad |\mathbf{r}_1| = 1,$$

$$\mathbf{r}_2 = \rho\cos\phi\cos\theta\,\mathbf{i} + \rho\cos\phi\sin\theta\,\mathbf{j} - \rho\sin\phi\,\mathbf{k}, \quad |\mathbf{r}_2| = \rho,$$

$$\mathbf{r}_3 = -\rho\sin\phi\sin\theta\,\mathbf{i} + \rho\sin\phi\cos\theta\,\mathbf{j}, \quad |\mathbf{r}_3| = \rho\sin\phi.$$

It is easy to check that the frame $\mathbf{r}_1, \mathbf{r}_2, \mathbf{r}_3$ is orthogonal. Hence the vectors $\mathbf{r}^k$ of the reciprocal basis are codirected with the $\mathbf{r}_k$:

$$\mathbf{r}^1 = \mathbf{r}_1,$$

$$\mathbf{r}^2 = \frac{1}{\rho}(\cos\phi\cos\theta\,\mathbf{i} + \cos\phi\sin\theta\,\mathbf{j} - \sin\phi\,\mathbf{k}),$$

$$\mathbf{r}^3 = \frac{1}{\rho\sin\phi}(-\sin\theta\,\mathbf{i} + \cos\theta\,\mathbf{j}). \qquad \square$$

A.3   Nabla in Curvilinear Coordinates

In Cartesian coordinates the first differential of a function f is

$$df = f_x\,dx + f_y\,dy + f_z\,dz$$

$$= \left(\mathbf{i}\,\frac{\partial}{\partial x} + \mathbf{j}\,\frac{\partial}{\partial y} + \mathbf{k}\,\frac{\partial}{\partial z}\right) \cdot (\mathbf{i}\,dx + \mathbf{j}\,dy + \mathbf{k}\,dz)$$

$$= \nabla f \cdot d\mathbf{r}.$$

Let us do a similar trick for a function f depending on u^1, u^2, u^3:

$$df = \frac{\partial f}{\partial u^1}\, du^1 + \frac{\partial f}{\partial u^2}\, du^2 + \frac{\partial f}{\partial u^3}\, du^3$$

$$= \frac{\partial f}{\partial u^k}\, \delta^k_m\, du^m$$

$$= \frac{\partial f}{\partial u^k}\, \mathbf{r}^k \cdot \mathbf{r}_m\, du^m.$$

Denoting

$$\nabla f = \frac{\partial f}{\partial u^k}\, \mathbf{r}^k$$

and taking into account that

$$d\mathbf{r} = \mathbf{r}_m\, du^m$$

we get

$$df = \nabla f \cdot d\mathbf{r} \quad \text{where} \quad \nabla = \mathbf{r}^k\, \frac{\partial}{\partial u^k}. \tag{A.10}$$

A key observation is that $\nabla f \cdot d\mathbf{r}$ is an invariant quantity. In vector calculus we often deduce a needed expression in Cartesian coordinates and then represent it through objective vectors and dot and vector products. Then, to represent the expression in curvilinear coordinates, we represent each of the vectors in the curvilinear frame. Note that ∇f is one of the objective vectors.

By means of the operator ∇ we can define

$$\nabla \cdot \mathbf{F} = \mathbf{r}^k\, \frac{\partial}{\partial u^k} \cdot \mathbf{F} = \mathbf{r}^k \cdot \frac{\partial \mathbf{F}}{\partial u^k}, \tag{A.11}$$

$$\nabla \times \mathbf{F} = \mathbf{r}^k\, \frac{\partial}{\partial u^k} \times \mathbf{F} = \mathbf{r}^k \times \frac{\partial \mathbf{F}}{\partial u^k}, \tag{A.12}$$

for a vector function $\mathbf{F}$.

A.4 Covariant Derivatives

Now let us discuss the partial derivative of $\mathbf{F}$ with respect to a coordinate u^k when $\mathbf{F}$ is expanded in the frames $\mathbf{r}^k$ and $\mathbf{r}_k$. With

$$\mathbf{F} = F^k \mathbf{r}_k = F_k \mathbf{r}^k$$

we get

$$\frac{\partial \mathbf{F}}{\partial u^k} = \frac{\partial F^m}{\partial u^k}\mathbf{r}_m + F^m\frac{\partial \mathbf{r}_m}{\partial u^k} = \frac{\partial F_n}{\partial u^k}\mathbf{r}^n + F_n\frac{\partial \mathbf{r}^n}{\partial u^k}.$$

In a Cartesian basis the derivatives of the basis vectors are zero. But if the $\mathbf{r}_k$ are variable, we need to know the derivatives of the basis vectors.

Consider the expansion of $\partial \mathbf{r}_i/\partial u^k$. As the $\mathbf{r}_k$ constitute a basis in $\mathbb{R}^3$, we can write

$$\frac{\partial \mathbf{r}_i}{\partial u^k} = \Gamma^j_{ik}\mathbf{r}_j. \tag{A.13}$$

The expansion coefficients, known as *Christoffel coefficients*, can be found with use of the reciprocal basis $\mathbf{r}^k$. Dot multiplication by $\mathbf{r}^m$ yields

$$\frac{\partial \mathbf{r}_i}{\partial u^k}\cdot\mathbf{r}^m = \Gamma^j_{ik}\mathbf{r}_j\cdot\mathbf{r}^m = \Gamma^j_{ik}\delta^m_j = \Gamma^m_{ik}.$$

These coefficients are also denoted in another way:

$$\Gamma^m_{ik} = \left\{\begin{array}{c} m \\ ik \end{array}\right\} = \frac{\partial \mathbf{r}_i}{\partial u^k}\cdot\mathbf{r}^m. \tag{A.14}$$

If $\mathbf{r} = \mathbf{r}(u^1, u^2, u^3)$ is twice continuously differentiable we have

$$\frac{\partial \mathbf{r}_i}{\partial u^k} = \frac{\partial}{\partial u^k}\left(\frac{\partial \mathbf{r}}{\partial u^i}\right) = \frac{\partial}{\partial u^i}\left(\frac{\partial \mathbf{r}}{\partial u^k}\right) = \frac{\partial \mathbf{r}_k}{\partial u^i}$$

and

$$\Gamma^j_{ik} = \Gamma^j_{ki}, \tag{A.15}$$

a symmetry property that saves much labor.

Now we return to the task of finding $\partial\mathbf{F}/\partial u^k$. By (A.13),

$$\frac{\partial \mathbf{F}}{\partial u^k} = \frac{\partial}{\partial u^k}(F^i\mathbf{r}_i)$$

$$= \frac{\partial F^i}{\partial u^k}\mathbf{r}_i + F^i\frac{\partial \mathbf{r}_i}{\partial u^k}$$

$$= \frac{\partial F^j}{\partial u^k}\mathbf{r}_j + F^i\Gamma^j_{ik}\mathbf{r}_j$$

$$= \left(\frac{\partial F^j}{\partial u^k} + F^i\Gamma^j_{ik}\right)\mathbf{r}_j. \tag{A.16}$$

The parenthetical expression is denoted by

$$\nabla_k F^j = \frac{\partial F^j}{\partial u^k} + F^i \Gamma^j_{ik} \tag{A.17}$$

and termed the kth *covariant derivative* of the jth contravariant component of $\mathbf{F}$. It is easy to see that in a Cartesian basis $\Gamma^j_{ik} \equiv 0$, and $\nabla_k F^j$ becomes a usual partial derivative $\partial F^j / \partial x_k$.

Similarly

$$\frac{\partial \mathbf{F}}{\partial u^k} = \left(\frac{\partial F_j}{\partial u^k} - \Gamma^i_{jk} F_i \right) \mathbf{r}^j = \nabla_k F_j \, \mathbf{r}^j \tag{A.18}$$

where $\nabla_k F_j$ is the kth covariant derivative of the jth covariant component of $\mathbf{F}$.

The Laplacian operator ∇^2 is given by

$$\nabla \cdot (\nabla f) = g^{ij} \left(\frac{\partial^2 f}{\partial u^i \partial u^j} - \Gamma^k_{ij} \frac{\partial f}{\partial q^k} \right). \tag{A.19}$$

The reader is urged to compute the Christoffel coefficients for the cylindrical and spherical coordinate systems.

A.5 Integration

We present a couple of useful formulas associated with the principal theorems that can be written in curvilinear coordinates. For Gauss' theorem we have

$$\iiint_E \nabla \times \mathbf{F} \, dV = \iint_{\partial E} \mathbf{n} \times \mathbf{F} \, dS. \tag{A.20}$$

In vector form Stokes' theorem can be written as

$$\iint_S (\mathbf{n} \times \nabla) \cdot \mathbf{F} \, dS = \int_{\partial S} \mathbf{F} \cdot d\mathbf{r}. \tag{A.21}$$

Certain formulas of continuum physics are easier remembered, written down, and applied when phrased in compact vectorial form. Any such expression can be expanded into component form, in a chosen coordinate system, whenever required. To derive integration formulas in vectorial form, we normally start with the formulas in Cartesian coordinates. Among the relations that can be obtained in this way are the gradient theorem

$$\iint_{\partial E} f \mathbf{n} \, dS = \iiint_E \nabla f \, dV \tag{A.22}$$

and the curl theorem

$$\iint_{\partial E} (\mathbf{n} \times \mathbf{A}) \, dS = \iiint_E \nabla \times \mathbf{A} \, dV. \tag{A.23}$$

As extensions of Green's identities, we can state the vector Green's first identity

$$\iint_{\partial E} \mathbf{n} \cdot (\mathbf{A} \times \nabla \times \mathbf{B}) \, dS = \iiint_E [(\nabla \times \mathbf{A}) \cdot (\nabla \times \mathbf{B}) - \mathbf{A} \cdot \nabla \times \nabla \times \mathbf{B}] \, dV, \tag{A.24}$$

and the vector Green's second identity

$$\iint_{\partial E} \mathbf{n} \cdot (\mathbf{B} \times \nabla \times \mathbf{A} - \mathbf{A} \times \nabla \times \mathbf{B}) \, dS = \iiint_E (\mathbf{A} \cdot \nabla \times \nabla \times \mathbf{B} - \mathbf{B} \cdot \nabla \times \nabla \times \mathbf{A}) \, dV \tag{A.25}$$

for vector fields $\mathbf{A}, \mathbf{B}$.

Hints and Answers

Chapter 1

1.2. (a) $(x, -y)$. (b) $(-x, y)$. (c) $(-x, -y)$. (d) (y, x).

1.7. $d = [r_1^2 + r_2^2 - 2r_1 r_2 \cos(\theta_1 - \theta_2) + (z_1 - z_2)^2]^{1/2}$.

1.8. (a) $\mathbf{r} = \mathbf{i} - \mathbf{j} + 2\mathbf{k}$.

1.11. $|\mathbf{A} \times \mathbf{B}|$ is the area of the parallelogram determined by the vectors $\mathbf{A}, \mathbf{B}$.

1.18. $x(t) = t$ and $y(t) = mt + b$ with $-\infty < t < \infty$.

1.20. Use a plotting package.

1.21. (a) Parabola. (b) Hyperbola $x^2/a^2 - y^2/b^2 = 1$ since $\sec^2 t - \tan^2 t \equiv 1$.

1.22. Use plotting software, choosing reasonable values for r and R.

1.24. A general point on the line $y = mx + b$ is given by $(x, mx + b)$. Writing $d^2 = (x - x_1)^2 + (mx + b - y_1)^2$ and using differentiation, we find that d^2 is minimized at $x = \bar{x} = [x_1 - (b - y_1)m]/(1 + m^2)$. Evaluating d^2 at $\bar{x}$ and taking square roots, we get $\bar{d} = |mx_1 + b - y_1|/(1 + m^2)^{1/2}$.

1.25. (a) This curve is a *cardioid*.

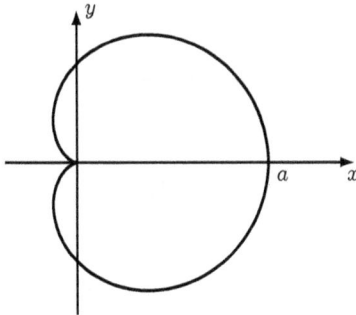

1.26. (a) Note that negative values of y cannot satisfy this equation; therefore the graph has only two "leaves" and they are both located in the range $0 \leq \theta \leq \pi$.

1.27. (a) $(r, \theta) = (a, \pm\pi/3)$.

1.28. $r^2 - 2rr_0 \cos(\theta - \theta_0) + r_0^2 = a^2$.

1.29. $y = ax^2 - b$, a parabola.

1.30. $x^2/a^2 - y^2/b^2 = \cosh^2 t - \sinh^2 t = 1$, a hyperbola.

1.31. Rewrite as $(x+1)^2/2 + (y-2)^2/3 = 1$ to see that the curve is a shifted ellipse.

1.32. (a) $y = c(x-b)/a + d$. (b) $y^2 = b^2 x/a$.

1.33. (a) $x = a\cos^4 t$ and $y = b\sin^4 t$ for $0 \leq t \leq \pi/2$. (b) $x = a\cos t$ and $y = a\cos t + a\sin t$ for $0 \leq t \leq 2\pi$. (c) $x = \cos^3 t$ and $y = \sin^3 t$ for $0 \leq t \leq 2\pi$. This curve is called an *astroid*:

1.37. (a) Helical spiral.

1.38. (a)
$$\tau = \frac{6\,\mathbf{i} + 6t\,\mathbf{j} + 3t^2\,\mathbf{k}}{(36 + 36t^2 + 9t^4)^{1/2}}.$$

(b)
$$\tau = \frac{2\,\mathbf{i} + 3t\,\mathbf{j} + 6t^4\,\mathbf{k}}{(4 + 9t^2 + 36t^8)^{1/2}}.$$

1.39. (a) Half the region inside an ellipse. (c) A triangular region (d) The region outside an ellipse.

1.41. (a) $3x - 2y - 2z = 18$. (b) $x/a + y/b + z/c = 1$. This is the so-called intercept form of the equation of a plane.

1.42. In component form the equation reads

$$\left(x - \frac{A_x}{2}\right)^2 + \left(y - \frac{A_y}{2}\right)^2 + \left(z - \frac{A_z}{2}\right)^2 = \frac{A_x^2 + A_y^2 + A_z^2}{4}.$$

1.43. (a)

$$(x - a)^2 + (y - b)^2 + (z - c)^2 = (a - x_1)^2 + (b - y_1)^2 + (c - z_1)^2.$$

(b)

$$\left(x - \frac{a+d}{2}\right)^2 + \left(y - \frac{b+e}{2}\right)^2 + \left(z - \frac{c+f}{2}\right)^2$$
$$= \frac{1}{4}[(a - d)^2 + (b - e)^2 + (c - f)^2].$$

1.44. (a) ellipsoid. (b) cone. (c) elliptic cylinder. (d) parabolic cylinder.

1.45. Cylinder along the z-axis with lemniscate cross section.

1.48. Some second-degree algebraic equations describe geometric objects that are not surfaces. For example, $x^2 + y^2 = 0$ describes only the coordinate origin. The equation $x^2 - y^2 = 0$ describes the pair of lines $y = \pm x$. The equation $x^2 + y^2 = -1$ has no real solutions and hence no graph at all.

1.49. $(x^2 + y^2)^{1/2} = |z| \tan \alpha$, a cone.

1.50. The ellipse

$$\frac{x^2}{(1/\sqrt{2})^2} + \frac{(y - 1/2)^2}{(1/2)^2} = 1.$$

1.51. See the sketch below.

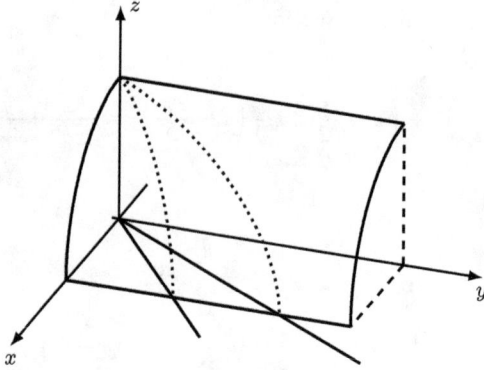

1.52. (a) The elliptic cylinder can be parameterized as

$$x = e\cos t, \quad y = \sin t \qquad (0 \le t \le 2\pi).$$

The equation of the plane yields

$$z = \frac{d}{c} - \frac{a}{c}x - \frac{b}{c}y = \frac{1}{c}(d - ae\cos t - b\sin t).$$

So the curve of intersection is given by

$$\mathbf{r} = e\cos t\,\mathbf{i} + \sin t\,\mathbf{j} + \frac{1}{c}(d - ae\cos t - b\sin t)\,\mathbf{k}.$$

For example, if $a = 3$, $b = 2$, $c = 1$, $d = 4$, and $e = 2$, it is

$$\mathbf{r} = 2\cos t\,\mathbf{i} + \sin t\,\mathbf{j} + (4 - 6\cos t - 2\sin t)\,\mathbf{k}.$$

(b) The curve of intersection is given by $\mathbf{r} = t\,\mathbf{i} + (1 - t)\,\mathbf{j} + [t^2 + (1 - t)^2]\,\mathbf{k}$ for $0 \le t \le 1$. The sketch below shows the plane (solid line) cutting the parabola (dashed line) in the curve (dotted).

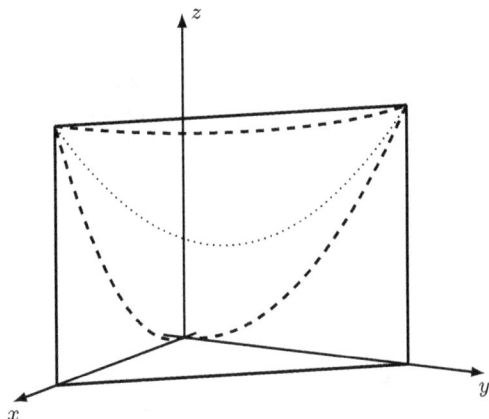

(c) The curve of intersection is given by $\mathbf{r} = t\,\mathbf{i} + (1-t^2)^{1/2}\,\mathbf{j} + (1-t^2)^{1/2}\,\mathbf{k}$ for $0 \le t \le 1$. The sketch below shows the cylinder $x^2 + y^2 = 1$ (solid line) cut by the cylinder $x^2 + z^2 = 1$ (dashed line) in the curve (dotted).

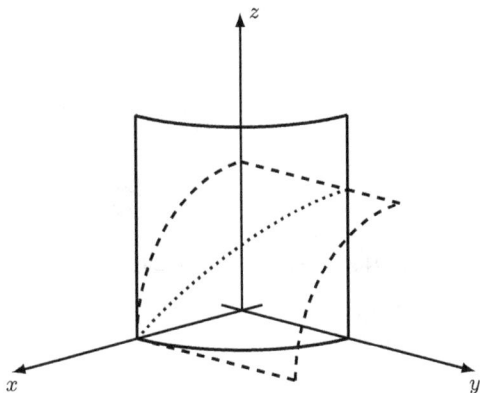

1.53. Make cross-sectional sketches for the cases $v = 0$, $u = 0$, and $u = \pi$.

1.54. Use a plotting package.

1.55. (a) $x^2/a^2 + (y^2 + z^2)/c^2 = 1$; (b) $(x^2 + y^2)/a^2 + z^2/c^2 = 1$; (c) $x^2/a^2 - (y^2 + z^2)/c^2 = 1$; (d) $(x^2 + y^2)/a^2 - z^2/c^2 = 1$; (e) $x^2 + y^2 = 2pz$; (f) $x^2 = y^2 + z^2$.

1.57. (b)
$$\mathbf{r}'(t) = v_0 \cos\theta_0\,\mathbf{i} + (v_0 \sin\theta_0 - gt)\,\mathbf{j}.$$

(d)
$$y = (\tan\theta_0)x - \frac{gx^2}{2(v_0 \cos\theta_0)^2}.$$

1.58. (a) Parabolic trajectory $y = \frac{1}{2}\eta E_0 (x/u_0)^2$. (b) Straight line trajectory $x = \eta E_0 z/g$.

1.59. It is helical motion, with the axis of the helix parallel to the z-direction. To see this more clearly, we can study the projection of the motion onto the xy-plane. We take the first two equations and eliminate t:

$$x^2 = \left(\frac{u_{x0}}{\omega_c}\right)^2 \sin^2 \omega_c t, \qquad \left(y + \frac{u_{x0}}{\omega_c}\right)^2 = \left(\frac{u_{x0}}{\omega_c}\right)^2 \cos^2 \omega_c t,$$

$$\therefore \quad x^2 + \left(y + \frac{u_{x0}}{\omega_c}\right)^2 = \left(\frac{u_{x0}}{\omega_c}\right)^2 (\sin^2 \omega_c t + \cos^2 \omega_c t) = \left(\frac{u_{x0}}{\omega_c}\right)^2.$$

This is $x^2 + (y + r_c)^2 = r_c^2$ where $r_c = u_{x0}/\omega_c$. We call r_c the *cyclotron radius*. In the special case $u_{z0} = 0$, the resulting uniform circular motion is termed *cyclotron motion*.

Chapter 2

2.3. An example is the strip $\{(x,y) : -1 \le x < 1\}$.

2.5. (a) The region $y \le x$, which is the closed half-plane consisting of points that lie on or below the line $y = x$. (b) $x + y > 0$, an open half-plane. (c) The whole plane with the origin deleted. (d) The portion of the plane bounded by the hyperbola $x^2 - y^2 = 1$ and containing its foci. (e) The disk $x^2 + y^2 \le 2^2$.

2.6. (a) We need $-1 \le z/(x^2+y^2)^{1/2} \le 1$ or $-r \le z \le r$ where $r = (x^2+y^2)^{1/2}$. This is the region of space between the two halves of the cone $z^2 = x^2 + y^2$. (e) The region of space above the plane $x + y + z = 0$.

2.9. (a) $f(r,\theta,z) = 1 + r^2 z \sin\theta \cos\theta$. (b) $f(\rho,\theta,\phi) = 1 + \rho^3 \sin^2\phi \cos\phi \sin\theta \cos\theta$.

2.10. Both can be derived from the obvious fact that $(a - b)^2 \ge 0$.

2.11. The square neighborhood specified by $|x - a| < \varepsilon/\sqrt{2}$ and $|y - a| < \varepsilon/\sqrt{2}$ is contained in the circular ε-neighborhood of (a,b).

2.13. Set $y = cx$ to obtain expressions that are not independent of c.

2.14. (a) 1 and -1. (b) Both 0.

2.15. (a) The iterated limit

$$\lim_{x \to 0} \lim_{y \to 0} x \sin(1/y)$$

does not exist.

2.16. Since $f(t,0) = 0$ but $f(t,t) = 1/2$ for all nonzero t, the function $f(x,y)$ cannot be defined at $(0,0)$ in such a way that continuity results.

2.17. We can find that partial limits over various paths through point $(1,1)$ are zero. These include the paths (1) for $x = 1$ when $y \to 1$, (2) for $y = 1$ when $x \to 1$, and (3) for $x = y$ and $x \to 1$. These facts merely suggest zero as a candidate for a limit value. A proof could be attempted via the ε-δ definition of the double limit. But in this particular case it is simpler to use Theorem 2.3. Factoring the expression we get

$$\frac{(x^2 - 2x + 1)(y - 1)}{(x^2 - 2x + 1) + 2(y - 1)^2} = \frac{(x - 1)^2}{(x - 1)^2 + 2(y - 1)^2}(y - 1),$$

where for $(x,y) \neq (1,1)$

$$\left| \frac{(x - 1)^2}{(x - 1)^2 + 2(y - 1)^2} \right| \leq 1$$

and

$$y - 1 \to 0 \quad \text{as } (x, y) \to (1, 1).$$

So by Theorem 2.3 the needed limit is zero.

2.18. 3 by l'Hospital's rule.

2.19.

$$|f(x, y) - f(0, 0)| = \left| \frac{y^3}{x^2 + y^2} \right| \leq \frac{|y||x^2 + y^2|}{|x^2 + y^2|} = |y| \leq (x^2 + y^2)^{1/2}.$$

2.20. Let $\varepsilon > 0$ be given. Choose $\delta = \varepsilon/(|a| + |b|)$. For $|x - x_0| < \delta$ and $|y - y_0| < \delta$, we have

$$|ax + by - (ax_0 + by_0)| = |a(x - x_0) + b(y - y_0)|$$
$$\leq |a||x - x_0| + |b||y - y_0|$$
$$\leq (|a| + |b|)\delta$$
$$\leq \varepsilon.$$

2.21. (a)

$$|f(x, y) - 0| = \frac{x^2}{x^2 + y^2}|y| \leq |y| \leq (x^2 + y^2)^{1/2}.$$

(b)

$$\left| \frac{x + y}{x^2 + y^2 + 1} \right| \leq |x + y| \leq |x| + |y| \leq 2(x^2 + y^2)^{1/2}.$$

(c)

$$\left| y\cos\left(\frac{1}{y-x}\right) \right| \le |y| \le (x^2+y^2)^{1/2}.$$

(d)

$$\left| \frac{x^3+y^3}{x^2+y^2} \right| = \frac{|x^3+y^3|}{x^2+y^2} \le \frac{|x|^3+|y|^3}{x^2+y^2} = \frac{(x^2)^{3/2}+(y^2)^{3/2}}{x^2+y^2}$$

$$\le \frac{(x^2+y^2)^{3/2}+(x^2+y^2)^{3/2}}{x^2+y^2} = \frac{2(x^2+y^2)^{3/2}}{x^2+y^2} = 2(x^2+y^2)^{1/2}.$$

(e) $|xy| \le (x^2+y^2)/2$. (f) $|x|+|y| \le \sqrt{2}\sqrt{x^2+y^2}$.

2.22. (a) Any point on one of the coordinate planes $x=0$, $y=0$, or $z=0$. (d) Only the point (a,b,c).

2.24. (a) $\partial f/\partial x = yx^{y-1}$ and $\partial f/\partial y = x^y \ln x$. (b) $\partial f/\partial x = y/(x^2+y^2)$ and $\partial f/\partial y = -x/(x^2+y^2)$. (c) $\partial f/\partial x = 1/y - y/x^2$ and $\partial f/\partial y = -x/y^2 + 1/x$. (d) $\partial f/\partial x = a\cos(ax+by)$ and $\partial f/\partial y = b\cos(ax+by)$. (e) $\partial f/\partial x = \sin y\, x^{\sin y - 1}$ and $\partial f/\partial y = x^{\sin y}\cos y \ln x$. (f) $\partial f/\partial x = y+1/y$ and, $\partial f/\partial y = x(1-1/y^2)$. (g) $\partial f/\partial x = e^{xy}\sin y(y\cos x - \sin x)$ and $\partial f/\partial y = e^{xy}\cos x(x\sin y + \cos y)$.

2.25. (a) $2\ln 2$ and $-2\ln 2$. (b) 1 and $1/2$. (c) $3\sin 1$ and $\cos 1$. (d) Both $(\cos 1)/2$.

2.27. (a)

$$\frac{\partial f}{\partial x} = e^y + ye^x, \quad \frac{\partial f}{\partial y} = xe^y + e^x, \quad \frac{\partial^2 f}{\partial x^2} = ye^x, \quad \frac{\partial^2 f}{\partial y^2} = xe^y, \quad \frac{\partial^2 f}{\partial x \partial y} = e^x + e^y.$$

2.29. In each case, substitute into the equation and show that an identity is obtained.

2.30. (c) We have $f(0+\Delta x, 0+\Delta y) - f(0,0) = 0 + |\Delta x\, \Delta y|$ where

$$\frac{|\Delta x\, \Delta y|}{(\Delta x^2 + \Delta y^2)^{1/2}} \le \frac{\frac{1}{2}(\Delta x^2 + \Delta y^2)}{(\Delta x^2 + \Delta y^2)^{1/2}} = \tfrac{1}{2}(\Delta x^2 + \Delta y^2)^{1/2}.$$

Hence F is differentiable at $(0,0)$.

2.31. (a) $dz = \frac{2}{x^2+y^2}(x\,dx + y\,dy)$. (b) $dz = x^{y-1}(y\,dx + x\ln x\,dy)$. (c) $dz = \frac{1}{x^2+y^2}(-y\,dx + x\,dy)$. (d) $dz = \frac{y}{x}\,dx + \ln x\,dy$. (e) $dz = e^{-xy}[(1-xy)\,dx - x^2\,dy]$. (f) $dz = \frac{1}{y}\,dx - \frac{x}{y^2}\,dy$. (g) $dz = \cosh(x+y)(dx+dy)$.

2.32. We have

$$df = -\frac{2xz}{(x^2+y^2)^2}\,dx - \frac{2yz}{(x^2+y^2)^2}\,dy + \frac{1}{x^2+y^2}\,dz,$$

so the answers are (a) $df = -\frac{2}{25}\,dx - \frac{4}{25}\,dy + \frac{1}{5}\,dz$ and (b) $df = -\frac{1}{2}\,dx - \frac{1}{2}\,dy + \frac{1}{2}\,dz$.

2.33. $du = A\,dx + B\,dy + C\,dz.$

2.34. (a)
$$d^2 f = \frac{-2y^2\,dx^2 + 4xy\,dx\,dy - 2x^2\,dy^2}{(y - x)^2}.$$

(b)
$$d^2 f = \frac{2(y - x)(y + x)\,dx^2 - 8xy\,dx\,dy - 2(y - x)(y_x)\,dy^2}{(x^2 + y^2)^2}.$$

2.35. Use Leibniz's rule for integrals. (a) $dF = f(x + y)(dx + dy)$. (b) $dF = -f(x)\,dx + f(y)\,dy$.

2.37. (a) $\nabla f = yz\,\mathbf{i} + xz\,\mathbf{j} + xy\,\mathbf{k},$

$$\frac{\mathbf{i} + \mathbf{j}}{\sqrt{2}} \cdot \nabla f \bigg|_{(1,2,3)} = \frac{1}{\sqrt{2}}(yz + xz)\bigg|_{(1,2,3)} = \frac{9}{\sqrt{2}}.$$

2.40. (a) $\nabla \cdot \mathbf{A} = 2(xy + yz + zx)$ and $\nabla \times \mathbf{A} = -(y^2\,\mathbf{i} + z^2\,\mathbf{j} + x^2\,\mathbf{k}).$

2.46. (a)
$$\frac{df}{dt} = \frac{e^t(t\ln t - 1)}{t\ln^2 t}.$$

2.47. (a)
$$\frac{\partial f}{\partial u} = 2uv\cos(u^2 v), \qquad \frac{\partial f}{\partial v} = u^2\cos(u^2 v).$$

2.48. (a)
$$\frac{df}{dx} = \frac{e^x + (x^2 + 1)e^y}{e^x + e^y}.$$

(b)
$$\frac{df}{dx} = \frac{y - 2(x + 1)^2}{(x + 1)^2 + y^2}.$$

(c)
$$\frac{df}{dx} = \frac{x + ye^x}{\sqrt{x^2 + y^2}}.$$

(d)
$$\frac{df}{dx} = x^y\left(\frac{y}{x} + \frac{\ln x}{x}\right).$$

(e)
$$\frac{df}{dx} = \frac{1}{x^2 + 1}.$$

2.49.
$$\frac{df}{dx} = 2e^x \sin x.$$

2.52. (a) $dz = f'(xy)(y\,dx + x\,dy)$.

2.53. It will suffice to treat only the f term in u. Regarding f as a single variable function dependent on the variable z which is $z = x - ct$, we have by the chain rule
$$f_t(x - ct) = f'(x - ct)(x - ct)_t = -cf'(x - ct).$$
Similarly the second derivative is
$$f_{tt}(x - ct) = (-cf'(x - ct))_t = c^2 f''(x - ct).$$

The derivatives of f with respect to x are $f_x(x - ct) = f'(x - ct)$ and $f_{xx}(x - ct) = f''(x - ct)$. Direct substitution into the equation completes the task.

2.54. (a) $1 + v_1 v_2 t^2$.

2.55. (a) No. (b) No. (c) Yes. (b) No.

2.56. (a) $y' = (y^x \ln y - yx^{y-1})/(x^y \ln x - xy^{x-1})$. (b) $y' = (\sqrt{x} - 1)/\sqrt{x}$. (c) $y' = -y/x$. (d) $y' = y/x$.

2.57. (a) $y' = -x/y$. Directly from this we can get
$$y'' = -\frac{y(1) - xy'}{y^2} = -\frac{y - x(-x/y)}{y^2} = -\frac{y^2 + x^2}{y^3} = -\frac{a^2}{y^3}.$$

2.58. (a)
$$\frac{\partial z}{\partial x} = \frac{z}{x(z-1)}, \qquad \frac{\partial z}{\partial y} = \frac{z}{y(z-1)}.$$
(b)
$$\frac{\partial z}{\partial x} = \frac{2x}{1-2z}, \qquad \frac{\partial z}{\partial y} = \frac{2y}{1-2z}.$$

2.59. Set $(ax^2 + by^2 + cz^2)^{-1} = k > 0$ and rearrange as
$$\frac{x^2}{(1/\sqrt{ka})^2} + \frac{y^2}{(1/\sqrt{kb})^2} + \frac{z^2}{(1/\sqrt{kc})^2} = 1.$$

2.62. (a) $x_0 x/a^2 + y_0 y/b^2 + z_0 z/c^2 = 1$. (b) $bx + ay - e^{-ab} z = 2ab - 1$. (c) $[b\sin(ab)]x + [a\sin(ab)]y + z = 2ab\sin(ab) + \cos(ab)$. (d) $z = x + 1$.

2.63. (a) $x = 0$, $z = -y$. (b) $y = 0$, $z = 1-x$. (c) $a^2(x-x_0)/x_0 = b^2(y-y_0)/y_0 = c^2(z - z_0)/z_0$. (d) $(x - 1)/2 = (y - 1)/2 = (z - 2/a)/(-a)$.

2.64. The surface is the sphere $x^2 + (y - 1)^2 + z^2 = 3^2$, which clearly has two points where the normal to the tangent plane is along $(1, 2, 2)$. The equations of these planes have the form

$$1(x - x_0) + 2(y - y_0) + 2(z - z_0) = 0$$

where (x_0, y_0, z_0) are the coordinates of the points at which the tangent planes are sought. As we know, $\nabla f(x, y, z) = 2x\,\mathbf{i} + 2(y - 1)\,\mathbf{j} + 2z\,\mathbf{k}$ at point (x_0, y_0, z_0) is a normal to the tangent plane. So there exists λ such that $\nabla f(x_0, y_0, z_0) = \lambda(1, 2, 2)$:

$$2x_0 = \lambda, \qquad 2(y_0 - 1) = 2\lambda, \qquad 2z_0 = 2\lambda.$$

Substituting $x_0 = 1/2\lambda$, $y_0 - 1 = \lambda$ and $z = \lambda$ into the surface equation we get

$$\frac{1}{2^2}\lambda^2 + \lambda^2 + \lambda^2 = 9.$$

This yields $\lambda = \pm 2$, and for (x_0, y_0, z_0) we find the two triples $(1, 3, 2)$ and $(-1, -1, -2)$. The two tangent plane equations are

$$(x - 1) + 2(y - 3) + 2(z - 2) = 0 \quad \text{and} \quad (x + 1) + 2(y + 1) + 2(z + 2) = 0.$$

2.65. (a) $(-2, -7/2)$ and $(-2/3, -5/6)$. (b) $f(0, -2) = -2e^{-1}$, local minimum.

2.66. (a) $(0, 2)$ is the only critical point; it is a point of local minimum. (b) $(0, 0)$ is the only critical point; it is a point of local maximum where the function value is 0. (c) $(1/3, 1/3)$ is a point of local maximum. (d) $(4/3, 1/3)$ is a point of local minimum.

2.67. (a) $(0, 0)$ is a point of local minimum. (b) $(0, 0)$ is a critical point but not an extreme point.

2.68. (a) $(1/2, 1/2)$, point of minimum as seen by expressing f in terms x only as $[x^2 + (1 - x)^2]^{1/2}$. (b) $f(1, 1) = 2$, $f(-1, -1) = -2$. (c) $(1/2, 1/2)$, point of maximum as seen by expressing f as $x(1 - x)$. (d) $f(1/\sqrt{2}, 1/\sqrt{2}) = \sqrt{2}$, $f(-1/\sqrt{2}, -1/\sqrt{2}) = -\sqrt{2}$. (e) $f(1, 1) = 2$, point of minimum.

2.69. (a) $(1, 1)$ and $(-1, -1)$ are points of maximum. (b) $(1, 1)$ is a point of maximum; $(-1, -1)$ is a point of minimum. (c) $(1, 1)$ and $(-1, -1)$ are points of maximum; $(-1, 1)$ and $(1, -1)$ are points of minimum.

2.71. (a) Maximize $f(x, y, z) = xyz$ subject to $4x + 4y + 4z = 12a$. This can be done by the Lagrange multiplier approach or by use of the AM–GM inequality as

$$xyz \leq \left(\frac{x+y+z}{3}\right)^3 = a^3.$$

Equality holds if and only if $x = y = z$. (b) One approach is to use the AM–GM inequality. We have

$$x^2 y^2 z^2 = (x^6 y^6 z^6)^{1/3} \leq \frac{x^6 + y^6 + z^6}{3}$$

with equality if and only if $x = y = z$. But at points on the sphere this last condition means that $x = y = z = R/\sqrt{3}$. Hence the maximum value of the expression is $R^6/27$. (c) Let us minimize the squared distance $f(x, y) = (x - a)^2 + y^2$ subject to the constraint $y^2 - kx = 0$. Lagrange's equation $\nabla f = \lambda \nabla g$ gives

$$2(x - a) + \lambda k = 0, \qquad 2y - 2y\lambda = 0,$$

to be solved together with the constraint equation. Here the second equation, rewritten as $y(1 - \lambda) = 0$, shows that $y = 0$ or $\lambda = 1$. But $\lambda = 1$ substituted into the first equation gives $x = a - k/2 < 0$, which is incompatible with the constraint equation under the condition that $k > 0$. So we reject this possibility and take $y = 0$. The constraint equation then gives $x = 0$. Evaluating $f(0, 0) = a^2$, we see that the required minimum distance is a. (e) We don't need a picture: the problem is simply to maximize the volume $V = xyz$ under the restriction $2x + 3y + 5z \leq 2000$. Clearly x, y, z are positive and bounded, so the maximum exists. For $g(x, y, z) = 2x + 3y + 5z$, a linear function, V cannot attain its maximum at points where $g(x, y, z) = 2x + 3y + 5z < 2000$. So the maximum of V occurs on the boundary of the restriction $2x + 3y + 5z = 2000$. The Lagrange multiplier approach leads to the system of equations $\nabla f = \lambda \nabla g$ and $g(x, y, z) = 2000$, that is

$$yz = 2\lambda, \qquad xz = 3\lambda, \qquad xy = 5\lambda, \qquad 2x + 3y + 5z = 2000.$$

Equating expressions for λ from the first three equations and taking into account that the sides cannot be zero we get $x = 5z/2$ and $y = 5z/3$. Substituting these into the last equation we get

$$2\frac{5x}{2} + 3\frac{5x}{3} + 5z = 2000$$

from which we find z and next, x and y, finally obtaining $V_{\max}$. (h) The situation in the first octant is shown below.

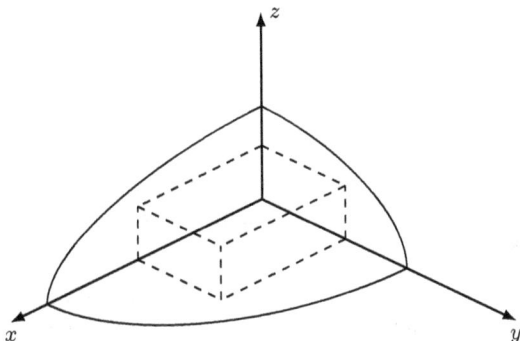

With $V = 8xyz$ we can apply the Lagrange multiplier method in a straightforward way. Another approach is to use the AM–GM–QM inequality, writing

$$\left(\frac{x}{a} \cdot \frac{y}{b} \cdot \frac{z}{c}\right)^{1/3} \leq \frac{1}{3}\left(\frac{x}{a} + \frac{y}{b} + \frac{z}{c}\right)$$

$$\leq \left[\frac{1}{3}\left(\frac{x^2}{a^2} + \frac{y^2}{b^2} + \frac{z^2}{c^2}\right)\right]^{1/2} = \frac{1}{\sqrt{3}}$$

which gives

$$8xyz \leq \frac{8}{3\sqrt{3}} abc.$$

Equality is attained if and only if $x/a = y/b = z/c$; making substitutions into the constraint equation, we find $x = a/\sqrt{3}$, $y = b/\sqrt{3}$, and $z = c/\sqrt{3}$.

2.72. (a) The maximum value is $1 + a$, attained at the points $(\pm 1, 1)$. The minimum value is $-a$, attained at $(0, -1)$. (b) The maximum value is 3, attained at $(0, 1)$. (c) The maximum value is $1/27$. The minimum value is 0. (d) The maximum value is $1/2$. The minimum value is $-1/2$. (e) The triangle is shown below.

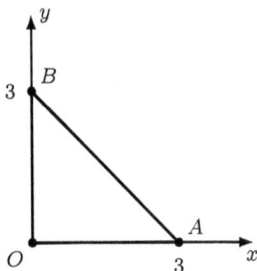

The critical points of f are solutions of the equation $\nabla f = \mathbf{0}$, that is

$$f_x(x, y) = 4y - 2y^2 + 2 = 0, \qquad f_y(x, y) = 4x - 4xy = 0.$$

The system has two solutions $x = 0, y = 1 + 2\sqrt{2}$ and $x = 0, y = 1 - 2\sqrt{2}$. Both points lie outside the triangle OAB. So the needed extrema must be achieved on the boundary, which consists of three segments OA, AB, and BO. Consider the max-min problem on each of them.

(1) On OA we have $y = 0$. We should find the extremal values of $g_1(x) = f(0, y) = 2x$ over $y \in [0, 3]$. So the minimum f is 0, taken at $(0, 0)$ and the maximum is 6 at the point $(6, 0)$.

(2) On AB we have $y = 3 - x$. Now the max-min problem is for $g_2(x) = f(x, 3 - x) = -2x^3 + 8x^2 - 4x$ over $x \in [0, 3]$. So first the critical points: $g_2'(x) = -6x^2 + 16x - 4 = 0$ brings us $(4 \pm \sqrt{10})/3$. Both values are inside $[0, 3]$. So the points of possible max-min of f on AB are

$$((4 - \sqrt{10})/3, (5 + \sqrt{10})/3), \quad ((4 + \sqrt{10})/3, (5 - \sqrt{10})/3), \quad (0, 3). \quad \text{(B.1)}$$

Note that the endpoint $(3, 0)$ has entered to OA.

(3) On BO we have $x = 0$, so we seek find possible max-min points of $g_3(y) = f(0, y) = 0$. The maximum and minimum are both equal to 0 and are taken at any point of BO.

So it remains to calculate f at the points (B.1) and select the max-min values. These values, 6 and 0, are max-min values of f on the triangle. We also should show at which points they are taken.

2.73. (a) The lone critical point $(0, 0)$ is not a point of local extremum. (b) The maximum value is 4, attained at points $(\pm 2, 0)$. The minimum value is -4, attained at points $(0, \pm 2)$. (c) The Lagrange system

$$2x + \lambda = 0, \qquad -2y + \lambda = 0, \qquad x + y = 1,$$

has no solution.

2.74. (a) Add the inequalities $x \le |x|$ and $y \le |y|$ to get $x + y \le |x| + |y|$. Then replace x by $-x$ and y by $-y$. Equality holds if and only if $x = 0$, or $y = 0$, or x and y have the same sign.

2.75. (a) The curve x^α lies below its tangent line through the point $(1, 1)$.

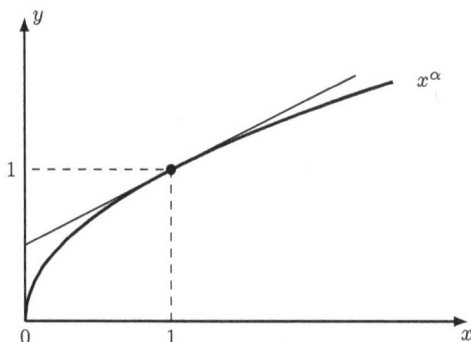

2.76. (c) Putting $\alpha = 1/p$, we get

$$x^{1/p} - \frac{x}{p} \le \frac{1}{q}.$$

Now put $x = A^p B^{-q}$ and multiply through by B^q. Here we have assumed that $B > 0$, but Young's inequality holds trivially if $B = 0$.

2.77. We show first that if $A_1, \ldots, A_n$ and $B_1, \ldots, B_n$ are any nonnegative numbers satisfying

$$\sum_{i=1}^n A_i^p = 1 \quad \text{and} \quad \sum_{i=1}^n B_i^q = 1,$$

then

$$\sum_{i=1}^n A_i B_i \le 1.$$

For this we apply Young's inequality to each pair of numbers A_i, B_i,

$$A_i B_i \le \frac{A_i^p}{p} + \frac{B_i^q}{q},$$

and sum over i:

$$\sum_{i=1}^n A_i B_i \le \frac{1}{p} \sum_{i=1}^n A_i^p + \frac{1}{q} \sum_{i=1}^n B_i^q = \frac{1}{p} + \frac{1}{q} = 1.$$

Finally, set

$$A_i = \frac{a_i}{\left(\displaystyle\sum_{i=1}^n a_i^p \right)^{1/p}}, \qquad B_i = \frac{b_i}{\left(\displaystyle\sum_{i=1}^n b_i^q \right)^{1/q}},$$

in order to obtain Hölder's inequality.

2.78. Write

$$\sum_{i=1}^{n}(a_i + b_i)^p = \sum_{i=1}^{n} a_i(a_i + b_i)^{p-1} + \sum_{i=1}^{n} b_i(a_i + b_i)^{p-1}$$

and apply Hölder's inequality to the two sums on the right:

$$\sum_{i=1}^{n}(a_i + b_i)^p \le \left[\sum_{i=1}^{n} a_i^p\right]^{1/p} \left[\sum_{i=1}^{n}(a_i + b_i)^{(p-1)q}\right]^{1/q}$$

$$+ \left[\sum_{i=1}^{n} b_i^p\right]^{1/p} \left[\sum_{i=1}^{n}(a_i + b_i)^{(p-1)q}\right]^{1/q}.$$

Since $(p-1)q = p$ and $1/q = 1 - 1/p$, we have

$$\sum_{i=1}^{n}(a_i + b_i)^p \le \left\{\left[\sum_{i=1}^{n} a_i^p\right]^{1/p} + \left[\sum_{i=1}^{n} b_i^p\right]^{1/p}\right\} \cdot \left[\sum_{i=1}^{n}(a_i + b_i)^p\right]^{1-1/p}.$$

Minkowski's inequality follows.

Chapter 3

3.1. The limiting procedure gives the expected answer of 4.

3.2. Partition $[0, 1]$ uniformly with $\Delta x \equiv 1/n$ so that the Riemann sums take the form

$$\sum_{k=1}^{n} f(\xi_k)\,\Delta x_k = \frac{1}{n}\sum_{k=1}^{n} f(\xi_k).$$

Taking the ξ_k all rational, we get 0 for these Riemann sums. Taking the ξ_k all irrational, we get

$$\frac{1}{n}\sum_{k=1}^{n} f(\xi_k) = \frac{1}{n}\sum_{k=1}^{n} 1 = \frac{1}{n}\cdot n = 1.$$

So the quantity

$$\lim_{\Delta x \to 0} \sum_{k=1}^{n} f(\xi_k)\,\Delta x_k$$

depends on our choice of the ξ_k.

3.3.

$$M = \int_{0}^{L} x(L - x)\,dx = \frac{L^3}{6}.$$

3.6. $I_1 = 27/2$. $I_2 = (e-1)/3$, by reversing the integration order. $I_3 = 1/24$.

3.7. (a) $(e-1)^2$. (b) $\pi/12$. (c) $\ln(4/3)$. (e) $2/3$

3.8. $1/8$.

3.9. (a) The region lies between the parabola $y = x^2$ and the line $y = x$ for $0 \le x \le 1$.

$$I = \int_0^1 \int_{x^2}^x f(x,y)\, dy\, dx.$$

(b) The region lies inside the upper half of the circle $x^2 + y^2 = 1$.

$$I = \int_0^1 \int_{-\sqrt{1-y^2}}^{\sqrt{1-y^2}} f(x,y)\, dx\, dy.$$

(c) The region lies between the x-axis and the line $y = x$ for $0 \le x \le 1$.

$$I = \int_0^1 \int_y^1 f(x,y)\, dx\, dy.$$

(d) The region lies between the x-axis and the parabola $y = x^2$ for $0 \le x \le 1$.

$$I = \int_0^1 \int_{\sqrt{y}}^1 f(x,y)\, dx\, dy.$$

3.10. (a)

$$\int_0^1 \int_{x/2}^x f(x,y)\, dy\, dx.$$

(b)

$$\int_0^1 \int_y^{y+1} f(x,y)\, dx\, dy.$$

3.11. (a)

$$\int_A xy\, dA = \int_0^1 \int_{x/3}^{3x} xy\, dy\, dx = \frac{10}{9},$$

$$\int_A xy\, dA = \int_0^{1/3} \int_{y/3}^{3y} xy\, dx\, dy + \int_{1/3}^3 \int_{y/3}^1 xy\, dx\, dy = \frac{10}{9}.$$

(b)

$$\int_A (x+y)\, dA = \int_0^1 \int_{x^2}^{\sqrt{x}} (x+y)\, dy\, dx = \frac{3}{10},$$

$$\int_A (x+y)\, dA = \int_0^1 \int_{y^2}^{\sqrt{y}} (x+y)\, dx\, dy = \frac{3}{10}.$$

(c) $1/3$.

3.12. (a) $\pi(e-1)$. (b) $\pi/3$. (c) The curve $xy = k$ in polar coordinates takes the form $r\cos\theta\, r\sin\theta = k$, or

$$r = \sqrt{\frac{k}{\sin\theta\cos\theta}}.$$

We have

$$I = \int_{\tan^{-1}(1/2)}^{\tan^{-1}(2)} \int_{\sqrt{\frac{1}{\sin\theta\cos\theta}}}^{\sqrt{\frac{2}{\sin\theta\cos\theta}}} (r\sin\theta)^2 r\, dr\, d\theta = \int_{\tan^{-1}(1/2)}^{\tan^{-1}(2)} \sin^2\theta \cdot \frac{3}{4\sin^2\theta\cos^2\theta}\, d\theta$$

$$= \frac{3}{4}\tan\theta\,\Big|_{\tan^{-1}(1/2)}^{\tan^{-1}(2)} = \frac{9}{8}.$$

3.13. (a) $\pi/8$. (b) $\pi(e-1)/4$. (c) $\pi/6$. (d) $\pi(2\ln 2 - 1)/4$.

3.14.

$$\int_0^\alpha \int_0^{a\cos\theta} r\sin^2\theta\, r\, dr\, d\theta = \frac{a^3 \sin^3\alpha(5 - 3\sin^2\alpha)}{45}.$$

3.15. In general, the area of a region bounded by $r = f(\theta)$ between the two rays $\theta = \alpha$ and $\theta = \beta$ is

$$A = \int_\alpha^\beta \int_0^{f(\theta)} r\, dr\, d\theta = \frac{1}{2}\int_\alpha^\beta [f(\theta)]^2\, d\theta.$$

For the present problem we get $A = (\pi + 2)/16$.

3.16. (a)

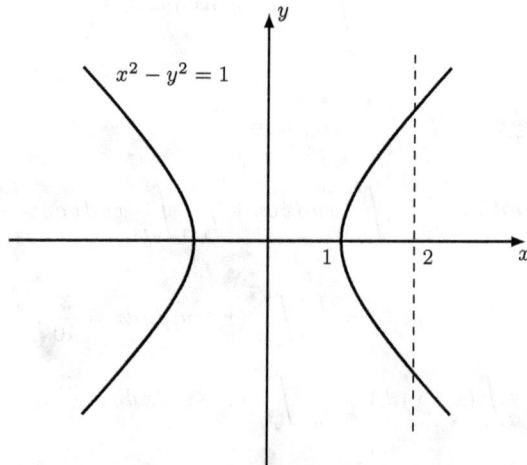

$$A = 2 \int_1^2 \int_0^{\sqrt{x^2-1}} dy\, dx = 2\sqrt{3} - \ln(2 + \sqrt{3}).$$

(b) First find the two intersection points, which are $(0,0)$ and $(2p, 2p)$.

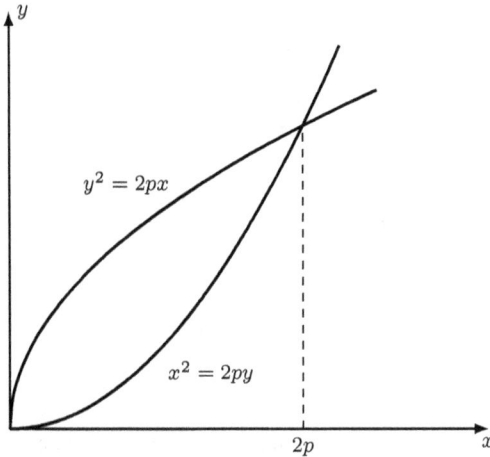

The enclosed area is

$$A = \int_0^{2p} \int_{x^2/(2p)}^{\sqrt{2px}} dy\, dx = \frac{4p^2}{3}.$$

(c) Type 1 and Type 2 calculations:

$$A = \int_0^1 \int_{1-x}^{1-x^2} dy\, dx = \frac{1}{6}, \quad A = \int_0^1 \int_{1-y}^{\sqrt{1-y}} dx\, dy = \frac{1}{6}.$$

(d)

$$A = \int_0^1 \int_{(x-1)^2}^{\sqrt{1-x^2}} dy\, dx = \frac{3\pi - 4}{12}.$$

3.17.

$$4 \int_0^a \int_0^{b\sqrt{1-\frac{x^2}{a^2}}} dy\, dx = \pi ab.$$

3.18.

$$\int_A dA = \int_{-1}^1 \int_{x^2}^1 dy\, dx = \int_0^1 \int_{-\sqrt{y}}^{\sqrt{y}} dx\, dy = \frac{4}{3}.$$

$$\int_A x^2\, dA = \int_{-1}^1 \int_{x^2}^1 x^2\, dy\, dx = \int_0^1 \int_{-\sqrt{y}}^{\sqrt{y}} x^2\, dx\, dy = \frac{4}{15}.$$

3.19. (a)
$$2 \int_0^\pi \int_0^{a(1+\cos\theta)} r \, dr \, d\theta = \frac{3\pi a^2}{2}.$$

(b)
$$\int_0^\pi \int_0^\theta r \, dr \, d\theta = \frac{\pi^3}{6}.$$

(c)
$$2 \int_0^{\pi/2} \int_0^{a\cos\theta} r \, dr \, d\theta = \frac{\pi a^2}{4}.$$

(d)
$$2 \int_0^\pi \int_0^{1-\cos\theta} r \, dr \, d\theta = \frac{3\pi}{2}.$$

3.20. (a) Express the circle in polar form as $r = 6\sin\theta$. Then

$$A = \int_{\tan^{-1}(1/2)}^{\pi/4} \int_0^{6\sin\theta} r \, dr \, d\theta \approx 1.99575.$$

(b) In the first quadrant the two circles intersect at $r = 1$, $\theta = \pi/6$.

$$A = 2 \int_{\pi/6}^{\pi/2} \int_1^{2\sin\theta} r \, dr \, d\theta = \frac{\pi}{3} + \frac{\sqrt{3}}{2}.$$

3.21. (a) $(a+1)b/3$. (b) $2/3$.

3.22. (a) $(\pi/2, \pi/8)$. (c) $(4a/(3\pi), 0)$. (d) $(5a/6, 0)$.

3.24. (a)
$$I_{xx} = \int_{-b}^b \int_{-a\sqrt{1-y^2/b^2}}^{a\sqrt{1-y^2/b^2}} y^2 \, dx \, dy = \frac{\pi a b^3}{4}.$$

(b) $\pi b a^3/4$. (c) $I = I_{xx} + I_{yy} = \pi ab(a^2+b^2)/4$.

3.25. The diagonal strip R is mapped onto the vertical strip $|u| \le 1$. In the sketch below, some pairings between points are also shown ($a \leftrightarrow A$, $b \leftrightarrow B$, etc.). The inverse transformation equations are $x = v$ and $y = v - u$.

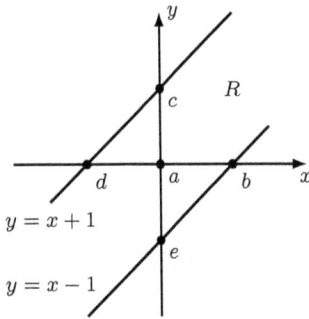

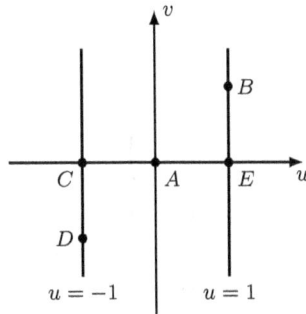

$y = x + 1$

$y = x - 1$

$u = -1$ $u = 1$

3.26. (a) $-e(e + 1)$. (b) e^2.

3.27. $\partial(x, y)/\partial(u, v) = ad - bc$ and $\partial(u, v)/\partial(x, y) = 1/(ad - bc)$.

3.29.

$$\int_0^1 \int_x^1 xy \, dy \, dx = \frac{1}{8} = 2 \int_{-1/2}^0 \int_{-v}^{v+1} (u + v)(u - v) \, du \, dv.$$

3.30. Region R is sketched at left below. Rewriting the boundary line equations as $y - x = 1$, $y + x = -1$, $y - x = -1$, $y + x = 1$, we are prompted to let $u = y - x$ and $v = y + x$. This transforms these four lines into $u = 1$, $v = -1$, $u = -1$, $v = 1$, respectively, which yield the boundary of S. Images of the corner points are indicated ($a \leftrightarrow A$, etc).

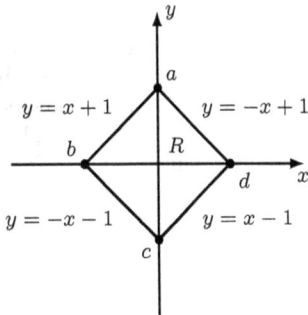

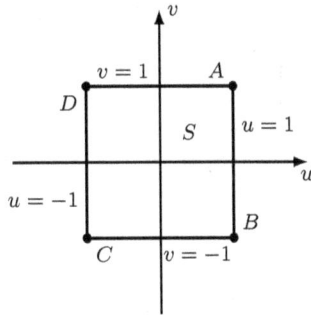

Solving the transformation equations for x and y, we get the inverse transformation equations $x = (v - u)/2$ and $y = (v + u)/2$. The Jacobians are

$$\frac{\partial(u, v)}{\partial(x, y)} = \begin{vmatrix} u_x & v_x \\ u_y & v_y \end{vmatrix} = \begin{vmatrix} -1 & 1 \\ 1 & 1 \end{vmatrix} = -2$$

and

$$\frac{\partial(x, y)}{\partial(u, v)} = \begin{vmatrix} x_u & y_u \\ x_v & y_v \end{vmatrix} = \begin{vmatrix} -1/2 & 1/2 \\ 1/2 & 1/2 \end{vmatrix} = -\frac{1}{2}.$$

Note that they are reciprocals.

3.31. The sketch below should be useful.

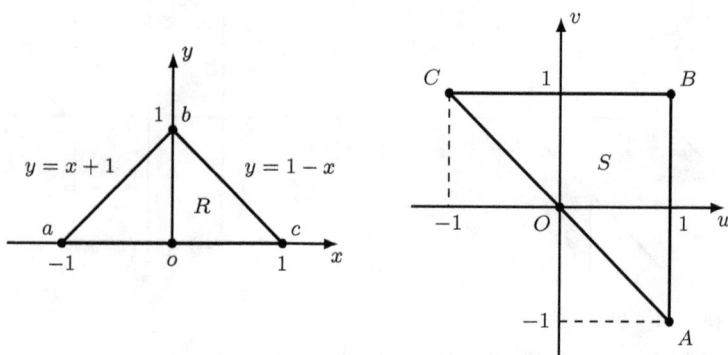

(a) Using the left-hand picture, we write

$$I = \int_0^1 \int_{y-1}^{1-y} x^2 y \, dx \, dy = \frac{1}{30},$$

$$I = \int_{-1}^0 \int_0^{x+1} x^2 y \, dy \, dx + \int_0^1 \int_0^{1-x} x^2 y \, dy \, dx = \frac{1}{30}.$$

(b) Let us determine S by mapping the boundary of R over to the uv-plane. Side ab of R lies along the line $y = x + 1$. Writing this as $y - x = 1$, we see that ab maps over to AB in the right portion of the sketch. Side bc of R lies along the line $y = 1 - x$. Writing this as $y + x = 1$, we see that bc maps over to BC. Side ca of R lies along the line $y = 0$. For this, the transformation equations give $u = -x$ and $v = x$, hence $v = -u$. So ca maps over to CA, and S is the triangle shown on the right. (c) Using the inverse transformation relations

$$x = \tfrac{1}{2}(v - u), \qquad y = \tfrac{1}{2}(v + u),$$

we find

$$J = \frac{\partial(x, y)}{\partial(u, v)} = -\frac{1}{2}.$$

Hence

$$I = \int_{-1}^1 \int_{-u}^1 \left(\frac{v-u}{2}\right)^2 \left(\frac{v+u}{2}\right) \cdot \left|-\frac{1}{2}\right| dv \, du = \frac{1}{30},$$

$$I = \int_{-1}^1 \int_{-v}^1 \left(\frac{v-u}{2}\right)^2 \left(\frac{v+u}{2}\right) \cdot \left|-\frac{1}{2}\right| du \, dv = \frac{1}{30}.$$

3.32. (a)

$$M = \int_0^{2\pi} \int_0^a (2a + r\cos\theta) r\, dr\, d\theta = 2\pi a^3.$$

(b)

$$\int_0^1 \int_0^1 (x^2 + y^2)^{1/2}\, dx\, dy = 2 \int_0^{\pi/4} \int_0^{\sec\theta} r^2\, dr\, d\theta = \frac{2}{3} \int_0^{\pi/4} \sec^3\theta\, d\theta$$

$$= \tfrac{1}{3}[\sec\theta \tan\theta + \ln|\sec\theta + \tan\theta|]\big|_0^{\pi/4}$$

$$= \tfrac{1}{3}[\sqrt{2} + \ln(\sqrt{2} + 1)].$$

3.33. 2 millicoulombs.

3.34.

$$M_y = \int_0^1 \int_{x^2}^{\sqrt{x}} x\, dy\, dx = \frac{3}{20}, \qquad M_x = \int_0^1 \int_{x^2}^{\sqrt{x}} y\, dy\, dx = \frac{3}{20},$$

and

$$M = \int_0^1 \int_{x^2}^{\sqrt{x}} dy\, dx = \frac{1}{3}.$$

Therefore $x_c = 9/20$ and $y_c = 9/20$.

Chapter 4

4.3. $I_a = 65/8$, $I_b = 195(e-1)/8$, $I_c = 3(e-1)$, $I_d = 0$, $I_e = a^6/48$.

4.4. (a) $1/720$, (b) $\pi a^4/4$.

4.5. $I = 2\pi/15$.

4.6. (a) $68/15$. (b) $\pi/2$. (c) $\pi/2$.

4.7.

$$\int_0^1 \int_0^x \int_0^{xy} dz\, dy\, dz = \frac{1}{8}.$$

4.8. (a) $1/6$. (b) $2\pi(1-\cos\alpha)a^3/3$. (c) $2\phi_0 a^3/3$. (d) $1/6$. (e) $2\pi abh(h^2+3c^2)/3c^2$. (f) $\pi z_0^3/3$. (g) $88/105$.

4.9. Use the change of variables $x = ar\cos\theta$, $y = br\sin\theta$. The equation of the cross-sectional ellipse becomes $r^2 + z^2/c^2 = 1$, and the Jacobian of the transformation is

$$\frac{\partial(x,y)}{\partial(r,\theta)} = \begin{vmatrix} a\cos\theta & b\sin\theta \\ -ar\sin\theta & br\cos\theta \end{vmatrix} = abr.$$

Then

$$V = 8 \int_0^c \int_0^{\pi/2} \int_0^{\sqrt{1-z^2/c^2}} abr \, d\theta \, dz = \frac{4}{3}\pi abc.$$

4.10. Use cylindrical coordinates. (a) $V = 4\pi a^{5/2}/5$. (b) $V = \pi a^{5/2}/5$.

4.11.

$$V = \pi \int_a^b f^2(x) \, dx.$$

For the parabola this gives $V = 2\pi ah^2$.

4.12. The plane tangent to the surface $F(x, y, z) = xyz - a^3 = 0$ at point $P = (x_0, y_0, z_0)$ has intercept form

$$\frac{x}{3x_0} + \frac{y}{3y_0} + \frac{z}{3z_0} = 1.$$

The required volume depends only on the product $x_0 y_0 z_0$ and can therefore be expressed purely in terms of a.

4.13. We can find the first-octant portion of the required volume and multiply by 8. The curve of intersection in the first octant was found in Problem 1.52. The projection of $\mathbf{r}(t)$ onto the xy-plane is $\mathbf{r}_{xy}(t) = t\,\mathbf{i} + (1-t^2)^{1/2}\,\mathbf{j}$, which traces out the first-quadrant portion of the circle $x^2 + y^2 = 1$. Hence D_{xy} for the triple integral is the quarter unit disk $\{(x, y): x^2 + y^2 \le 1, \ x \ge 0, \ y \ge 0\}$. The "ceiling" of the desired volume region is formed by the cylinder $x^2 + z^2 = 1$ over D_{xy}. We have

$$V = 8 \int_0^1 \int_0^{\sqrt{1-y^2}} \int_0^{\sqrt{1-x^2}} dz \, dx \, dy = \frac{16}{3}.$$

4.17. $M = 4\pi ka^5/5$.

4.18. $(x_c, y_c, z_c) = (3/4, 3/2, 2/3)$.

4.19. $z_c = 3h/4$.

4.20. $(1/4, 1/4, 1/4)$.

4.22. Suppose to the contrary that $f(x_0, y_0, z_0) \ne 0$. Without loss of generality we may assume that $f(x_0, y_0, z_0) > 0$. By continuity there is a neighborhood N of (x_0, y_0, z_0) and a number $\varepsilon > 0$ such that $f(x, y, z) > \varepsilon$ throughout N. Then

$$\iiint_N f(x, y, z) \, dV > \iiint_N \varepsilon \, dV = \varepsilon \cdot V(N) > 0,$$

a contradiction.

Chapter 5

5.1. (a)

$$\int_L x\,dy + y\,dx = \int_0^1 2x\,dx = 1.$$

(b)

$$\int_L x\,dy - y\,dx = \int_0^1 x^2\,dx = \frac{1}{3}.$$

(c)

$$\int_0^{\pi/2} (\cos t \cdot 3\sin t - 1)(-\sin t)\,dt + \cos^2 t \cdot 3\sin t \cdot 3\cos t\,dt = \frac{9}{4}.$$

(d) 38.

5.2. Since $\mathbf{r}'(t) = -\sin t\,\mathbf{i} + \cos t\,\mathbf{j} + \mathbf{k}$ we have

$$\int_C \mathbf{A}(\mathbf{r}) \cdot d\mathbf{r} = \int_C \mathbf{A}(\mathbf{r}(t)) \cdot \mathbf{r}'(t)\,dt = \int_0^{2\pi} [-t\sin t + \cos^2 t + \sin t]\,dt = 3\pi.$$

5.3. $\sqrt{3}(1 - e^{-2\pi})$.

5.9.

$$\int_L f(x,y,z)\,ds$$

$$= \int_a^b f(x, y(x), z(x))\,\{1 + [y'(x)]^2 + [z'(x)]^2\}^{1/2}\,dx,$$

$$\int_L \mathbf{F}(x,y,z) \cdot \mathbf{r}$$

$$= \int_a^b [P(x, y(x), z(x)) + Q(x, y(x), z(x))\,y'(x) + R(x, y(x), z(x))\,z'(x)]\,dx,$$

where $\mathbf{F} = P\mathbf{i} + Q\mathbf{j} + R\mathbf{k}$.

5.10. In (5.6) let the parameter t be θ. Write

$$x(\theta) = \cos\theta, \qquad y(\theta) = \sin\theta, \qquad z(\theta) = h(\theta).$$

Then

$$x'(\theta) = -\sin\theta, \qquad y'(\theta) = \cos\theta, \qquad z'(\theta) = h'(\theta),$$

and

$$[x'(\theta)]^2 + [y'(\theta)]^2 + [z'(\theta)]^2 = 1 + [h'(\theta)]^2.$$

Substitute into (5.6) to get

$$\int_L f(x, y, z)\, ds = \int_a^b f(\cos\theta, \sin\theta, h(\theta))\, \{1 + [h'(\theta)]^2\}^{1/2}\, d\theta.$$

5.11. (a) In (5.6) let the parameter t be θ. Write

$$x(\theta) = \sin[h(\theta)]\cos\theta, \qquad y(\theta) = \sin[h(\theta)]\sin\theta, \qquad z(\theta) = \cos[h(\theta)].$$

Then

$$x'(\theta) = \sin[h(\theta)](-\sin\theta) + \cos\theta\cos[h(\theta)]h'(\theta),$$
$$y'(\theta) = \sin[h(\theta)]\cos\theta + \sin\theta\cos[h(\theta)]h'(\theta),$$
$$z'(\theta) = -\sin[h(\theta)]h'(\theta),$$

and

$$[x'(\theta)]^2 + [y'(\theta)]^2 + [z'(\theta)]^2 = \sin^2[h(\theta)] + [h'(\theta)]^2.$$

Substitute into (5.6) to get

$$\int_L f(x, y, z)\, ds = \int_a^b f(\sin[h(\theta)]\cos\theta, \sin[h(\theta)]\sin\theta, \cos[h(\theta)])$$
$$\cdot\, \{\sin^2[h(\theta)] + [h'(\theta)]^2\}^{1/2}\, d\theta.$$

(b) With $f \equiv 1$ and $h(\theta) = \theta$ we have

$$\int_L 1\, ds = \int_{\pi/4}^{\pi/2} \{\sin^2\theta + 1\}^{1/2}\, d\theta.$$

This integral does not exist in closed form but can be approximated numerically.

5.12. (a)

$$L = \frac{(9a + 4)^{3/2} - 8}{27}.$$

5.13. (a)

$$s = \int_0^3 (1 + 4t^2 + 4t^4)^{1/2}\, dt = 21.$$

5.14. Taking $t = x$ as the parameter, describe the curve of intersection as $x(t) = t$, $y(t) = t^2/(2a)$, $z(t) = t^3/(6a^2)$.

5.15. Intuitively we have

$$ds = [(dr)^2 + (r\,d\theta)^2]^{1/2} = \left[\left(\frac{dr}{d\theta}\right)^2 + r^2\right]^{1/2} d\theta.$$

(a) 8.

5.16. $L = \sinh a$.

5.18. $(0, 2/\pi, 0)$.

5.19. $L = 8$.

Chapter 6

6.3. (a)

$$\int_0^{2\pi}\int_0^{\pi} (R\sin\theta)^2 R^2 \sin\theta\,d\theta\,d\phi = \frac{8\pi R^4}{3}.$$

(b) Use cylindrical coordinates:

$$I = \int_0^1\int_0^{2\pi}(\sqrt{1-\sin^2\phi} + \cos\phi)z\,d\phi\,dz = \int_0^{2\pi}(|\cos\phi| + \cos\phi)\,d\phi\int_0^1 z\,dz$$

$$= \frac{1}{2}\int_0^{2\pi}|\cos\phi|\,d\phi = 2\int_0^{\pi/2}\cos\phi\,d\phi = 2.$$

6.4. Recall that for a surface $z = g(x,y)$ defined for $(x,y) \in D_{xy}$, we have

$$\iint_S f(x,y,z)\,dS = \iint_{D_{xy}} f(x,y,g(x,y))(1+g_x^2+g_y^2)^{1/2}\,dA.$$

(a) 0. (b) 1/8. (c) 3π.

6.5. (a)

$$A = \frac{(a^2+b^2+1)^{1/2}x_0^2}{2k}.$$

(b)

$$I = \frac{(a^2+b^2+1)^{1/2}[2k(a+1)+b+1]x_0^3}{6k^2}.$$

6.6. We have

$$\mathbf{r}_u \times \mathbf{r}_v = \begin{vmatrix} \mathbf{i} & \mathbf{j} & \mathbf{k} \\ \cos v & \sin v & 1 \\ -u\sin v & u\cos v & 0 \end{vmatrix}, \quad |\mathbf{r}_u \times \mathbf{r}_v| = \sqrt{2}u,$$

and

$$A = \int_0^{2\pi} \int_0^1 \sqrt{2}u \, du \, dv = \pi\sqrt{2}.$$

6.7. (a) $3\pi\sqrt{14}$. (b) $2\pi[(a^2+1)^{3/2}-1]/3$. (e) The sketch below shows the plane (solid line) cutting the cone (dashed line) in a curve (dotted).

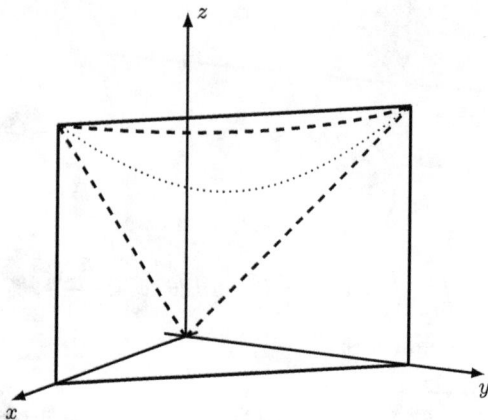

The relevant domain in the xy-plane is the triangle D_{xy} bounded by the coordinate lines and the line $x+y=1$. Differentiation gives

$$z_x^2 = \frac{x^2}{x^2+y^2}, \qquad z_y^2 = \frac{y^2}{x^2+y^2},$$

and we get

$$S = \int_0^1 \int_0^{1-x} \sqrt{2} \, dy \, dx = \frac{1}{\sqrt{2}}.$$

6.8. Parameterizing the surface by

$$\mathbf{r}(u,v) = u\,\mathbf{i} + v\,\mathbf{j} + (d/c - au/c - bv/c)\,\mathbf{k}$$

we find

$$|\mathbf{r}_u \times \mathbf{r}_v| = [(a/c)^2 + (b/c)^2 + 1]^{1/2}.$$

The required area is

$$A = \frac{1}{2}\frac{d}{a}\frac{d}{b}\frac{(a^2+b^2+c^2)^{1/2}}{c}$$

as expected.

6.9. Use

$$A = \iint_{D_{xy}} (1 + z_x^2 + z_y^2)^{1/2} \, dx \, dy.$$

(a) Integrate in polar coordinates. The answer is $A = \pi(5^{3/2} - 1)/6$. (b) $A = \sqrt{3}/2$. (c) $A = \pi(2^{3/2} - 1)/6$.

6.10. For part (a), think of finding the area of a strip having length $2\pi|f(x)|$ and width $\{1 + [f'(x)]^2\}^{1/2}$.

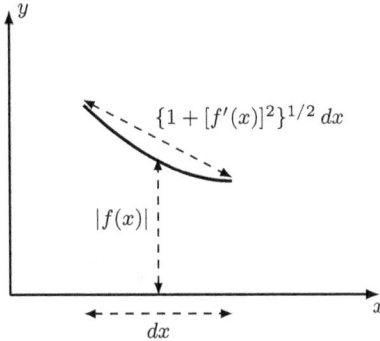

So

$$\text{(a)} \quad A = 2\pi \int_a^b |f(x)|\{1 + [f'(x)]^2\}^{1/2}\, dx,$$

$$\text{(b)} \quad A = 2\pi \int_c^d |g(y)|\{1 + [g'(y)]^2\}^{1/2}\, dy,$$

$$\text{(c)} \quad A = 2\pi \int_e^f |h(z)|\{1 + [h'(z)]^2\}^{1/2}\, dz.$$

6.11. We can find dy/dx from $2y\, dy = 4a\, dx$. The answer is

$$A = \frac{8\pi}{3}\sqrt{a}[(h+a)^{3/2} - a^{3/2}].$$

6.12. Use part (c) of Problem 6.10. The answer is $2\pi a^2[1 + 4\pi/(3\sqrt{3})]$.

6.13.

$$S_y = 2\pi a\left[a + \frac{b^2}{\sqrt{a^2 - b^2}}\sinh^{-1}\left(\frac{\sqrt{a^2 - b^2}}{b}\right)\right].$$

6.14. (a) 1. (b) 0. (c)

$$\int_0^{2\pi}\int_0^{\pi/2}\cos\phi\, a^2\sin\phi\, d\phi\, d\theta = \pi a^2.$$

Chapter 7

7.1. (d) 2/3.

7.2. (a)

$$\int_L P\,dx + Q\,dy = \int_0^2 2y^2\,dy - \int_0^1 2x\,dx = \frac{13}{3},$$

$$\iint_D \left(\frac{\partial Q}{\partial x} - \frac{\partial P}{\partial y}\right) dA = \int_0^1 \int_0^2 (4xy^2 - x)\,dy\,dx = \frac{13}{3}.$$

(b) 1/3.

7.3. $P = x^2 y$, $Q = xy^2$,

$$I = \iint_D \left(\frac{\partial Q}{\partial x} - \frac{\partial P}{\partial y}\right) dA = \int_{-1}^1 \int_{-1}^1 (y^2 - x^2)\,dy\,dx = 0.$$

Alternatively

$$I = \int_L P\,dx + Q\,dy$$

$$= \int_{-1}^1 x^2(-1)\,dx + \int_{-1}^1 (1)y^2\,dy - \int_{-1}^1 x^2(1)\,dx - \int_{-1}^1 (-1)y^2\,dy$$

$$= 0.$$

7.5. (a) We have

$$\int_{\partial S} \mathbf{F}\cdot d\mathbf{r} = \int_{\partial S} y\,dx + z\,dy + x\,dz = \int_0^1 (1)\,dy - \int_0^1 (1)\,dx - \int_0^1 (1)\,dy = -1$$

and

$$\nabla \times \mathbf{F} = \begin{vmatrix} \mathbf{i} & \mathbf{j} & \mathbf{k} \\ \dfrac{\partial}{\partial x} & \dfrac{\partial}{\partial y} & \dfrac{\partial}{\partial z} \\ y & z & x \end{vmatrix} = -\mathbf{i} - \mathbf{j} - \mathbf{k}, \qquad d\mathbf{S} = \mathbf{k}\,dS$$

so that

$$\iint_S \nabla \times \mathbf{F}\cdot d\mathbf{S} = -\iint_S dS = -1.$$

(b) 1/2. (c) 0.

7.6. We have $\nabla \times \mathbf{A} = -(\mathbf{i} + \mathbf{j} + \mathbf{k})$. Dotting this with the unit normal to the plane, we get $\nabla \times \mathbf{A}\cdot\mathbf{n} = -(\mathbf{i} + \mathbf{j} + \mathbf{k})\cdot(\mathbf{i} + \mathbf{j} + \mathbf{k})/\sqrt{3} = -\sqrt{3}$. Since the area enclosed by the intersection (a circle) is πR^2, the answer is $-\sqrt{3}\pi R^2$.

7.7. Let $\mathbf{A} = (x - y)\,\mathbf{i} + (x + y)\,\mathbf{j} + (x - z)\,\mathbf{k}$. Then $\nabla \times \mathbf{A} = 2\,\mathbf{k}$, so $\nabla \times \mathbf{A} \cdot \mathbf{n} = 2$ where $\mathbf{n}$ is unit normal to the planar surface enclosed by C. The answer is 8π by Stokes' theorem.

7.8. $3/2$.

7.9. (a) $1/2$. (b) $1/24$.

7.10.

$$\int_0^4 \int_0^{4-x} \int_0^{8-2x-2y} dz\,dy\,dx = \frac{64}{3}.$$

7.11. Use the divergence theorem. (a) Since $\nabla \cdot \mathbf{A} = 3$, one needs simply $3/8$ multiplied by the volume of the whole ellipsoid. The answer is $\pi abc/2$. (b) $3\pi a^2 h$. (c) π.

7.12. $4\pi a^3$.

7.13. In the divergence theorem set $\mathbf{A} = f\,\mathbf{i}$:

$$\iint_{\partial E} f\,\mathbf{i} \cdot d\mathbf{S} = \iiint_E \nabla \cdot (f\,\mathbf{i})\,dV.$$

Then use the facts that $\nabla \cdot (f\,\mathbf{i}) = \mathbf{i} \cdot \nabla f$ and $\mathbf{i}$ is a constant to write

$$\mathbf{i} \cdot \left(\iint_{\partial E} f\,d\mathbf{S} - \iiint_E \nabla f\,dV \right) = 0.$$

Repeat to obtain analogous equations with $\mathbf{i}$ replaced by $\mathbf{j}$ and $\mathbf{k}$, respectively. This shows that the parenthetical quantity is perpendicular to any vector, hence is the zero vector.

7.14. Set $\mathbf{A} = \nabla f$ in the divergence theorem.

Index

applications of, 181
change of variables in, 176
cyclic substitution in, 176
estimates for, 167
linearity of, 167
over general bounded domain, 166
two-dimensional object, 26

unit tangent, 12

vector function, 5
vector(s)
  acceleration, 13

position, 1
unit, 14
velocity, 11
vector-operator, 67
velocity, 11
virtual displacement principle, 237

wave equation, 60, 107
weak solution, 237
Weierstrass theorem, 55
work, 194

Young's inequality, 112

www.ingramcontent.com/pod-product-compliance
Lightning Source LLC
Chambersburg PA
CBHW050635190326

41458CB00008B/2283